T0138282

Reckoning with Matter

Reckoning with Matter

Calculating Machines, Innovation, and Thinking about Thinking from Pascal to Babbage

MATTHEW L. JONES

The University of Chicago Press
Chicago and London

The University of Chicago Press, Chicago 60637
The University of Chicago Press, Ltd., London
© 2016 by The University of Chicago
All rights reserved. Published 2016.
Printed in the United States of America

25 24 23 22 21 20 19 18   2 3 4 5

ISBN-13: 978-0-226-41146-0 (cloth)
ISBN-13: 978-0-226-41163-7 (e-book)
DOI: 10.7208/chicago/9780226411637.001.0001

Library of Congress Cataloging-in-Publication Data

Names: Jones, Matthew L. (Matthew Laurence), 1972– author.
Title: Reckoning with matter: calculating machines, innovation, and thinking about
    thinking from Pascal to Babbage / Matthew L. Jones.
Description: Chicago ; London : The University of Chicago Press, 2016. | Includes
    bibliographical references and index.
Identifiers: LCCN 2016015339 | ISBN 9780226411460 (cloth: alk. paper) |
    ISBN 9780226411637 (e-book)
Subjects: LCSH: Calculators—History. | Computers—History. | Technology—History.
Classification: LCC QA75.J66 2016 | DDC 510.28/4—dc23 LC record
    available at https://lccn.loc.gov/2016015339

♾ This paper meets the requirements of ANSI/NISO Z39.48-1992 (Permanence of Paper).

To my parents, Wally and Pru,
and my sister, Lindsay

I shall not stop to examine either the Wisdom or the Morality of these several Propositions; my Object in this part of this Logical Investigation, being simply to shew by means of specifick Examples, in what Manner, a <u>Piece of Mechanism</u> can delve into the <u>Mind of Man</u> and discover the inmost Recesses of the <u>Human Heart</u>.

CHARLES STANHOPE, C. 1800

Contents

		Introduction	1
1		Carrying Tens: Pascal, Morland, and the Challenge of Machine Calculation	13
	FIRST CARRY	Babbage and Clement Mechanize Table Making	44
2		Artisans and Their Philosophers: Leibniz and Hooke Coordinate Minds, Metal, and Wood	56
	SECOND CARRY	Babbage Gets Funded	88
3		Improvement for Profit: Calculating Machines and the Prehistory of Intellectual Property	97
	THIRD CARRY	Babbage Claims His Property	122
4		Reinventing the Wheel: Emulation in the European Enlightenment	126
	FOURTH CARRY	Babbage Confronts Prior Art	157
5		Teething Problems: Charles Stanhope and the Coordination of Technical Knowledge from Geneva to Kent	162
	FIFTH CARRY	Babbage's Collaborators Emulate	200
6		Calculating Machines, Creativity, and Humility from Leibniz to Turing	212
	FINAL CARRY	Epilogue	239

Acknowledgments 251
Conventions 255
Abbreviations 257
Notes 259
References 293
Index 321

Introduction

This scientific invention has been, at all times, and particularly in the last two centuries, an object of serious but fruitless meditations by eminent men.

ADVERTISEMENT FOR ARITHMOMETER OF
THOMAS DE COLMAR, 1870[1]

Inventing a calculating machine would be so undemanding, the young Charles Babbage noted to himself in 1822, that he'd have attention to spare for reflecting upon the process itself. Babbage explained that he wished "to watch . . . the progress of the mind in the pursuit of a mechanical invention to arrest if possible the perishable traces of its course." Thereby he could "communicate to the world what is of far higher value than the most ingenious machine"— namely, "the art by which it was contrived." The relative triviality of the mechanical problems would enable a profound inquiry into the creative mind: "The comparatively small exertion of intellectual labor required in mechanical discovery" leaves "the mind at liberty to attend to the progress of its own operations."[2] Inventing the machine proved far more taxing than Babbage had envisioned. Half a lifetime of labor left no completed machines.

Babbage was the latest in a two-hundred-year line of philosopher-engineers whose hubris about the triviality of mechanical invention was bound to their desire to grasp, improve, and rationalize that process—what Babbage called Francis Bacon's "hope for the discovery of a philosophical theory of invention."[3] Making calculating machines always proved more taxing than their inventors envisioned. This book studies failed efforts to contrive calculating machines, alongside their creators' reflections on the inventive process. In the accounts of their efforts that follow, a descending bassline of philosophical hubris checked again and again over two centuries accompanies the melodies and motifs of schematic ideas and clever contrivance, of second-order reflection on the nature of inventive labor, of the coordination of mind and hand, of the labor history of invention, and of the prehistory of intellectual property and of creative genius.

Nearly all the calculating machines considered here, from those of Pascal and Leibniz to those of Stanhope and Babbage, were failures in the eyes of their makers and other beholders. No mechanical calculating machine was manufactured in more than a handful of copies before the second half of the nineteenth century.[4] Nearly two centuries after the first attempts at mechanizing calculation, contemporaries mocked the machines as costly and ineffective. "Many of these instruments," one reviewer remarked in 1832, "must be viewed as mathematical toys, and mere objects of curiosity. In the present state of numerical science, the operations of arithmetic may all be performed with greater certainty and dispatch by the common method of computing by figures, than almost by any mechanical contrivance whatsoever." Less highfalutin devices were another matter altogether: "We must except the scale and compasses, the sector, and the various modifications of the logarithmic line with sliders, all of which are valuable instruments, and may often be employed with singular advantage. The chief excellence of these instruments consists in their simplicity, the smallness of their size, and the extreme facility with which they may be used." In sharp contrast, these "qualities which do not belong to the more complicated arithmetical machines, and which . . . render the latter totally unfit for common purposes, even if they performed the operations more conveniently and more readily than they do."[5] Calculating machines did not enter into wide use, notably among actuaries and accountants, in Western Europe and the United States until the 1870s at the earliest—and not without continuing skepticism.[6]

This book studies the complicated, failure-prone, largely unused machines in the period before they became everyday commodities, from the attempts of Pascal in the 1640s through Babbage's efforts in the 1820s–1840s. Copies and counterfeits, as well as profoundly innovative and improved machines, proliferated throughout the period, though none was manufactured in any number.[7] Through the optic of these failed technical artifacts, this book peers into diverse forms of technical life—social arrangements of practitioners, legal conceptions of the ownership of work and of ideas, philosophical conceptions of knowledge and of skill. *Reckoning with Matter* brings to light the concrete processes of imagining, elaborating, testing, and building key components for calculating machines. I look at examples of inventive activity and the challenges faced by natural philosophers of coordinating material production with essential knowledgeable craftspeople. Philosophers, engineers, and craftspeople wrote about their distinctive competencies, about technical novelty, and about the best way to coordinate different sorts of technical practitioners. Their diverse written accounts authorized partitions of glory and lucre; they helped promote and attack incipient notions of "intellectual property"; they reveal central features of aesthetic and legal debates of the early-modern period. I look at conceptions of invention right up to the instauration of modern patent regimes and the solidification of the concept of Romantic genius. Considering the makers of calculating machines highlights the varied early-modern ways of understanding and rewarding creative making that drew upon earlier creations, in the moment before imitation came to be seen as antithetical to original creation rather than integral to the inventive process. These conceptions of creativity and making are often more incisive—and more honest—than those still dominating our own legal, political, and aesthetic culture. Ultimately, *Reckoning with Matter* uses these machines, with their dense accumulations of documentation, as recording instruments that track major contingencies of European early modernity, from its economic history to its visions of creative activity itself.

Carry, Numeracy, and the Debased Status of Computation

Threaded through this study is the story of efforts to mechanize a skill ubiquitous in postindustrial society: the process of carrying ones in addition. If Jane has 889 apples and Sven gives her 111, she has 1,000 apples. Using a discipline of mind and hand central to primary education, most of us perform this operation by adding 9 to 1, writing 0 below the 1, and performing a carry by putting a 1 above the 8 in the tenths place.[8] Adding the 8 and two 1s in the tenths place produces another carry, duly marked in the hundreds place. Ditto for the hundreds.

	1	$1\,1$	$1\,1\,1$	$1\,1\,1$
889	889	889	889	889
+ 111	+ 111	+ 111	+ 111	+ 111
	0	00	000	1000

The first carry itself causes a carry, which in turn causes another carry—this is called the propagation of carry. We could do this all day, until we got bored or lost attention or found a child to do it for us. The difficulty of making a mechanical—or electronic—apparatus perform this corporeal-mental operation reliably, securely, cheaply, and speedily is the warp running through all the chapters that follow. In every case, philosopher-inventors underestimated the challenge of mechanizing carry.

Relatively few people in early-modern Europe possessed this skill or equivalent ways of performing addition, such as using an abacus. In his 1673 pamphlet for his calculating machines, Samuel Morland appended a short introduction to the four operations of basic arithmetic, so "to render them plain and obvious to the meanest capacities." He included a table to explain the decimal system: "For everyplace to the left, is ten times the value of the next place to the right."[9] Numeracy was highly unevenly distributed: probably more people in the thriving commercial centers of Italy and the Netherlands could add and subtract than in France and Britain, and more people in cities than in the country.[10] The growing sophistication and density of commercial transactions, combined with the expansion of schooling, prompted more people to become capable of basic sums.[11] Men and women in seventeenth-century Western Europe needed help performing arithmetic. Printers, instrument makers, and mathematicians were happy to provide it in the form of introductory texts, facilitating instruments, and tables of all sorts. Simple multiplication, let alone the more difficult task of figuring interest, proved far beyond the skills of even the educated and commercial orders.[12] Despite their ever-greater importance in commercial transactions, navigation, and taxation, the ability to perform sophisticated arithmetic, algebra, and accounting operations was the province of a narrow stratum of commercial and governmental elites.

Reckoning belonged primarily to commercial orders and not to aristocracies of arms or aristocracies of the mind. Even natural philosophers and geometers typically viewed reckoning askance, as a mechanical activity—a mental one, to be sure—low on the totem pole of knowledge. Algebra suffered from the taint of the mechanical: René Descartes insisted that his new geometry, reworked with his form of algebra, had nothing to do with the mechanical "logistic" of the commercial classes—a claim as revealing as it was untrue.[13] Over the course of the seventeenth and eighteenth centuries, despite its commercial associations, algebra and its heir, algebraic analysis, became essential to higher mathematics. When George Berkeley attacked the foundations of calculus, he described its algorithms as an appropriate tool for a commercial mind, but no succor for a properly metaphysical one.[14]

The debased, mechanical reputation of computation contributed greatly to the iconoclastic edge of Thomas Hobbes's notorious reduction of thinking to reckoning: "By Ratiocination, I mean computation. Now to compute is either to collect the sum of many things that are added together, or to know what remains when one thing is taken out of another. Ratiocination, therefore, is the same with addition and subtraction."[15] The most anachronistic recasting of this claim as an intimation of artificial intelligence correctly captures the fact that Hobbes was seeking to make the mental more mechanical. Almost none of his contemporaries followed Hobbes, except for Leibniz, and then only in a modified form. Even the most stalwart advocates of syllogistic logic saw little to recommend in Hobbes: whatever logical reasoning was, and no matter how formal, it was something greater than the debased mechanical

procedures of commercial arithmetic. With a few exceptions, the philosophical love affair with symbolic reasoning had to wait a few centuries to bloom.[16]

In recent years, many philosophers and philosophically inclined scientists
have speculated about the possibility of reducing all mental operations to
mechanical computation. The same penchant should not be projected into
the early-modern period. Given that computation was accorded such a lowly
status among mental activities, the mechanization of reckoning, however impressive, was rarely seen as powerful evidence that the mind could be reduced
to mechanism. For mechanical computation to threaten an incorporeal mind
seriously, something like Hobbes's reduction of thought to calculation was
required. Few in the seventeenth century accepted such a reduction—fewer still
in the eighteenth century, despite the growth of materialism and algebraic
analysis alike. Like our contemporaries who seek to demonstrate domains
of human competence not replicable by an algorithmic machine, most early
moderns—including the most godless materialists—saw thinking as something irreducible to calculation.

Historical Virtues of Failure

Mechanizing the carrying of tens was a profound technical challenge. The
perceived, and very real, failure to create machines that successfully mechanized carry is crucial to my history. Bespoke machines were extremely hard
to build—skilled labor was difficult to find; materials were nonstandardized.
Manufacturing the machines in any large number proved even more difficult.
No one succeeded in commercializing a mechanical calculating machine
until after 1850, and only by convincing potential adopters to overlook the
problems with the machines. The process of attempting to make calculating
machines tempered philosophical hubris in revealing ways. The challenges
of building machines make visible the array of technical practices, precision
instruments, managerial skills, interpersonal competencies, and legal structures
required for the production of calculating machines as commodities in the
marketplace. The failure of machines to carry robustly rests upon the failure
of inventors to produce communities capable of bringing machines into being. Among the chief culprits in the failure to reduce machines to practice—to
turn them into readily manufactured goods—was the inability of philosopher-
inventors to coordinate skilled artisans and to cooperate with them.[17] Without strong ties to artisans and craftspeople, philosophically trained inventors
lacked stable factual knowledge about materials and competencies in working
those materials. Without sufficiently robust social relations, they lacked adequate knowledge, propositional and factive, about the qualities of the materials

they needed. Materiality was social: to be undersocialized often led to designs not being materialized.[18] To be sure, there were limitations in the precision machinery available; yet those limitations were as much a consequence of the failure to coordinate as of technology itself.

Innovation, Collective Invention, and Hylomorphism

As in Babbage's era, concerns about invention preoccupy, rightly or wrongly, our own debates over the proper organization of our political economy of imperial competition. Such debates suffuse current popular and academic discussion of industrial competition, intellectual property, and educational systems.[19] Who is innovating, and what provides people with incentives to undertake such risks? How can we organize formal and vocational education to encourage innovation? How does one stay competitive? Do patents and copyrights help or hinder technological developments? For several genera-tions, academic historians have stressed the dangers of attending primarily to innovations and inventions when looking at technologies. Such a focus down-plays the omnipresent technologies, big and small, that structure everyday life. To concentrate too narrowly on "innovation" tends to mask the collective nature of most inventions and the evolutionary quality of most technologi-cal work. Heroic narratives of individual inventors, easy to find in the popular press, rightly do not dominate recent academic historical literature, Thomas Edison and Elmer Sperry being justified exceptions. The gulf between our common modern conception of technology as invention and its more every-day and collective reality is itself a fact of some importance to the development of our technological society: the "innovation-centered view of technology, of science, and of knowledge is deeply institutionalized" in modern culture.[20] However problematic it is as a generalization about the nature of most tech-nology in history, the innovation-centered view has been built into a range of institutions that seek to promote creativity, codified into legal systems protect-ing intellectual property, and rationalized within accounts of genius central to aesthetics and the everyday working of markets in art, literature, and the sciences. A vision of technology centered on radical innovation cannot be taken either as simply superstructural or as false consciousness. It matters, if for no other reason, because most people think it does.

To concentrate narrowly on invention leads one to treat individual inven-tors as uprooted, both from the social conditions of production and from the evolutionary development from the antecedent technical solutions used in any new contrivance or technical system.[21] In recent years, in contrast, the concern with innovation has become coupled, in scholarly and popular discussion, with a deep sense of the collective and cumulative quality of cultural production—something absent from much popular discussion of technology until the recent stark reemergence of attention to the salience of collective authorship.[22] Among programmers, savvy users, hobbyist makers, and legal activists alike, the social and collective qualities of the making of things and of knowledge have come to the fore. The awareness of the power of collective making structures new forms of collaboration, as in open-source programming, and has resulted in efforts to transform legal protections better to encourage innovation through reuse or remixing, much contested by "legacy" stakeholders, to celebrate rather than to denigrate collective work.[23] The craftsmanship of computer programming, in particular, has helped focus scholarly and nonacademic attention on the social conditions of innovation and on the central place of imitation within innovation. Collective invention is not new, but public awareness has been heightened as of late.[24] In numerous popular business books, in the interventions of Lawrence Lessig, and in the novels of Cory Doctorow alike, imitation is seen no longer as antithetical to innovation but as integral to it—music to the ears of the social historians of technology, if not to legacy intellectual property partisans.

Following the making of calculating machines, along with the accompanying debates about how best to make them, allows us to glimpse the growth, questioning, and annealing of a fundamental divide in the history of technology: the division between a determinist, cumulative model and a heroic, highly individualistic model.[25] Built upon an evolutionary and analogical account, the determinist model downplays the importance of individual creativity or a need for particular genius. The history it frames is a contextual one of gradual change, not of inspired, punctuated leaps. The heroic model insists upon the need for superior minds to break the deadening hand of authority and precedent. The "linear model" of technology, where science drives technology, is one variant of this model; modish Silicon Valley speak of "disruption" is another.

Reckoning with Matter eschews these two poles of writing the history of technology: the collective, determinist account of inventive activity and the individualist, heroic, creative account. The history of calculating machines fits poorly into the top-down understanding of technical innovation as driven by science-informed savants liberated from the routines of artisanal practice; the history fits no better within a bottom-up conception of innovation as primarily an analogical process drawing upon existing technologies. Both determinist and heroic models of technical change, especially once given legal and philosophical frameworks, are important for grasping the debates around technical innovation then and now.

Throughout the early-modern period, a third way of understanding cultural production was ubiquitous. This widely disseminated conception included no stark division between imitation and absolute originality, no bifurcation between social and individual making, and no absolute split between design and production. The refusal of these divisions has long been the dominant self-understanding of most actual craftsmanship, if rarely of formal aesthetic reflection.[26] The ancient category of *techne*, or skilled making, captured the braiding of imitation and originality, the social and the individual, and design and production. Inventors were not yet geniuses detached from social and intellectual context.

From 1600 until the early 1800s, Western European philosophy, aesthetics, and legal theory underwent a fundamental transformation from this vision of *techne* toward a bifurcation of design, understood as the production of ideas, from implementation, understood as the process of materializing those ideas. Calculating machines did not cause this shift. Early-modern calculating machines, with their rich attendant documentation, bring into sharp relief key aspects of the contingent development of this shift, in practices of making, in legal protection, and in philosophical accounts of creation, imitation, and fabrication.

Mindful Hands and Handy Minds

Artisans and craftspeople were integral to the process of making calculating machines: the pages that follow celebrate and document the contributions of named artisans—Ollivier, Joseph Clement, Philippe Vayringe—and anonymous makers alike. I do not, however, claim "authorship" for their contributions retrospectively. Authorship is the wrong category, for it involves a division between ideation and materialization that precludes an appreciation of the interplay of mind, hand, and material in making.[27] Attributing any category of property primarily understood as intellective to early-modern artisans is to do a singular injustice, for it requires a too narrowly mentalistic understanding of creation that systematically denies the substantiality of their work. This refusal of authorship is meant not to deprive craftspeople of their ideas but to deny the possibility of a facile separation of those ideas from concrete making. To recover authorship is to choose the legal and aesthetic framework that excludes a lion's share of the contributions of craftsmanship. Historically, efforts to draw on the mentalistic categories of philosophers worked against, not for, makers.

The recovery attempted here is not one animated by nostalgia for an artisanal culture we have lost but rather one that seeks to illuminate the extraordinary dynamism of an early-modern artisanal culture looking to integrate theoretical knowledge and new tools on its own terms. This dynamism runs parallel to calls for natural philosophy to become more involved, more connected, with artisanal knowledge of materials.[28] Neither of these pictures reveals a monotonic course of history toward deskilling and top-down design, toward a linear model of technological development. To view these dynamisms is to grasp alternative futures of technological and scientific development.

Artisans did not just possess tacit knowledge and corporeal skill while philosophers had theory and propositional knowledge. Considerable recent scholarship using the conceit of the "mindful hand" serves to discard the wonted anachronistic divide between "science" and "technologies" in favor of studying the multitude of sites and networks where knowledge and thing-making came together. Such studies illustrate the diverse ways of combining more formal knowledge, craft knowledge (traditional as well as new), and practices of production.[29] Such efforts at integration were challenging and prone to failure. *Reckoning with Matter* stresses the difficulties of minding the hand as well as manualizing the mind: difficulties of skill, social organization, and reward systems, as well as natural philosophy.

The book tracks one genealogy of the gulf between theory and practice later to be reified by the new categories of science and technology. Blaise Pascal and Gottfried Leibniz regarded themselves as engaged in minding the hand and manualizing the mind. Both philosophers, in the end, were rather poor at both; the less famous figures of Samuel Morland, Philipp Hahn, and Charles Stanhope were, on the contrary, quite good at them. Babbage's early hubris gave way to a studied production of great skill, uniting the design of practical contrivances with profound theoretical inspiration.

Many early-modern natural philosophers dreamed of perfecting a divide between theory-inspired design and mechanical implementation, between formally knowledgeable philosophers and mechanics. In such a conception of making, which Tim Ingold calls "hylomorphic," ideas precede materialization, as form precedes matter.[30] Such a division of design from production had little reality in the world of actually making artifacts. The division could only materialize through new practical tools that made hylomorphism a real possibility as a strategy for design and for the social organization of production. Insofar as that division was partially realized in the nineteenth century, it was a result of the tools produced through a culture committed to the coordination of theory and practice, of scientific practitioners, artisans, and engineers—a culture of "emulation." Emulating craftspeople produced the

machines necessary for precision production; emulating craftspeople likewise produced the means of formal drawing that helped stabilize a mentalistic
conception of invention possible within modern legal systems. Alongside the
radical reorganization of labor, these parallel technologies made possible the
realization of the hylomorphic conception of invention, the process of deskilling, and the creation of a modern productive order. It authorized a writing
of two sorts of histories: those that presuppose the hylomorphic model and
those that reject the salience of ideas outright. Neither suffices.

Competing over Expertise: The Persona of the Inventor-Philosopher

In defending his calculating machine in the 1640s, Blaise Pascal denounced two
parties that ignorantly slandered his machine: mere artisans, lacking theory,
and mere philosophers, lacking actual knowledge of material nature. To bolster his machine, he proffered himself as a new sort of philosopher-inventor-
engineer, capable of uniting theoretical knowledge to practical knowledge of
the material world and those who could work it—a new sort of persona important to the French crown and its often depleted treasury.[31] The "new philosophers" of the seventeenth century, from Francis Bacon and René Descartes
to Robert Boyle and Gottfried Leibniz, all claimed that the superiority of their
accounts of nature rested upon their deeper involvement in knowing matter
and their competencies in transforming it.[32] Among themselves, natural philosophers debated the relationship of theory to craft knowledge and the levels
of certainty possible to human beings; they differed on the proper form of engagement needed to gain knowledge of the material and disputed the salience
of a spiritual or mental realm; they differed on the importance of mathematics
to knowledge of nature. If they came to no agreement on how minded the
hand or handy the mind ought to be, they collectively underscored the need
for a new convergence of theory and skill. These different configurations of the
handiness of mind belong to different sorts of natural philosophical personae.

Many natural philosophers offered themselves, along with their new social forms, academies, and scientific societies, as new and superior experts
necessary for princely magnificence and material progress alike.[33] Few succeeded in winning over monarchs or representative bodies in a sustained
way. No consensus on the technical expertise essential to the state emerged,
despite the rhetoric of the new academies of sciences. Government officials
knew not to whom they ought to turn; their skepticism toward theoretically
oriented natural philosophers is well known (and entirely justified). Much of
the history of calculating machines involves philosopher-inventors attempting

to validate their variant of the technical persona through the material production of a challenging machine. Calculating machines served, for a number of them, as a material proxy to motivate a choice of technical expert. The difficulty of producing the machines allowed them to be used as means for authenticating experts. This gambit succeeded on occasion: the effort lets us glimpse the disorder of—and ignorance around—the sources of technical expertise from 1600 until the mid-1840s.

Calculating Machines and Modern Computers

Drawing a genealogy from modern digital electronic computers directly to early-modern mechanical calculating machines would be a wildly anachronistic, post-hoc effort to valorize these little-used machines, mostly deemed failures. Some years ago, Michael Mahoney signaled the dangers of creating a teleological history of the general-purpose stored program computer: "The dual nature of the computer is revealed in its dual origins: hardware in the sequence of devices that stretches from the Pascaline to ENIAC, software in the series of investigations that reaches from Leibniz's combinatorics to Turing's abstract machines." Nothing organically connected these histories until the 1930s: "They belong to different histories, the electronic calculator to the history of technology, the logic machine to the history of mathematics and they can be unfolded separately without significant loss of fullness or texture. Though they come together in the computer, they do not unite. The computer remains an amalgam of technological device and mathematical concept, which retain separate identities despite their influence on one another."[34]

The history of mechanical calculation and the history of logic both figure in *Reckoning with Matter*. In his work, so essential for unscrambling the plural histories of the computing, Mahoney placed too great a caesura between mechanisms and mathematical logic. The two histories of calculating machines and logic intersected time and again—not of necessity, but contingently. Not until the mid-twentieth century, however, did they come together robustly, as a way of transforming automatic calculating engines into universal machines. Focusing on the genealogy of mechanical and electromechanical calculation up through the first electronic digital computers upsets the apparent quantum jump in the novelty of the early stored-program machines. Studying the challenge of creating stable and secure mechanisms for carrying tens mitigates against an idealist history of computing, one in which Turing's concept of the general-purpose computer is understood as the wellspring of modern computation.[35]

Roadmap to Computation

The chapters in *Reckoning with Matter* move from concrete stories of labor, technique, and wages, to industrial espionage and intellectual property, and, ultimately, to philosophical speculation; in form and content, they weave together the history of technology with the history of philosophy without sacrificing the autonomy of either. In covering most of the major known machines from 1600 to 1830, the book offers no catalog; while looking closely at the machines themselves, with a particular focus on carry mechanisms, the book does not describe the machines in much technical detail, easily found elsewhere. Offering prehistories of computing machines, of deskilling, of intellectual property, and of artificial intelligence, the chapters follow a roughly chronological path, punctuated by five forays into the nineteenth century.

Set in the seventeenth century, the first two chapters study the often-frustrated efforts of Blaise Pascal, Gottfried Wilhelm Leibniz, and Samuel Morland to create calculating machines, with a focus on their interactions with skilled artisans. Turning to the prehistory of intellectual property, the third chapter examines these three efforts to create calculating machines within early-modern systems for protecting and encouraging manufactures and, indirectly, invention. Moving into the eighteenth century, the fourth chapter documents the effervescent "emulation" of calculating machines across Europe in the wake of Leibniz's infamous failure to produce a well-functioning model. Focusing on a remarkably well-documented Enlightenment example, the fifth chapter tracks the creation of a series of calculating machines by the English nobleman Charles Mahon, the third Earl Stanhope. Stanhope's materialized design practice emerged from late eighteenth-century ways of forming materials, of coordinating different practitioners, of representing forms and matter. Shifting to intellectual history, the final chapter looks at the significance—and, often, the insignificance—of calculating machines in eighteenth-century reflection on the nature of thinking, mathematics, and original creation. Five times the narrative jumps chronologically to a "carry" section that recounts the story of Babbage's Difference Engine through the thematic lens of the preceding chapter. A sixth and final carry concludes with the commercial success and scientific importance of calculating machines in the late nineteenth and twentieth centuries.

Carrying Tens: Pascal, Morland, and the Challenge of Machine Calculation

In a letter of September 1646, Pierre Petit, engineer to the king, wrote a short history of techniques for facilitating arithmetical procedures. The most important development since logarithms was Blaise Pascal's arithmetical machine, a "piece invented truly with as much good fortune and speculation as its author has wit and knowledge" (see figure 1.1). Petit tempered his praise: "Since it comprises a number of wheels, springs and movement, and since it requires the head and hands of a good watchmaker to understand it and put it into practice, as well as the skill and knowledge of a good Arithmetician to be able to use it, I fear that its use will never become common." He accurately predicted that the machine would end up "in cabinets and great libraries," not in government bureaus and trading houses as Pascal had hoped.[1] Contemporaries echoed Petit's concerns. One collector of gossip noted that only one artisan in Rouen could build the machine, and only with Pascal present.[2] The projector and sometime fugitive Sir Balthasar Gerbier concluded his 1648 description of Pascal's machine by noting that "a man must first be exact jn Arithmeticke before he can make use of this Instrument, which cost 50 pistols." He deemed the machine "a Rare Invention farre saught, and deare baught: putt them jn the Storre house was the Prince of Orange wont to saye and lett us proceede on the ordinary readdy way."[3]

In setting forth the impracticalities of Pascal's machine, Petit stressed its dependence on two forms of skill: that of the arithmetician and that of the watchmaker. The machine did not obviate the skills involved in doing arithmetic. Making something as complex as a calculating machine capable of performing addition and subtraction depended likewise on superior artisanal skills. The design of the machine was not sufficiently simple to allow ordinary artisans to produce it easily in a merely imitative, or a fairly machinelike,

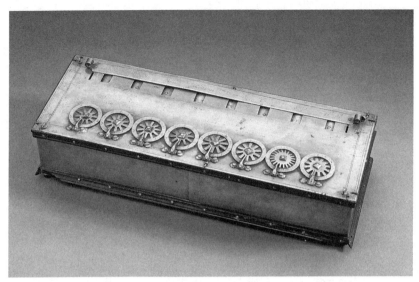

FIGURE 1.1. Machine arithmétique de Pascal à chiffres plus sous et deniers, 1645. Inventory no. 19600-0001. Closed view. (© Musée des arts et métiers-CNAM, Paris. Photo: J-C Wetzel.)

fashion—what David Pye calls the "craftsmanship of certainty." Its production, like its devising, required creativity and improvisation—the "craftsmanship of risk."[4] The superior skills and creative problem solving necessary to build Pascal's machine meant that ordinary watchmakers or other artisans could not produce the machines in any standardized fashion. As a consequence, Pascal's machines were too dear for his envisioned markets of merchants, government tax officials, and architects.[5]

Using the examples of the calculating machines of Blaise Pascal and Sir Samuel Morland from the seventeenth century, this chapter investigates the skills necessary to calculate and the skills necessary to design and build calculating machines. I intertwine a discussion of the major technical challenges involved in producing calculating machines with evidence about the nature of the contributions of the artisans involved in producing those machines. After analyzing Samuel Morland's detailed description of the work involved in producing his multiplying instrument, I look at Pascal's polemical portrayal of artisanal knowledge and skill. In Pascal's account, his mastery of arithmetic and of designing machines supplanted the need for such skills in users and the artisans producing the machines. Pascal and Morland treated the subject of the artisanal production of their machines very differently. In writing to patrons, Morland celebrated the world-class artisans who constructed his machines: he gave their names and detailed their labors. Pascal left his artisans nameless and depicted their labor as skillful but imitative. When advertising his machine, Morland directed prospective buyers to "Mr. Humphry Adamson" who was "the onely Workman" who could produce the machine "with that exactness that is absolutely necessary for such Operations."[6] In contrast, Pascal advised "the curious" to visit Gilles Personne de Roberval, professor of mathematics at the Collège de France, at his lodgings.[7] Morland brought together his buyers and his producers; Pascal insulated his buyers from his artisans.[8]

The history of calculating machines reveals how often such tools have been mistakenly understood to be proxies, things capable of substituting for human beings; that history likewise shows how often genuine proxies, skilled artisans, have been mistakenly understood to be tools. By misattributing human activity, it is easy to disregard the skills necessary to use calculators and computers, just as it is easy to disregard the skills necessary to produce them.[9] Such misattributions ease the dividing of intellectual from manual work and help justify the social hierarchies attendant upon that division.[10]

The study of artisanal knowledge, tacit knowledge, and skill has been central to the history and philosophy of science and technology for many years.[11] In this chapter and the next, I use the empirical case of early-modern calculating machines to clarify the distinct sorts of knowledge captured by those terms and to show the richness implicit in that array of meanings.[12] The artisanal knowledge and skills involved in making calculating machines include the following:

1. Propositional knowledge, gained through long-term experience working with materials. Sometimes such knowledge is articulated, though rarely in the formal language of elite natural philosophers. Such knowledge need not be consciously cognized in linguistic terms at all. Examples might include properties like the ductility and malleability of different metals or the springiness of springs.
2. Discernment, or the acuity of senses in making judgments about perceptions, such as gauging the size of barrels or the quality of cheese or wine or metals.[13]
3. Dexterity in doing work with hands, gained after long-term experience with materials, proceeding often without conscious mental reflection but with profound manual perspicacity.[14]
4. Knowledge of the social world where other artisanal knowledge and skills can be found—essential knowledge in an age of deep imperfection of information.

In practice, the first three sorts of artisanal knowledge are often tightly bound together in the workshops of today as well as in the early-modern past. Dexterity in metalwork, for example, constantly involves using a powerful discernment about the qualities of various metals. Like many of their recent counterparts, numerous early-modern philosophers worked hard to deprecate the independent importance, if not existence, of artisanal knowledge.[15]

Itemized Tasks

On the second of April 1669, the English inventor Samuel Morland wrote to Charles Stuart, Duke of Lennox, with a plea. Morland hoped soon to have "an Employm!" of the "Instrument for Accompts & Numbers" he had loaned the duke. He asked the duke "either to let mee haue it again" or "to let mee have so much Money for it as will pay for ye making of another."[16] While we do not know whether the duke returned the calculating machine, a few years later Morland sent a similar machine, made in the same period, to a more important duke, Cosimo III of Florence (see figure 1.2).[17]

Morland's multiplication machine is a variation of Napier's bones given a circular form and a degree of automation. Like the bones, the instrument allows one to perform multiplication by reducing it to a series of additions of single-digit numbers.[18] Skill in using Napier's bones to aid in multiplication was

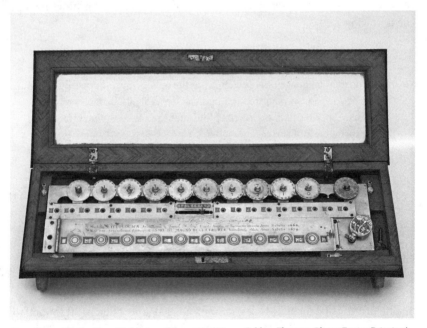

FIGURE 1.2. Morland's multiplying machine, 1666. (Museo Galileo–Florence. Photo: Franca Principe.)

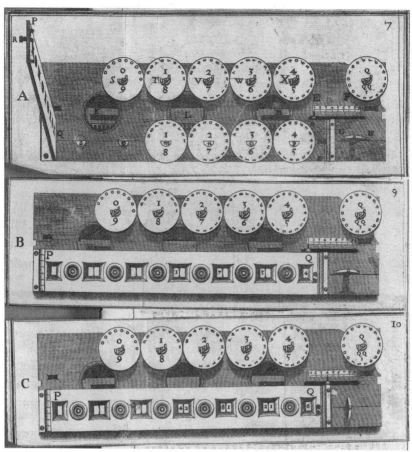

FIGURE 1.3. Morland's multiplying machine, from Samuel Morland, *The Description and Use of Two Arithmetick Instruments* (London, 1673), plates A7, B9, C10. *A*, open; *B*, preparing to multiply; *C*, multiplication by 4. (Courtesy of Houghton Library, Harvard University, Hyde 90W-163.)

common among mathematical practitioners of Morland's time. Morland's machine automates this process.

The machine, shown in figure 1.3A, comprises a large number of disks engraved with numbers (S, T, V, W, X), posts to store and use the disks, a platform with a key (GH) to drive the internal mechanism, a pointer and a numbered line (EF), and a gate (AP) with viewing holes, shown open in figure 1.3A and closed in figures 1.3B–C. To use the machine to multiply, say, 1734 by 4, the user selects the disks for 1, 7, 3, and 4. Moving left to right, one places one disk on the first, one on the second, another on the third, and a last disk on the fourth pinion on the machine (labeled p, o, n, m). The gate is then closed; only the digits 1, 7, 3, and 4 are visible, as in figure 1.3B. The user then turns GH, moving the pointer along EF to whatever number he or she wishes to multi-ply by—in this case, 4 (see figure 1.3C). In the windows, a series of numbers are now visible: 42, then 81, then 21, then 6. The numbers in each window need to be added to produce the final result: 4 + 2 = 6, 8 + 1 = 9, 2 + 1 = 3, 6 = 6, so 1734 × 4 = 6936. Multiplying a number with more than a single digit, such as 44, requires the user to repeat the entire process and then add the two results obtained. To aid in this, Morland paired this multiplying machine with an add-ing machine to record the results.

Morland's asking price for the luxurious machine was a princely £67 7s. 6d.; in comparison, the sum total of yearly wages for all the Duke of Len-nox's servants at his residences in London and Cobham came to £889.[19] Lest the duke question this steep asking price, Morland included a breakdown of costs. This unusual document lists the stages and types of work involved in making the machine as well as the names of the three artisans involved.[20] The cost breakdown Morland supplied to the Duke of Lennox reveals the variety of tasks and the materials involved in producing such a machine in a proto-industrial age.

A Note of what ye Great Arithmet: Instrument will cost.

1. The Multiplying Part

The 55 sylver plates came by weight before they were filed &c	6-00-00
To Mr Blondeau for twice cutting & planishing, & making small centers	2-15-00
To Mr Sutton for piercing, diuiding, grauing, waxing, & polishing them at 3–6 per plate	9-12-06
For plating ye Sylver, dividing & graving ye tabulating plates	1-10-00
Grauing and gilding y^e title plate	1-00-00
Mr Fromantle demands for making up y^e Multipluing-part & finding brass, & gilding as that is	25-00-00

[2.] The Adding part

The Adding part will cost	20-00-0
The Case	1-10-0
sum	67-7-6[21]

FIGURE 1.4. Morland's multiplying machine, detail, 1666. (Museo Galileo–Florence. Photo: Franca Principe.)

In listing his expenses, Morland explained that he would not charge for the "great deal of trouble I shall have to follow & direct workmen." The duke was indeed getting a good deal. Morland hired the best artisans in England in their respective trades, among the best in all of Europe.

Morland's machine required a large number of disks; the exemplar in Florence has fifty-five silver disks and seventeen silvered brass disks (see figure 1.4). The cost breakdown details many of the stages required to convert silver into the numbered disks necessary for the machine. To create the general form of each disk, the first artisan, Blondeau, had to cut each disk out of metal, "planish" or flatten it, and then put a center in it. Among the finest makers of coins in early-modern Europe, using the most recent technical innovations, Pierre Blondeau had received the protection of Richelieu before being brought to Cromwell's Commonwealth in 1649 to help with the English coinage.[22] An associate of Samuel Hartlib and his circle, he published a series of works defending the use of new mechanized methods for minting coins: "Monie coined with the hammer," he explained in 1653, "cannot bee made exactly round, nor equal in weight and bigness, and is often grossly marked."[23] He probably flattened Morland's disks in a screw press, a mechanized technique that he had used to revolutionize coin making in Britain.

The second artisan, Sutton, next performed a large number of steps, at 3s. 6d. per plate: he had carefully to divide each disk into precise angles and then engrave the numbers at appropriate angles on each disk. The premier engraver of mathematical diagrams in England, Henry Sutton was a maker of especially fine metal and paper mathematical instruments.[24] His death caused considerable concern among mathematicians, as there appeared to be no one else in England so skilled in the extremely precise engraving necessary for mathematical diagrams and instruments.[25]

The third artisan, Fromantle, built the internal mechanism, a rack-and-pinion device where the key rotates the disks a precise number of degrees and moves the pointer to the appropriate number. The Dutchman John Fromantle came to England to make pendulum clocks after Christiaan Huygens's design.[26] He and his family were famous for their fine clocks, which immediately became collectors' items. Morland hired Fromantle to manufacture other new inventions. He produced a "waywiser," or odometer, after Morland's design for the Duke of Lennox's carriage.[27]

Each of Morland's artisans was a famous innovator in his craft; two of them were expatriates who had brought needed skills and innovations to England.[28] They were not "simple" artisans capable only of reproducing extant works using repetitive techniques gained through habitual activity. The cost breakdown gives no sense of how fully Morland had specified his machine or what discretion he left to his innovative artisans to work out the details of their parts of the machine. In the chapters that follow, we will see the wide latitude early-modern inventors gave their artisans to fill out and often determine the technical details of their calculating machines. A clever inventor of all sorts of devices, Morland knew how to select artisans who would help bring his works to fruition. His numerous projects over the course of his life demonstrate that he understood the utility of coordinating the skills of others in order to refine and develop his designs. Morland recognized his limits and knew to ask for help (albeit sometimes too late). He was a savvy user of the expertise of others: he drew on the skills of the mathematician John Pell when he needed advanced mathematical help;[29] he drew on the skills of the engine maker Isaac Thompson when he needed to turn his model of a steam engine into a workable product; late in life, he solicited the help of the future archbishop of Canterbury to vet his religious writing.[30] He certainly took advantage of the skills of Blondeau, Sutton, and Fromantle in designing and implementing his multiplying machine. Morland's knowledge of the social world of skilled workmen made the materialization of his machine possible. Morland was, to be sure, the machine's inventor—the one who first envisioned it in general terms and brought together the knowledge needed to design and materialize it. All three of his skilled craftsman, however, shared in creating the device in an implemented form.

Automating Carry: From Morland to Pascal

Morland's multiplication machine was paired with a machine to aid in addition (see figure 1.5). The latter's lower asking price of twenty pounds reflects its relative simplicity; it is a pocket-sized device, of which several examples still exist. Unlike the more ambitious machines of Wilhelm Schickard, Pascal, and Leibniz, Morland's instrument cannot automatically perform carries.[31] Leibniz remarked that Morland seemed "to have wanted to avoid" performing carries "in order to have nothing to do with teethed wheels," as he was "too skillful" to miss seeing how it could be done. Since machines like Morland's did not mechanically perform carries, Leibniz reckoned, they were little more than curiosities. Since Pascal's machines could not perform carries well, they too were amusements, not useful tools: "Addition and subtraction hardly become easier with such machines than they are with a pen: these Machines are more for curiosity than for real use. This does not prevent them from being charming."[32] Making a machine capable of robustly performing carries was no small task. To see why, we need to analyze the steps involved in carrying in order to isolate the skilled human behaviors the machines were intended to facilitate or to replicate.

The competencies involved in doing arithmetic using arabic numerals remained rare in late seventeenth-century England, France, and Germany, if probably more widely distributed in Italy and the Netherlands.[33] In his best-selling *Mathematical Compendium*, the noted pedagogue and engineer Sir Jonas Moore offered only a feeble explanation of addition and subtraction, before remarking, "If any Gentleman, especially Ladies, that desire to look into their disbursements, or layings out, and yet have not time to practice in numbers, they may from Mr. *Humphrey Adamson* . . . , have those incomparable Instruments, that will shew them to play Addition and Subtraction in l. s. d. [pounds, shillings, and pence] and Whole Numbers, without Pen, Ink, or help of Memory; which were the invention of that worthy Person, and Ornament of his country, Sir *Samuel Morland* Baronet."[34] Rather than explaining the basic rules and skills for arithmetic, Moore's mathematical compendium simply sent readers to purchase Morland's machines.[35]

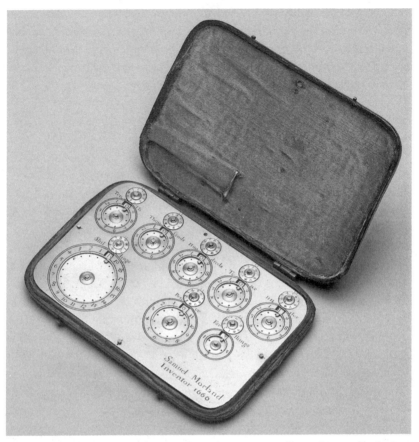

FIGURE 1.5. Morland's adding machine, 1666. Inventory no. 1876-0538. (Science Museum, London.)

Morland's booklet of 1673 advertising his machines doubled as an arithmetical primer: "For the better understanding of the Arithemetical *Instruments*, I shall endeavor so to explain and demonstrate the reason of the Operations of *Addition, Subtraction, Multiplication, Division*, and Extraction of the Square, and Cube-Roots, as to render them plain and obvious to the meanest capacities."[36] He worked from the assumption that his reader had no knowledge of either roman or arabic numbers. Having explained the numerals, he offered the following "Precept for ADDITION of Integers in Plain Numbers": "Having placed the Unites of the respective Progressions in Ranks and Files; then begin and add together the Unites of the right-hand-File, setting down the sum underneath, if it be under 10. but if it is just ten, set down 0. and carry 1. to the next place; and if above 10. set down the excess in the first place, for every 10. an Unite."[37] In this rather unclear manner, Morland specified the general rules for carrying numbers in ordinary addition. If two digits add up to less than ten, write the sum in the first column; if they add up to ten, write zero in the first column and carry a one to the next digit; if they add to more than ten, write the amount greater than ten in the first column and carry a one to the next digit. Morland did not specify or illustrate how to keep track of numbers carried.

The primary function of Morland's machine for addition was to automate part of these rules: the machine keeps track of the number to be carried over to the next digit. The machine replaces the practice of keeping track of carries in memory or on paper. The convention today includes a manual practice of marking a one above the tens column, a reminder to add a ten. Consider a straightforward example of addition:

$$\begin{array}{r} 136 \\ +\ \ 75 \\ \hline \end{array}$$

Since $6 + 5 = 11$, a 10 has to be carried to the next column:

$$\begin{array}{r} 1\ \ \ \\ 136 \\ +\ \ 75 \\ \hline 1 \end{array}$$

$1 + 3 + 7 = 11$, so a second 10 has to be carried in a similar manner:

$$\begin{array}{r} 1\ 1\ \ \\ 136 \\ +\ \ 75 \\ \hline 211 \end{array}$$

Note that the skills involved in performing addition are learned through something like drill; developing such skills means replacing the conscious use of rules with an unconscious, nearly automatic following of them. Morland's adding machine automates the habitual registering of accumulated carries, whether in memory or by noting them above the next digit. The machine does not perform the addition of the carried quantity in the next column automatically, so it replaces only one part of the practice of doing addition.

Let's see how a user would perform the same arithmetic problem on Morland's machine. The user of the machine first dials 136 into the hundreds, tens, and ones columns using a stylus:

Main addition dials   0 0 1 3 6

To add the 75, the user first turns the ones dial on the right five notches. Each time the entry dial reaches ten, a tooth connected to the dial advances the smaller carry dial just above the main digit dial. So we get the following:

Small carry dials   1
Main addition dials   0 0 1 3 1

The carry is automatically recorded but not automatically added to the tens column. To add the 7, the user turns the tens dial seven notches; the small carry dial above the tens place automatically advances one. So the state of the machine after adding the 5 and then the 7 is

Small carry dials   1 1
Main addition dials   0 0 1 0 1

To get the final result, the user, starting from the right, turns the tens dial the amount registered on the ones carry, so the zero becomes 1, and then turns the hundreds dial the amount shown on the tens carry. The result is

Small carry dials   0 0
Main addition dials   0 0 2 1 1

In adding a long series of numbers, these carried amounts will often be greater than ten; Morland advertised a machine with two sets of carry dials, which would eliminate the need to stop to perform carries partway through a long series of additions.

Using Morland's machine may seem no easier than the familiar method with pen and paper. Morland's device was advertised as most useful for adding long series of figures, as in accounting, where the accumulation of many carries becomes unwieldy and subject to error. The machine allows one to replace many of the skills involved in pen-and-paper arithmetic with skills involved using the carrying dials.

Pascal's goals for his machine some twenty years earlier were more ambitious.[38] In a pamphlet announcing and defending his calculating machine, Pascal set out the problems inherent in the practice of ordinary arithmetic with pen and paper: with a pen, "one is at every moment required to carry or to borrow the necessary numbers, and how many errors slip into these carries and borrowings, unless one has a very long habit [in performing these operations] and, moreover, a profound attention that tires the mind very quickly."[39] More than simply aiding the memory during the reasoning process, Pascal's machine promised to perform all the carries involved in addition itself and thereby to relieve memory and mind. It "makes up for the shortcomings of memory, and without carrying and borrowing, it does itself whatever he desires, without requiring him to think."[40] The user would need only to input the digits to be added or subtracted and would not need to worry at all about keeping track of the carried or borrowed numbers, whether by pen or through auxiliary dials.

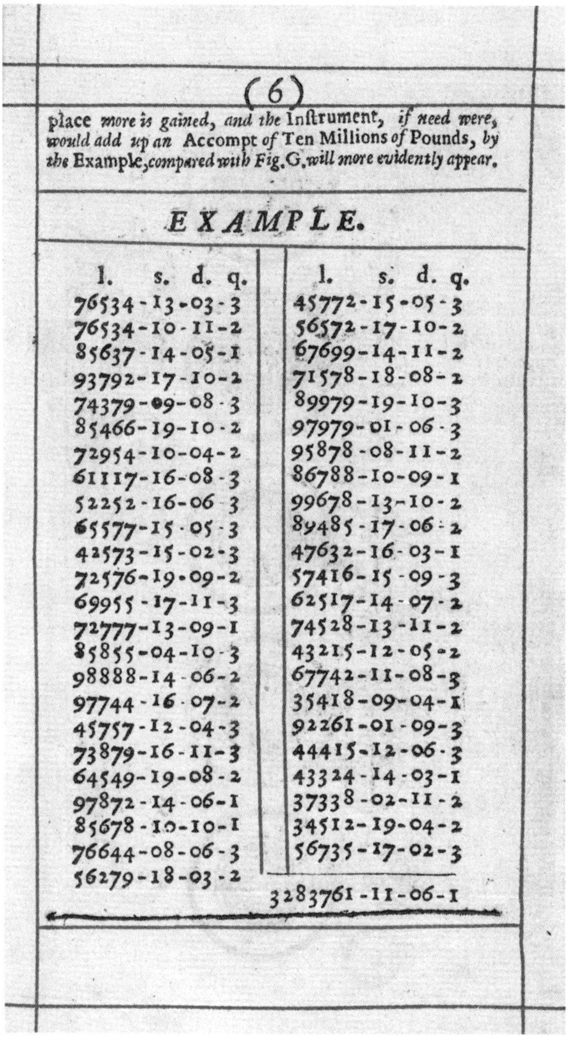

(6)

place *more is gained, and the* Inſtrument, *if need were, would add up an* Accompt *of* Ten Millions *of* Pounds, *by the* Example, *compared with* Fig.G, *will more evidently appear.*

EXAMPLE.

l.	s.	d.	q.		l.	s.	d.	q.
76534	13	03	3		45772	15	05	3
76534	10	11	2		56572	17	10	2
85637	14	05	1		67699	14	11	2
93792	17	10	2		71578	18	08	2
74379	09	08	3		89979	19	10	3
85466	19	10	2		97979	01	06	3
72954	10	04	2		95878	08	11	2
61117	16	08	3		86788	10	09	1
52252	16	06	3		99678	13	10	2
65577	15	05	3		89485	17	06	2
42573	15	02	3		47632	16	03	1
72576	19	09	2		57416	15	09	3
69955	17	11	3		62517	14	07	2
72777	13	09	1		74528	13	11	2
85855	04	10	3		43215	12	05	2
98888	14	06	2		67742	11	08	3
97744	16	07	2		35418	09	04	1
45757	12	04	3		92261	01	09	3
73879	16	11	3		44415	12	06	3
64549	19	08	2		43324	14	03	1
97872	14	06	1		37338	02	11	2
85678	10	10	1		34512	19	04	2
76644	08	06	3		56735	17	02	3
56279	18	03	2					
					3283761	11	06	1

FIGURE 1.6. Example of a long series of additions, from Samuel Morland, *The Description and Use of Two Arithmetick Instruments* (London, 1673), p. 6. (Courtesy of Houghton Library, Harvard University, Hyde 90W-163.)

Making a dependable, easy-to-use tens carry for multiple digits proved to be a difficult engineering problem requiring careful design, calibration, and accuracy in the production of the parts involved. Consider this problem:

$$999999$$
$$+ \qquad 1$$

In performing this calculation by hand, following the standard practice, we would probably write the following:

$$
\begin{array}{r}
1\,1\,1\,1\,1\,1 \\
9\,9\,9\,9\,9\,9 \\
+ \qquad\quad 1 \\
\hline
1\,0\,0\,0\,0\,0\,0
\end{array}
$$

After the first digit, each carry causes a carry on the next digit. Such a result posed no problem for Morland's machine, since the user would have to turn each of the dials individually to propagate the carry: first the tens, then the hundreds, then the thousands, and so forth. Making a machine that could automatically perform this series of carries simply by adding 1 on one dial turned out to be tricky.

Early-modern inventors faced two major problems in attempting to create a machine capable of accurately and consistently performing carries: first, providing sufficient force to propagate carries over a series of digits from right to left, and second, converting analog motions into digital results.[41] In a machine that performs carries automatically, something needs to provide the force necessary to propagate all the carries throughout the machine, from the ones column to the column of the highest digit. One approach was to make the effort of turning a dial in the first column provide force sufficient to turn as many of the subsequent wheels as needed to perform additional carries. Such a mechanism was used in the first known calculating machine designed to automate carry—that of the astronomer, geographer, and Hebraist Wilhelm Schickard—described in a 1624 letter from Schickard to Johannes Kepler. The only known example was apparently destroyed in a fire (see figure 1.7).[42] Schickard added an extra tooth to the gear for each digit: when a given digit reached 10, its extra tooth turned a secondary gear that then turned the next higher digit in the machine. (If a carry was required beyond the capacity of the machine, a bell would ring.) Such a mechanism is still used in some odometers and gas meters. In this sort of mechanism, turning the ones dial in the case of 999999 + 1 requires considerably more force than turning the same dial in the case of 100000 + 1. Adding 999999 + 1 requires sufficient force to make the first gear turn the next six dials and their associated mechanisms; adding 100000 + 1 only turns one dial and its associated mechanisms one click. So much force is required in the first case that it would likely break the ones dial.[43] Let's call this the *sufficient-force problem*. Solving this problem in a more robust manner involves finding a way for the machine to have some reserve of force adequate for propagating carries and some means for that force to be renewed automatically in the course of using the machine.

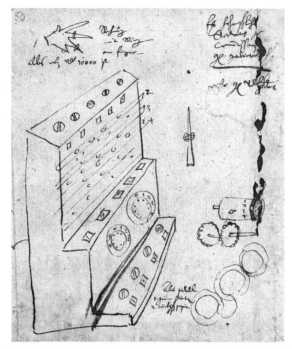

FIGURE 1.7. Wilhelm Schickard, Sketches and notes on calculating machine. (Württembergischen Landesbibliothek, Cod. hist. qt. 203.) Note the diagram of the carry tooth on the middle right.

Pascal's solution to the sufficient-force problem was the so-called *sautoir*.[44] The *sautoir* is essentially a fork that is raised as the digit on the machine approaches 10 and then falls, pulling down the mechanism of the next digit one stop. In figure 1.8, taken from Denis Diderot's article on the machine in the mid-eighteenth-century *Encyclopédie*, the *sautoir* is the forklike *34567*.

With this clever mechanism, Pascal claimed, "it is as easy to make one thousand and ten thousand wheels move all at once . . . as it is to make a single one move." The *sautoir* solved the sufficient-force problem.[45] In anachronistic terms, the principle behind the mechanism was to use the potential energy created by raising the weight of the *sautoir* to a sufficient height to advance the next digit one click when the *sautoir* falls. To ensure that the lower digits of the machine have sufficient force to set in motion a propagation of carries, the *sautoirs* of the lower digits are heavier than those of the higher digits. Finding weights appropriate for a well-functioning machine must have been a difficult task indeed. Contemporaries recognized the *sautoir* and its underlying principle as Pascal's great contribution.

The mathematics professor tasked with selling Pascal's machine, Gilles Personne de Roberval, noted "a thousand columns will move as easily as two";

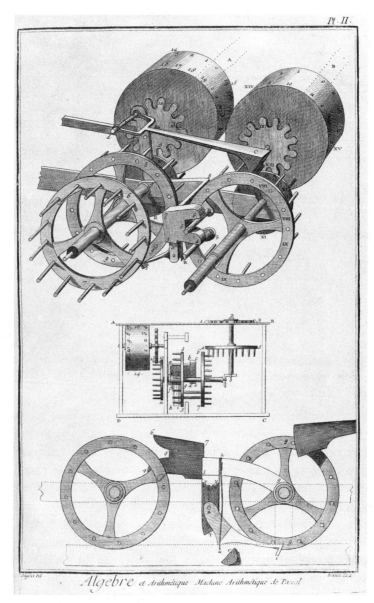

Pl. II.

FIGURE 1.8. Mechanism of Pascal's arithmetical machine, from Denis Diderot and Jean d'Alembert, eds., *Encyclopédie, ou, Dictionnaire raisonné des sciences, des arts et des métiers* (Geneva: Chez Briasson, 1751–1772), art. "Arithmétique." (Courtesy of Rare Books and Manuscript Library, Columbia University.)

The *sautoir* is the fork-like *34567*. The wheel *VIII-VIII-IX* moves clockwise as a number is added to its digit. As the wheel moves, the peg *rs*, mounted on that wheel, pushes the *sautoir 34567* up; a second mounted peg *RS* continues to push the *sautoir* up until *RS* reaches the edge of the *sautoir* at point *4* and lets it fall. When it falls, the catch bar *1* pushes the wheel *89* in a clockwise direction one stop. This design allows the wheels of the machine to turn in only one direction, so the machine cannot accomplish subtraction simply by being driven backward.

that is, Pascal's machine solved the sufficient-force problem. More than that, he added, its wheels "will also stop correctly, each in its proper place"; that is, Pascal's machine kept all the wheels at discrete stops corresponding to digital numbers.[46] To see this second problem involved in automating arithmetical carry, consider a mechanical odometer marking the epochal moment when a car reaches a million miles. As the odometer turns over, the nines become zeroes—but not quite, at least not strictly speaking. Most of the digits will be neither at 9 nor at 0: they will all be somewhere in between for a short period of time. Since the odometer is measuring a continuous quantity—distance traveled—this is not usually a problem; in fact, it might be understood to offer useful information. In a machine whose purpose is to add noncontinuous quantities, like the whole numbers, any such intermediate positions of the wheels would become a major difficulty. Adding any two counting numbers must produce another counting number: 8 or 9, not 8.7999999. A human being easily corrects for such an error, knowing that the result should really be read as 8 or 9, not the intermediate quantity the machine in fact shows. We make such judgment calls all the time in using electronic calculators. Dividing 7 by 11 and multiplying by 11 will produce 6.99999996 on many simple electronic calculators. Without consciously reflecting upon it, we compensate for such a result, understanding it as a 7; we do so without judging the calculator to be inherently untrustworthy.[47] A machine cannot make the judgment call involved in such a correction. Were one only ever to use the machine to perform the addition of two numbers, this problem would not be particularly severe: the human user would automatically correct the result. Imagine, however, adding a column of numbers, as in Morland's example, in figure 1.6. Over a long series of additions, even small errors would add up considerably: with each addition, the results of the machine would become increasingly erroneous. A human user could not easily correct the result, except by correcting it manually at each and every step. Let's call this the *keeping-it-digital problem*.[48] Solving the keeping-it-digital problem requires some way for the machine to maintain its dials exclusively on digital values, 0 to 9, and to correct automatically for or prevent the production of any intermediate results internally. Pascal's solution to this problem involved springs and a toothed lever (*c* in figure 1.8) to force the dials into discrete positions, whether or not the *sautoir* was engaged in performing carries.[49] Every digital calculating machine from Pascal's to the electronic computers of the twentieth century must keep it digital.

To perform carries adequately by the standards of their makers, these machines needed some source of force to allow propagation and some way of

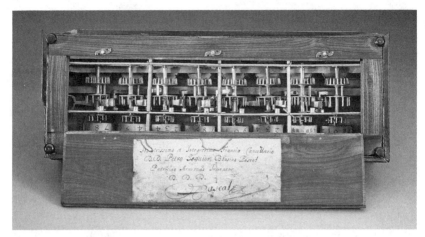

FIGURE 1.9. Machine arithmétique de Pascal à chiffres plus sous et deniers, 1645. Inv. 19600-0001. Open view. (© Musée des arts et métiers-CNAM, Paris. Photo: J-C Wetzel.)

consistently and accurately making analog motions report as digital results. As best we can tell from his notes and reconstructions, Schickard's machine was hard to use when multiple carries propagated and often showed intermediate results. Contemporaries judged Pascal's machine to have solved both problems. But for all its ingenuity, Pascal's solution to these problems precluded his machine from performing subtraction directly: as Roberval explained, he "did not know how to find the way to make his machine go from left to right, as well as right to left, on the same numbers."[50] Many critics felt this crippled the utility of the machine, even though subtraction could easily be performed indirectly using complements. They also noted that Pascal's use of weights required the machine to be used on level surfaces only.[51]

Solving the sufficient-force and keeping-it-digital problems with configurations of gears, springs, and weights was challenging conceptually and practically. These gears, springs, and weights had to be carefully specified and produced if the instrument was to perform according to the wishes of its inventors and the needs of its potential clients (see figure 1.9). The great historian of scientific instruments Anthony Turner explains that the robustness of the construction of the chief wheels of Pascal's machine required them to be cut from a single piece of brass rather than by having pins inserted into another piece. The process was laborious, performed "by taking a cylinder of brass of diameter equal to that required for the wheel, mounting it to a lathe and turning out its center until only a wall equal in thickness to that needed for the pins was left. Cutting and filing away the unwanted metal between them formed the individual pins."[52] Once their basic disposition was established, the various parts had to be carefully adjusted and constantly maintained. Nearly all reports on Pascal's machines testify to the care necessary to keep the machine in working order and even to use the machine when properly adjusted. And as Leibniz noted, engineers of Morland's caliber deliberately eschewed performing carries mechanically.

Philosophers hoping to build calculating machines needed more knowledge of the material world than was then available in formal natural philosophy. Devising mechanical solutions to the sufficient-force and keeping-it-digital problems required considerable knowledge of the material properties of gears, springs, and weights. Philosophers needed the help of people familiar with how to procure and produce springs of various elasticity and weights with given qualities. Watchmakers, above all, had considerable skill and experience in designing and using mechanisms with the properties of the various materials available.[53] While Morland's cost breakdown revealed the artisans and the labors involved in producing the machine, the precious document offers little indication about their role in the design and perfecting of the machine. We have no independent evidence concerning the role artisans played in designing and perfecting Pascal's machine.

A Polemical Picture of Theory, Labor, and Skill

In the mid-1640s, the young Blaise Pascal found his effort to produce and sell calculating machines obscured by two potential "clouds": one emanated by philosophers and mathematicians, the other by artisans. To dispel these clouds, Pascal laid out his distinctive competencies in a pamphlet dedicated to a patron, Chancellor Séguier. The first cloud came from people claiming "that this machine could be made more simply." Contemporaries criticized Pascal's machine as clever but too complicated. Such claims, he argued, "can be made only by certain minds who genuinely have some knowledge of mechanics and geometry, but who, *being unable to join the one to the other*, and the two together to the physical world, flatter themselves or deceive themselves with their imaginary concepts." Such philosophical figures convince "themselves that many things are possible that are not, as they possess only an imperfect theory of things in general." Their abstract knowledge about nature "is not sufficient for them to anticipate, in particular, all the inconveniences that come along" in the process of implementing a machine.[54] No contemporaneous general philosophical or mathematical account could possibly provide sufficient practical knowledge to design new devices or to judge them. The knowledge of such inconveniencies arises only through attempting to implement imagined designs and ameliorating their faults.

Contrasting himself to these philosophers deracinated from the world, Pascal claimed he combined theoretical and practical domains of knowledge.[55] He detailed the arduous process of gaining that competency: he learned to apply theory to practice in designing and attempting to produce a wide variety of calculating machines. Although his first attempt "satisfied many," he remained unhappy, so he built a second, then a third machine, which attained yet more approbation. Nevertheless, "continuing to perfect it still, I found reasons to change it, seeing in every [design], a difficulty in operating, or a rudeness in movement, or a propensity for becoming too easily corrupted by weather or travel." He celebrated his patience in constructing "up to fifty models, each different, some in wood, others in ivory and ebony, others in copper, before having accomplished the machine I now am revealing." Despite its many parts, the machine was now "so solid" that, given "the experience" of all his models, he could assure his potential customers and the reading public that no travel could upset the machine.[56] Numerous contemporary accounts claim that Pascal destroyed his health in his years of working on the machine.[57]

Insisting on the need to combine theory and practice was hardly unusual in mid-seventeenth-century France. Contemporary engineers, often in the employ of the king, stressed that they alone possessed the ability to unify theoretical knowledge of nature with practical knowledge in using and modifying it.[58] So, too, did master artisans keen to separate themselves from ordinary laborers, often their subordinates.[59] Pascal adopted the claims of competency derived from a grounding of theory in practice from engineers and elite craftsman. And then he decried artisans as simple imitators.

Artisans created the second cloud threatening the reputation of Pascal's invention: "bad copies of this machine that could be produced by presumption of artisans."[60] As Jean-François Gauvin argues, Pascal's denunciation of counterfeiters echoed similar denunciations made by clockmakers.[61] French clockmakers worked hard to fight off the constant threat of counterfeit clocks, often made by members of other guilds. The worst offenders, the goldsmiths, were only too happy to craft beautiful but nonfunctional simulacra of clocks. Some clockmakers likewise happily produced beautiful but nonfunctional simulacra of calculating machines.[62]

According to Pascal, the success of artisans in copying machines led them to believe that they possessed genuine creative ability and the theoretical knowledge necessary to guide it: "The more they are excellent in their art, the more it is to fear that their vanity" will lead them to believe "that they are capable of undertaking and executing new works by themselves, when they are ignorant of the principles and rules for those works." Pascal created a parallel between his inabilities with those of artisans:

> It is not in my power, with all the theory imaginable, to execute myself my own design without the aid of a worker who possesses perfectly the practice of the lathe, file and hammer, that are necessary to reduce the pieces of the machine into the measures and proportion I have prescribed to him using the rules of theory. Likewise it is equally absolutely impossible to all simple artisans, no matter how skilled in their art, to put a new piece into perfection, when that new piece has complicated movements, . . . without the aid of someone who gives them the measure and proportion of all the pieces . . . using the rules of theory.[63]

While Pascal could not physically manufacture something based on a fully specified design, he could specify the design of a complex machine using theory and knowledge of the properties of materials. The misfit of theoretical designs with the material world informs the perfecting of technical designs, but only the philosophically informed engineer, not the artisan, appears in Pascal's account to be able freely to recognize, compensate, and overcome such misfit by reference to theory. Whereas artisanal skill reveals the limitations and useful properties of different materials, an experienced philosopher such as Pascal recognizes those limitations or useful properties and adapts his design accordingly. The work of physically producing the machine is collaborative; the work of designing it is not.

Pascal claimed the ability to make adjustments and design modifications produced in the course of making, so central to artisanal practice, as part of his distinctive competencies.[64] He reduced craftspeople to instruments for implementing designs. In his creation narrative, Pascal explained that he produced his many designs always under the guiding aid of theory. In contrast, artisans "work through groping trial and error, that is, without certain measures and proportions regulated by art." They "produce nothing corresponding to what they had sought, or, what's more, they make a little monster appear, that lacks its principal limbs, the others being deformed, lacking any proportion."[65] In the Aristotelian and Horatian category Pascal invoked, a monster is precisely a material thing lacking a unifying form. Even as they experiment with new designs, savants always maintain the unity of the design through their command over theory and art; artisans acting alone merely modify pieces willy-nilly without regard to the whole. In the process of inventing, Pascal claimed, savants can autonomously regulate themselves and others to direct the production of new designs of unified machines and the improvement of current designs; artisans cannot regulate themselves to produce unified machines autonomously.[66]

In Pascal's account, the dependable success of artisans within their own domain of production reduces to their manual skills. These skills are produced through the repetition of actions under the direction of savants possessing theory and are nothing but the making habitual of practice. With new inventions, artisans must be regulated "until practice has made the rules of theory so common that [the rules] have finally been reduced into art, and until continual exercise has given to the artisans the habit of following and practicing these rules with confidence."[67] The nuanced Aristotelian notion of habit, still current in Pascal's time, allowed for corporeal habits, such as playing a flute, to be corporeal and still be tied to reflection on the theoretical grounds of the activity. In retaining all creative and theoretical activity for himself, Pascal denied the artisans any theoretical or free agency in their habitual activity.[68] However much they attempted to conceive, they could only misconceive. When they tried to innovate, they upset the epistemic, technical, and social order and, in so doing, produced only monsters or abortions that were nonfunctional disunities.

In arguing that artisanal labor should be noncreative, Pascal collapsed a well-understood legal distinction concerning the variety of artisans and laborers. In an important treatise written a few years later, Pascal's friend, the lawyer Jean Domat, distinguished the two sorts of "professions" to be found in certain mechanical arts: a higher form "that joins the art of inventing works that are exquisite in their genre to the industry of the hand" from "those who, with little or no invention, work on what others have invented." These two classes of artisans could be found in painting, sculpture, architecture, and "mechanics."[69] Such hierarchies among different levels of craftspeople figured in legal treatises, aesthetic writing, and every practice of compensation—as evident in Morland's bill for the duke. Pascal lumped all artisans together into the noninventive class, at least for complicated machines. Any invention requiring theory remained beyond the capacities of artisans.

Pascal was writing from a position of weakness. He needed the artisans more than they needed him; he was in no position to dictate stringent terms to them.[70] His polemical account of invention was to aid his quest for a royal privilege to grant him a monopoly on calculating machines in France as well as the state-sanctioned means for maintaining such a monopoly. This privilege would "suffocate, before their birth, all these illegitimate abortions that could be engendered otherwise than by the legitimate and necessary alliance of theory and art."[71] Addressing Chancellor Séguier, he sought control over the production of calculating machines that was eluding him. In 1649, he received just such a privilege, which echoed Pascal's litany of the dangers posed by artisans producing machines outside his control and provided the coercive controls Pascal wanted.[72]

Pascalian Defects and Imitation

As Pascal bitterly denounced counterfeiters and demanded royal intervention, his friend and middleman in distributing the machines, Gilles Personne de Roberval, was secretly setting out his own design for a calculating machine in a manuscript labeled, appropriately, "for me alone."[73] Roberval was inspired by what he saw as the primary defect of Pascal's machine: its input dials could be turned only in one direction.[74] Pascal's effort—and failure— spurred imitators further afield. Some aimed at radical simplification of the goals and the mechanism; others sought better mechanisms for the same goals. Some acknowledged Pascal's as a model, while others claimed complete independence. Some kept their inventions secret; others sold them; and still others published them, especially as learned periodicals proliferated in the early eighteenth century. Some worked from rumor, some from direct experience with the machines, and some from published accounts. Copying and reinvention animated the quest to mechanize calculation, from Pascal to the early twentieth century.

In May 1673, the great Jansenist theologian Antoine Arnauld wrote to Pascal's nephew concerning something he had seen after dinner one day. A "little clockmaker who, having seen a machine of M. Pascal, has perfected it so that it is incomparably easier." Its dials "turn one way or the other"—it could do subtraction directly; it could help with multiplication; it was much smaller (see figure 1.10).[75]

When the Parisian clockmaker in question, one sieur Grillet, discussed his new calculating machine of the 1670s, he echoed the by then standard complaints about Pascal's machine, all the while praising its ingenuity: "This machine, whose extremely beautiful effect cannot be denied, is extremely complex, and has this inconvenience: one must use it holding it horizontally, because of the weights that are the principal part of it, which cannot produce their effect if the machine is in any other position." Grillet zeroed in on the key difficulty with Pascal's machine: the weights involved in the delicate carry mechanism required the machine to be carefully calibrated and positioned just right. The central technical breakthrough of the machine prevented it from seeming useful to his contemporaries. Its ability to automatically perform carries was overshadowed by its size, complexity, and difficulty in use. Grillet explained, "The quantity and size of the pieces it requires make it extremely large and

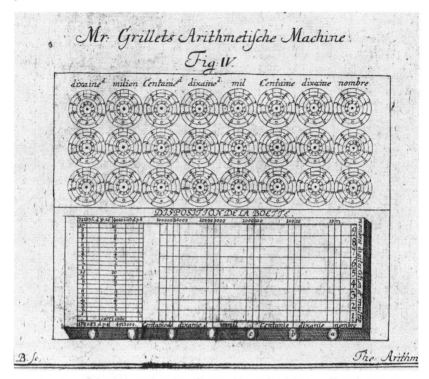

FIGURE 1.10. Grillet's Arithmetical Machine, from Jacob Leupold, *Theatrum arithmetico-geometricum: Das ist: Schau-Platz der Rechen- und Meß-Kunst* (Leipzig, 1727), plate VI. (Courtesy of Rare Books and Manuscript Library, Columbia University.)

heavy, and prevent that one could carry it around. Whence it comes that this machine will not be able to serve for ordinary use, and can only ever be a curiosity in a cabinet."[76] Pascal's old friend Arnauld evidently agreed with the little clockmaker: the Pascal machine was far too bulky, expensive, and finicky to be in common use.

Despite his claims to the contrary, contemporaries judged that Pascal had not provided a robust, accurate, dependable, portable solution to the problem of automating carries. Simpler machines, made by master artisans as well as the likes of Samuel Morland, might not automatically perform carries, but they had the virtue of working on their reduced terms. Not attempting to be proxies for the doing of arithmetic, these machines worked simply as tools to aid someone performing calculations. Grillet praised the simplicity of his mechanism and its appropriateness for the average user: "The Machine I have invented has none of these difficulties: it has neither drums nor weights, nor this great quantity of wheels that one sees in Monsieur Pascal's."[77] Whereas Pascal's instrument would require users to develop new, peculiar skills, particularly to perform subtraction, Grillet's instrument simply supplemented already extant arithmetical skills, as did Morland's roughly contemporaneous machine.[78]

More counterfeits—or improvements, or emulations—sprung forth in Poland. Its recently installed reforming monarchs—especially its Francophile queen, Louise-Marie de Gonzague—had Pascal's machines ordered from Roberval: "The Pascal you sent gives me great joy. The king never wants to permit the removal of the one that I put in his chamber."[79] Tito Livio Burattini, an Italian engineer in the service of the Polish crown, created a calculating machine that was sent as a gift to the Medici prince in Florence. Maker of lenses, sometime reformer of the Mint, and a projector of a "flying dragon," Burattini was a classic example of an emulator, someone skilled in "perfecting discoveries already made," who reduced two major technological innovations—the pendulum clock and the Pascal machine—to pocket-sized devices.[80] Burattini pioneered what would become a standard pattern: attempting to perfect a calculating machine became one important way early-modern engineer-savants tried to demonstrate their technical competence to sovereigns, especially modernizing sovereigns of peripheral centers, from Warsaw to Darmstadt to Venice and Lunéville. Burattini was celebrated—at least locally—for improving machines ready at hand and for reverse engineering. Experts capable of such technology transfer—and often industrial espionage—were precious to early-modern polities, however difficult they were to distinguish from mountebanks and "projectors" offering spurious technical and financial schemes.[81]

Efforts to emulate—to imitate and improve—Pascal's machine continued throughout the eighteenth century. Acquaintance with Pascal's efforts varied widely. The May 1733 issue of the *Journal des Savants* described an "arithmetical machine of M. de Boistissandeau" that had been featured in the latest volume of the *Mémoires de l'Académie des sciences*. The review added only one detail: "He imagined it without knowing that of M. Pascal."[82] Eighteen years later, in August 1751, the *Journal des Savants* carried another notice about a new calculating machine presented to the *Académie*, that of Jacob Rodrigues Pereira, a Portuguese-Spanish Jewish linguist and teacher living in France who became famous for his work on the education of deaf-mutes.[83] Pereira invented with considerable knowledge of previous efforts, the journal explained. Pereira could read about the machines of Pascal, Perrault, Lepine, and Boistissandeau in the printed catalog of instruments approved by the *Académie*:[84] "M. Pereira had the opportunity to see all these machines at will in the various depots where they are conserved. All Savants and curious can equally see them*; it appears that M. Pereira examined them with great care."[85] The asterisk referred to a note mapping out the Parisian geography of calculating machines that Pereira had likely inspected.

Around 1751, acknowledging the sources of Pereira's inspiration was not a form of denigration. Awarding him his due required no mental gymnastics to assert an ex nihilo quality of creation or to deny his knowledge of previous efforts. The geography of machines in Paris stemmed from a thriving culture of mechanical and scientific making in the first half of the eighteenth century, where machines and instruments were collected, displayed, and available for inspection to encourage further invention and elaboration.

One of Pereira's key patrons and connections to the crown, Michel Ferdinand d'Albert d'Ailly, duc de Chaulnes, possessed one of the foremost collections of scientific instruments of the time.[86] In this "cabinet," Pereira inspected a machine made by one Levi, as well as a Pascal machine itself. To encourage improvements to machines, the duc opened his collection to visitors as a resource for inspecting machines "at leisure."[87] In the middle of the eighteenth century, spaces such as the duc's cabinet provided an important means for overcoming barriers to the transmission of knowledge about machines, instruments, and manufacture; direct manipulation and inspection of the instruments could compensate for skills poorly communicated in print or otherwise kept secret. A major innovator in using micrometers with microscopes and in devising cutting machines for wheels, the duc de Chaulnes was committed to disclosure, either through inspection or in print, to encourage the perfecting of technology through collective improvement. He described his machine tools for dividing lines in great detail.[88] He contributed to a midcentury culture of fruitful imitation and improvement in Paris.

In refusing to describe the machine, the report about Pereira's new calculating machine hinted at the superiority of physical examination and manipulation of machines possible in spaces such as the duc's cabinet: "To understand the play of these machines, the reader needs the help of figures and perhaps seeing first hand the machines themselves, the various parts involved in their connection to the whole, in order to fix his imagination and to develop the degree of necessary knowledge of the properties and, so to speak, the functions of all the parts making up each of these machines."[89] Tactile experience of machines, made possible in spaces such as the cabinet of the duc de Chaulnes, was fundamental for a fuller understanding and future elaboration and improvement.[90]

The report on Pereira's machine in the *Journal des Savants* was immediately followed by a modest notice of the publication of the first volume of a new encyclopedia, edited by Denis Diderot and Jean Le Rond d'Alembert.[91] Pereira's machine was produced thanks to one eighteenth-century solution to the transmission of information and skill; the new volume of the encyclopedia, and its plates, which took another decade to appear, embodied an alternative practice for the transmission of technical knowledge. The *Ency-clopédie* deployed a new form of description and depiction that would seek to render direct inspection and handling of the machines less central for the transmission of technique.[92]

In his introduction to the *Encyclopédie*, Diderot explained the process necessary for this new form of the transmission of knowledge. He denied that artisans had conscious knowledge of what they did. Most artisans could work forty years "without knowing anything about these machines." Diderot had to take on the "annoying and delicate" role of a "midwife of minds" to articulate the skill of artisans as propositional knowledge. He was forced to undertake an apprenticeship—to imitate and thus to embody skills in order to become capable of articulating them. To produce articulated propositional knowledge, Diderot and his colleagues first had to gain a maker's knowledge through physical imitation: "It was necessary in many cases to obtain the machines, to construct them, to put one's hands to work, to make oneself, so to speak, into an apprentice, and to make bad works in order to teach others to make good ones."[93] The physical imitative labor of the encyclopedist and his draftsman produced propositional content through new forms of disclosure—a new descriptive and pictorial technology that would eliminate much of the need for imitative corporeal work and spaces for tactile experience, such as the duc de Chaulnes's cabinet.[94] The *Encyclopédie* was parasitic upon—indeed, made possible by—the culture of artisanal apprenticeship and imitation that it ultimately sought to supplant.

Six hundred pages later in the first volume, Diderot exemplified these new technologies of disclosure with his virtuoso description of Pascal's calculating machine: "I am certain, that it will appear . . . nearly as difficult to understand Pascal's machine," based on the earlier publications, "as to imagine another arithmetical machine." Absent the creation of proper techniques for disclosure, reinvention was easier than discerning the nature of a device. In technical matters, insufficient forms of disclosure could be worse than no disclosure. In his article, Diderot promised to "expose the mechanism of Pascal's machine in a manner so clear, that no effort of the mind to grasp it will be needed."[95] He explained the practice of depiction followed by his artists in depicting machines: "The machines and tools were sketched; nothing was omitted that could be shown distinctly to the eyes. Whenever a machine merits more detail, either because of its importance or the multiplicity of its parts, a more singular drawing replaces the whole composition."[96]

Providing the means for making available the content of artisanal knowledge and skills, the new descriptive and depictive technologies of the *Ency-clopédie* would overcome technical ignorance and, in principle, eliminate the need for an imitative mind capable of recreating the invention of another. Propositional knowledge of materials, the knowledge of how to work with tools, and the social worlds of production would all be made intelligible and visible to readers. With such a procedure for specifying the essence or substance of the invention, a "hylomorphic" model of invention and production, one that severs design from material production, becomes a far more thinkable and convincing account of technical production. With Diderot's depiction, the machine's design could be seized all at once.

In the 1784 edition of Pascal's complete works, the editor Bossuet reprinted Diderot's description and reprinted the impressive plates, which remain standard today. Pascal's polemical account, with its vague invocations of the various ways Pascal had sought to mechanize carry, was now, for the first time, accompanied by detailed description and schematic representation of its mechanism. Pascal's polemical depiction of artisans became plausible only through descriptive technologies created a century after him.[97]

The cabinet of the duc de Chaulnes and the *Encyclopédie* offered radically different eighteenth-century solutions to lowering "barriers" to the communication of technique necessary for its improvement.[98] The artisanal culture of emulation provided elite philosophers with a model for coordinating theory and practice, mind and hand, to their mutual improvement. This industrial enlightenment coexisted in a complex contrapunctual relationship with the better known high Enlightenment committed to the deskilling of artisans and the movement toward a hylomorphic model of design and creation.

Providing practical and effective ways of mechanizing carries depended on integrating the skills, competencies, and innovations of artisans, not reducing them to simple part-making machines. Creating these solutions required forms of coordination and management of labor and skill that involved cooperation and exchange, not absolute subordination. Inventors were not sovereigns, at least not if they wanted to succeed in the eyes of their contemporaries.

Wrong Imputations from Machines

Turning to religion late in life, Samuel Morland penned a short Christian apologetic, the *Urim of Conscience*. Early in the text, he draws on engineering, machines, and skilled labor in his defense of the Christian faith to set forth an unusual form of the design argument. Morland focuses on the difficulties in accounting for human skilled activity within a broadly mechanistic framework. He notes that he can "imagine the Soul of a Man, whilst his Body stands up right, to be a spiritual Engineer, and to be seated in the Brain, as in its Watch-Tower, and there to make use of each hand and arm to lift up a

ponderous Weight, to the perpendicular height of 6 or 12 Inches."[99] Accounting for the mechanisms by which human beings self-consciously perform
simple activities seems at least thinkable to Morland: one can imagine how an
intention to lift something could be converted into the range of the mechanical impulses that direct the body to perform general actions in a general way.
Far less easy to comprehend is how the commands of the "spiritual engineer"
become converted into the vast array of small actions necessary to perform
that action. Our bodies act in ways that far outstrip our conscious direction of
our action: "I do not at all apprehend, neither have I any *Idea*, or Imagination,
by what secret power the Soul contracts or dilates the Muscles, how it elabo-
rates, and sends forth the Animal Spirits." To illustrate the point, he offers for
"the Contemplation of the most skilful and subtil *Mechanick*, or Philosopher,
in the world" a series of remarkable feats, everyday as well as extraordinary,
of laborers and tricksters:

1. Of a Porter, taking up great and ponderous [burden] from off the ground,
   and heaving them on his shoulders.
2. Of a Waterman, who, upon a Wager, pulls in his Oars with both his Arms
   towards his body, and, at the same time, thrusts from him with his Thighs
   and Legs,
   [. . .]
4. Of one, who distorts all the Parts, Members and Joints of his body, so as to
   make it appear in many Different Figures, and strange Shapes.[100]

In this argument from design, Morland drew upon his grasp of the complexity, necessity, and often-subconscious nature of skilled activity to testify to
the complexity of the human machine. A human engineer could produce a
machine capable of performing relatively straightforward acts. Only a super-
human engineer could produce a machine of sufficient complexity to perform unconsciously such an array of activities. Contemporaries such as Boyle
made similar arguments; Morland's version of the argument focused less on
the challenges of mechanizing reason than on the challenges of mechanizing
skilled and agile human activity.

Contemporaries reckoned that Pascal's machine involved the mechanization of reason. His friend, the poet D'Alibray, wrote:

> Counting was the action of a reasonable man,
> And now we see your inimitable art,
> Gives this ability to the lowest of minds.
> This art requires neither reason nor memory,
> Thanks to you everyone practices it without difficulty or glory,
> So that everyone owes you both its glory and effect.[101]

Pascal's sister likewise explained, "This work was considered a new thing in nature, as it reduced into a machine a science that resides entirely in the mind, and having found the means to perform all the operations with perfect certainty, with no need for reasoning."[102] Like some advocates of artificial intelligence, these observers took the ability to create a machine capable of performing the behaviors of arithmetic as tantamount to creating a machine capable of reasoning, albeit in a limited way. Forgetting the skills necessary to make the machine perform its arithmetical behaviors and to correct its sometime defects, they saw it as a proxy for a reasoning being, not a tool to be used by one.

The misattribution of agency concerned Pascal. In his *Pensées*, Pascal offered a more nuanced view of the relationship between the calculating machine and human reason: "The arithmetical machine produces effects that approach thinking more than anything animals do. But it does nothing concerning which it could be said that it has a will, as is the case with animals."[103] However much it approaches thinking, the calculating machine captures only a small part of the phenomena of living beings, including humans.

Pascal was not arguing that animals have the faculty of willing their actions; rather, he was claiming that animals perform numerous sorts of actions that appear to stem from choice, from free actions of the will. Just before his remarks on the machine, he cryptically notes, "The history of the pike and the frog, of Liancourt: they do this always, and never otherwise, nor [do they do] anything else partaking in the mind." In the story, likely taken from the famous *Compleat Angler*, a frog tears out the eyes of a pike.[104] The story illustrates anthropomorphism gone mad. A fisherman and a bishop

> saw a frog, when the Pike lay very sleepily and quiet by the shore side, leap upon his head; and the frog having expressed malice or anger by his swolne cheeks and staring eyes, did stretch out his legs and imbraced the Pikes head, and presently reached them to his eyes, tearing with them, and his teeth those tender parts; the Pike, moved with anguish, moves up and down the water, & rubs himself against weeds, and what ever he thought might quit him of his enemy; but all in vain, for the frog did continue to ride triumphantly, & to bite and torment him till the Pikes strength failed; & then [the frog sunk] with the Pike to the bottom of the water; then presently the frog appeared again at the top and croaked, and seemed to rejoice like a Conqueror.[105]

Expanding Pascal's telegraphic claim, we might say that the apparent spontaneity of the frog is only apparent; its actions make us believe that the frog can think, emote, and choose actions other than those built into its mechanical makeup. Thinking thus would be a mistaken hypostatization, much like mistakenly using the apparent horror of the vacuum as grounds for claiming that there exists a genuine horror of the vacuum in nature, or using the reason-like behavior of a machine as grounds for imputing reason to it. As Pascal insisted time and again, the attribution of reason, intention, and emotion to nature is a dangerous mistake. The point was a good Cartesian one. The nonhuman stuff of the world, animals and calculating machines included, are simply machines: we ought not to mistake their effect.

1   First Carry

Babbage and Clement Mechanize Table Making

In his idiosyncratic autobiography *Passages from the Life of a Philosopher*, Charles Babbage explained, "The most important part of the Analytical Engine," his envisioned general-purpose calculating engine, "was undoubtedly the mechanical method of carrying tens."[1] Solving the problem of carrying tens figured prominently in the quest to construct Babbage's Difference Engines, a series of partially built and envisioned automated machines for mechanizing and securing the calculation and printing of numerical tables—crucial military, commercial, and scientific technologies well into the second half of the twentieth century.[2] The values of many kinds of tables can be approximated or computed exactly through repeated sets of additions, making it possible for a suitable adding machine to compute those values.

A Difference Engine comprised in large part an interlinked set of adding machines capable of long series of additions for the computation of tables; a robust carry mechanism was essential.[3] Begun in the early 1820s and then granted considerable state support, Babbage's effort to build the engine collapsed by 1834 and left behind a series of plans, a number of demonstration parts, and one beautifully constructed demonstration machine one-seventh the size of the envisioned engine (see figure C1.1). Amid the collapse of the Difference Engine project, Babbage turned to a more ambitious machine: his projected, but never built, Analytical Engine. This engine was to be more general-purpose, capable of being programmed to evaluate a wide variety of arithmetical and algebraic formulae.

Mechanizing carry posed many challenges: intellectual ones, as it presented a profound design challenge; technical ones, since creating carry mechanisms pushed the limits of then extant precision engineering technologies; and social ones, in that building something within those technologies required working

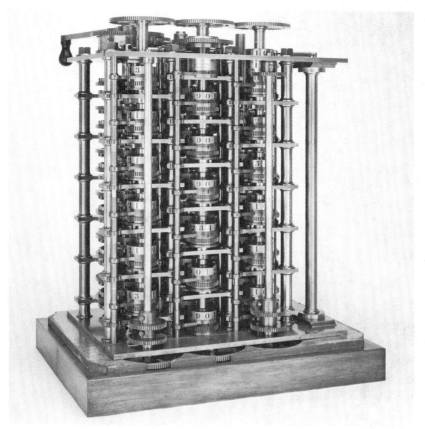

FIGURE C1.1. Babbage's Difference Engine No. 1, 1824–32. Inventory no. 1862-0089. (Science Museum, London.)

out relationships among an inventor, a chief engineer and draftsmen, and a team of workmen. Finding a practical solution to the problem of carry involved gaining knowledge of materials and the means of working materials—but above all, it meant finding and then working closely with machinists and artisans possessing knowledge of materials and how to work them. Only by integrating the process of invention and implementation could carries—and their mechanisms—come to be propagated expeditiously.

The Difference Engine

Consider the process of squaring whole numbers. Make a table of the natural numbers up to 6 and then their squares.

First write the out the differences between each pair of the squared values (3, 5, 7, 9, . . .), and then write out the differences between the first set of

TABLE C1.1. Computation of square numbers using second differences

n	n^2	1st Diff.	2nd Diff.
1	1		
		3	
2	4		2
		5	
3	9		2
		7	
4	16		2
		9	
5	25		2
		11	
6	36		

differences. The members of the second set of differences are all constant—in this case, all 2s. A cursory examination of the table suggests one could compute the squares entirely through repeated addition, with no need to multiply directly. So $4^2 = 16 = 7 + 9 = 7 + 5 + 3 + 1 = (3 + 2 + 2) + (3 + 2) + 3 + 1$. A polynomial of a degree n can, in general, be computed using the n-th differences.

Automating procedures devised for rooms full of human computers, Babbage's Difference Engine computes series of functions through addition alone.[4] The engine automates the process of adding differences, starting with some constant. The machine first adds the constant second difference (2 in our example) to the last calculated first difference (5)—this produces the next first difference (7). It then adds this new first difference to the last computed value for the function (9) to produce the next computed value (16). Or, in the case of a polynomial of degree 2, for any n, $f(n) = f(n - 1) + \text{difference}_1(n - 1) + \text{difference}_2(n - 1)$ for n greater or equal to 2.

Babbage realized that the machine could perform these additions in parallel: while it was adding 7 to 9 to get 16, it could simultaneously be computing the addition of the next second difference, 2, to the next first difference, 9, to get 11.[5] While the machine was, in its basic design, limited to constant differences, Babbage experimented with procedures for computing with variable differences in a process he called "eating its own tail."[6] Since many important functions can be approximated as polynomials—as in Taylor series—the Difference Engine would compute all the values between values put into it.[7] The machine would then automatically typeset all the computed values on plates to allow the printing of tables. Human beings would need neither to perform the computations nor to set the type in the production of mathematical tables.

By June 1822, Babbage had a demonstration piece of the calculating apparatus, now called "Difference Engine 0," about which we know little.[8] He showed the engine to progressively larger groups of spectators as he began to push for government funding to advance the project. A Dr. Wallastson "spent some time in examining and in working it; it made no mistake and 'did as It was bid'[;] the sum total of his remarks were thus expressed 'All this is very pretty but I do not see how it can be rendered productive.'"[9] Others dismissed its functioning, not just its potential utility. By 1823, Babbage sought a better construction: "My model is so badly made that I am discontented with it."[10]

A working demonstration piece of his developed design for the calculating part, called "Difference Engine 1" (DE1), was produced by 1833; a monument to early nineteenth-century British engineering, the fully functional machine is on display in the Science Museum in London. (The printing mechanism was never completed.) He radically revised and simplified plans for the machine in 1846–48. Not built in his lifetime, "Difference Engine 2" (DE2) was constructed at the Science Museum in London at the end of the twentieth century, largely following his detailed plans.[11]

After the ceasing of efforts to build the Difference Engine in 1834, Babbage sought to create a far more ambitious "analytical" engine that would automate computation using the standard range of arithmetical operators and using a set of data entered via punch cards. The machine would reduce the computation of those operators to repeated addition, performed in a central processing "Mill." No analytical engine was ever completed, though Babbage and his chief draftsman, Charles Jarvis, left voluminous technical drawings and notebooks.[12] An ambitious effort, "Plan 28," is currently under way to attempt to construct an Analytical Engine.[13]

Sequential Carry

In Babbage's earliest surviving manuscript concerning the development of his Difference Engine, dated 1822 but likely from 1821, he reflected upon adding using linear rods versus circular gears for the carry mechanism. Performing a carry—a movement from 9 to 0—on a linear rod would require a jump; the movement from 9 to 0 would be different from all other movements, a "breach of continuity." Observing this breach "induced" Babbage to set out a crucial design principle: "<u>Always to prefer a circular motion to any other when its immediate \object/ relates to number</u>."[14]

This earliest manuscript offered a proof of concept that a Difference Engine *would be* possible if carry could be mechanized. In proving the possibility of performing addition mechanically, Babbage was initially rather cavalier about carry, noting that a "figure wheel" moving from 9 to 0 could have such a protruding lever: "The use of this lever is, that every time the figure wheel passes from 9 to 0 it may act on *some mechanism which* shall cause the next adjacent figure wheel to advance one step."[15] Like every projector of calculating machines, Babbage soon realized that "some mechanism" is far from trivial to create. His insouciance about the carry mechanism did not last long once Babbage began trying to construct machines.

Babbage's several Difference Engines used, as best as we can reconstruct, a variety of carry mechanisms.[16] As every inventor and fabricator of calculating machines discovered, considerable trial and error was required. Babbage's existing machine and drawings, all probably from the early 1830s, use "a memorandum taken by the machine of a carriage to be made:"[17] a motion from 9 to 0 moves a lever that signals a carry to be made *after* the primary additions are complete. An elegant spiral axis with arms arrayed helically—called by an early supporter "a mechanical arrangement of singular felicity"—powers the sequential run of carries that have been flagged, or in the idiolect of the machines, "warned," and allows for carries to cause additional carries.[18] Babbage retained a form of this helical arm in his designs for the second Difference Engine (DE2); the action of these helical arms produces parallel beautiful waves, which can be seen when the machine is demonstrated at the Science Museum in London or, until recently, the Computer History Museum in Mountain View, California. Neither was novel, as we will see: warning mechanisms went back to German polymath Gottfried Wilhelm Leibniz, and the spiral arm went back to the third Earl Stanhope, as was recognized later in the century. Babbage may have created both independently. Or not. We don't know.

As Babbage moved from mentally envisioning to actually producing machines from 1820 to 1833, he became ever more interested in ensuring that the machines would remain "secure" and precise. Above all, they needed to keep their results digital. Ideas of possible ways of adding and carrying came easily; doing so robustly and securely was the challenge, as he noted in 1837: "The great difficulties in constructing the calculating engine have in general arisen from certain conditions I had laid down as desirable for the perfect security and certainty of its action."[19] In the extant fragments of the first Difference Engine, for example, crucial components in the addition mechanism are prevented from taking up nondiscrete values—kept digital—with a spring-loaded roller mechanism that keeps the scalloped "unbolting wheel" from

FIGURE C1.2. Charles Babbage and Joseph Clement, calculating wheels with spring roller for "keeping it digital." (Science Museum, London.)

any intermediate, nondigital positions (see figure C1.2).[20] The various security mechanisms, Lardner wrote, mean "that the machine will either calculate rightly, or not at all."[21] These mechanisms, seen then and now as a central part of the genius of the machine, were collective products.

Hylomorphism and the Polemical Account of Labor

Once he had secured a £1,500 grant from the Treasury to fund the construction of a full-sized Difference Engine in 1823, Babbage set out to industrial north England "to see everything to which I can get access that relates to machinery."[22] His research resulted in his *On the Economy of Machinery and Manufactures* of 1832, among the nineteenth century's most important economics texts. In its twenty-seventh chapter, "On Contriving Machinery," Babbage drew up a division between the design of a machine and its implementation. He extolled cerebration before construction: "It is possible to construct the whole machine upon paper, and to judge of the proper strength to be given to each part as well as to the framework which supports it, and also of its ultimate effect, long before a single part of it has been executed. In fact, all the contrivance, and all the improvements, ought first to be represented in the drawings."[23] Just after he stated this hylomorphic framework, a dream of the instantiating a fully drawn design in matter, he insisted on the need for a journey through material implementation: "On the other hand," he noted, "there are effects dependent upon physical or chemical properties for the determination of which no drawings will be of any use. These are the legitimate objects of direct trial." He gave several examples: "It is impossible by any drawings to solve difficulties such as these, experiment alone must determine their effect."[24] In his autobiography, he explained, in reference to the analytical engine, "Draftsmen of the highest order were necessary to economize the labour of my own head; whilst skilled workmen were required to execute the experimental machinery to which I was obliged constantly to have recourse."[25] The process of making up the very design of the machine was collaborative, as was creating pieces of it. Mere imitation was not in question.

Clement, Security, and Nonhylomorphic Design

In the wake of the death of his father and wife in the course of a year, Babbage traveled to Italy in 1827 on a grand tour to help overcome his depression. He left his friend Herschel to supervise the work of his engineer, Joseph Clement, on the Difference Engine. Progress had seemed extremely slow since 1823 as Clement worked on preliminaries to creating the parts of the machine. Clement spent his time devising and building new precision instruments for making the parts and creating the series of drawings necessary to envision a fully specified Difference Engine.

Herschel's letters and notes let us glimpse the state of work in Clement's shop. Herschel describes the making of drawings; the building of new precision machining instruments, especially the lathe; and the making of patterns for parts. "In the work-shop patterns are making for a good number of wheels but the preparation of <u>tools</u> is still the main business."[26] Planning was all well and good, but Herschel was unsatisfied: "With regard to the <u>work</u>, I confess I see very little progress. They are now at work on a long screw 6 or 7 feet long . . . and Clement has been occupied also in making parts of the framework and carrying part &c. to try his plan of making the axes work in sheaths."[27] Babbage understood the value of the preparatory work better than Herschel: "The accounts you give of the progress of the machine are by no means discouraging. As to the work done on the drawings it is a species of work which makes but little show." Babbage worried, however, about the ownership of the tools Clement was spending so much time building.[28] Debates over ownership and pay soon contributed to the collapse of the Difference Engine project.

FIGURE C1.3. Proposal by Clement to increase "security" of engine. Herschel to Babbage, 18.12.1827. Royal Society HS 2.218. © Royal Society.

Although confused and telegraphic, Herschel's reports to Babbage illustrate Clement's part in the work of bringing "some mechanism" capable of addition and carry into being.[29] Even if he was skeptical of Clement's application and focus at times and increasingly worried about his motives, Babbage knew him to be an excellent draftsman and master mechanic. He had hired Clement not as a mere implementer but precisely as someone knowledgeable and skilled at creating actual mechanisms from ideas and sketches and as someone capable of the free, creative craftsmanship of risk, not just the imitative craftsmanship of certainty.[30] "I left Clement," he wrote Herschel, "so well acquainted with all the mechanical actions of the machine that I have not the least fear of his making such changes as he may think necessary."[31] Clement had the freedom—indeed, the charge—to transform radically the design in the course of development. Babbage had confidence in the ability of Clement and his subordinates to improve their working design and fully expected them to do so.

The suggestions and contrivances produced by Clement and his team take numerous forms but largely focus on improving the "security" of the machine. This "security" involved keeping the various gears digital—that is, keeping them at integer units, not intermediate positions. Many of Clement's suggestions figure centrally in contemporary and later accounts of the contrivances necessary to ensure the accuracy of the machine. In several letters, Herschel described, somewhat poorly, a number of proposals for improving the security mechanisms of the machine:

> Another thing he has made drawings of is a plan for fixing the axes firm in their places till the wheels come into their proper places to be bolted. This he says was to have been done by rollers but he prefers a more decisive species of detent [see figure C1.3] thus to be driven up into notches by springs and which are to be (like the former contrivance) at one end of a lever, the other being commanded by the teeth of a ratchet wheel fixed on the bolting axis thus [see figure C1.4] B is the spring forcing the detent in. A is the spring playing in a

plane perpendicular to the paper, carrying a stud, its office to hold the detent back when it ought not to go in.[32]

In the weeks following, Clement continued to work on this fixing mechanism. Not just sketching a changed individual contrivance, he updated the master plans and elevations of the engine to include the new holding devices: "An old drawing 4 feet by 2 with much new work done on it, Viz: plans and Elevations of the System of Calculating and bolting axes, and the manner of holding the alternate ones locked on the new scheme described in my last where the inclined plane principle is used to hold the wheels fast jammed thus in their proper places" (see figure C1.5).[33]

Replying from Italy, Babbage commended Clement: his "plan for locking the axes in their places until the wheels are ready to be bolted seems to me better than rollers which latter I never intended and know not how they got into the drawings."[34] Despite these developments, rollers won out. The demonstration machine in London retains just such rollers—perhaps a compromise that Babbage ultimately was unhappy with. Remarkably, Babbage denied authorship of one of the signal security mechanisms of the first Difference Engine. His design for Difference Engine 2 abandoned these in favor of a new locking mechanism, in which a wedge running vertically along each set of figure wheels inserts itself into the figure wheels to preclude all intermediate values. Both the first and second Difference Engines were designed to jam—and to stop—if the wheels were so far from discrete values that the operation was certainly compromised.[35] These contrivances—a collective product of Babbage, Clement, and perhaps other unnamed workmen—proved essential

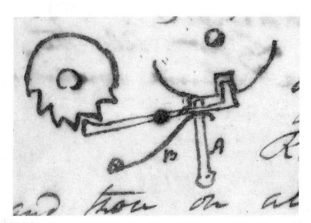

FIGURE C1.4. Proposal by Clement to increase "security" of engine. Herschel to Babbage, 18.12.1827. Royal Society HS 2.218. © Royal Society.

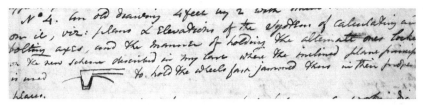

FIGURE C1.5. Proposal by Clement for a new scheme for locking. Herschel to Babbage, 12.2.1828. Royal Society HS 2.219. © Royal Society.

in maintaining state support for the design and construction of the engine (as discussed below).

The possibility of losing Clement terrified Babbage. The drawings did not, in fact, fully describe the machine: "Into his head I have for several years been conveying all my ideas on the subject of the machine and he is consequently in full possession of them." Clement had the ideas in his head—and their practical projection on paper. But those drawings were not readily obvious to anyone else: "Much of that labor is now fixed in drawings which it would require much time to make another person fully understand . . . still much remains in his mind ready to be produced."[36] Babbage's ally, the Duke of Somerset, captured the embodiment of knowledge about the machine in congratulating Babbage when it seemed he had found a replacement for Clement: "I am very glad to hear that you have found another Draftsman who understands the machine; for when you had but one, your situation must have been one of great anxiety. It was like having a single copy of a work, which, if lost, would be irretrievable."[37] The single copy comprised the ideas, the drawings, and the chief engineer.

Anticipation

Babbage created the various paper versions of his analytical engines by drawing upon all the splendors and miseries of designing and implementing the first Difference Engine. Since his experiences with the previous Difference Engines underscored the importance of contrivances to ensure the security of the calculations, Babbage put such contrivances at the heart of his later designs. In attempting to improve dramatically the design of his first Difference Engine, Babbage became worried above all about the time required to perform carry, as befit a practical economist much worried with factory organization: "The difficulty did not consist so much in the more or less complexity of the contrivance as in the reduction of the *time* required to effect the carriage." Babbage came to a breakthrough on paper—a new vision of effecting carry by teaching the machine to look ahead: "At last I came to the conclusion that I had exhausted the principle of successive carriage." He continued in an anthropomorphic vein: "I concluded also that nothing but teaching the engine to foresee and then to act upon that foresight could ever lead me to the object I desired, namely to make the whole of any unlimited number of carriages in one unit of time."[38] Babbage's assistant apparently "seriously thought [Babbage's] intellect was beginning to become deranged."[39] Babbage explained what it meant to anticipate: "If the mechanism which carries could be made to forsee* that its own carriage of a ten to the digit above when that digit happens to be a nine would at the next step give notice of a new carriage then a contrivance might be made by which, acting on that knowledge, it should effect both carriages at once."[40] Carries would effectively be performed before the additions that would cause them. The envisioned mechanism was complex and would have been heavy, large, and a profound challenge to manufacture, requiring extensive and costly hand finishing.[41] By December 1837, Babbage had an impressive series of stereotype plates made up to illustrate the functioning—monuments to his draftsmen's and collaborators' development of technical illustration.[42] He distributed copies of these plates, though he never published them.

The complexity of ensuring robust and expeditious carry led Babbage to the most radical and modern-appearing quality of the analytical engine: its innovative architecture. He explained that the sheer amount of machinery necessary for carriage "led me to separate the machine into two parts—one (the mill) in which operations were performed[,] the other (the store) in which quantities used and produced were inserted."[43] The analytical engine was thus to have a "mill" (a central processing unit) distinct from its "store" (memory).

While discussing the idea of anticipation, Babbage clarified his understanding of mechanizing operations of reason: "In substituting mechanism for the performance of operations hitherto executed by intellectual labour it is continually necessary to speak of contrivances by which certain alterations in parts of the machine enable it to execute or refrain from executing particular functions. The analogy between these acts and the operations of mind almost forced upon me the figurative employment of the same terms." He was careful to underscore that he was not suggesting that the machine in fact had knowledge in any meaningful sense: "For instance, the expression 'the engine *knows*, etc.' means that one out of many possible results of its calculations has happened and that a certain change in its arrangement has taken place by which it is compelled to carry on the next computation in a certain appointed way."[44] Sequential carry mimicked a human technique for performing addition. The foresight of the anticipating carry had no obvious analog in human arithmetic (though it can be understood as a property of addition). While the proposed machine did not "think," Babbage's machine, with its ability to forsee carries, outstripped the cognitive capacities of its human creator.

In October 1695, Leibniz sketched an emblem to celebrate his calculating machine, which he never brought to perfection. Hovering above his machine, the words "Greater than man" pointed to the essential lesson of his invention: its accuracy and speed in calculation surpassed those of its creators. "We wonder," Leibniz wrote, "at the divine reason built into this thing. / Greater than man is a machine built by human hands."[45]

Artisans and Their Philosophers: Leibniz and Hooke Coordinate Minds, Metal, and Wood

"... the arithmetical instrument that puts all the labor of the mind into wheels."[1]

In the early 1690s, Gottfried W. Leibniz had to disappoint a potential buyer. Since "the present Monarch of China infinitely loves arithmetic," the Jesuit missionary Claudio Grimaldi had a "great desire" to have a copy of Leibniz's machine to give to the Kangxi emperor.[2] Leibniz could not, alas, "satisfy him so quickly," as he did not have a machine on hand, and the machine had far too "many gears" to be replicated with any speed.[3] Noted for his facility with the abacus and his knowledge of mechanical devices, the Kangxi emperor got his calculating machines anyway, likely built by the skilled artisans in the joint Jesuit-Chinese workshops within the Forbidden Palace.[4] The machines, which aided addition and subtraction, have a construction apparently unlike any of the known European designs.[5] However inspired by rumors about European machines, they were almost certainly designed and built in workshops in China.[6] As part of their effort to secure access to the Chinese emperor, the Jesuits sent clockmakers to build and repair clocks, automata, and other sophisticated mechanical devices.[7] These skilled Jesuits trained Chinese clockmakers, and they labored together in workshops. This organized group of skilled and innovative artisans likely constructed a series of calculating machines.

Like their counterparts in Europe, the Chinese machines were possible thanks to accumulations of artisanal capital: artisans, their knowledge, and their competencies. Unlike their counterparts in Europe, the machines were produced in organized workshops under imperial control. Absent such organization, building machines required finding artisans and coordinating them. The history of the development of calculating machines is a history of the coordination and organization of mechanical innovation, artisanal skill, and autonomy. One recent study of early-modern technology notes, "Artifacts

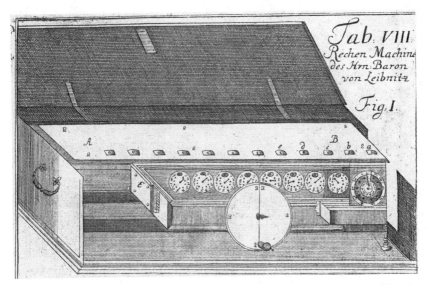

FIGURE 2.1. Diagram of exterior of Leibniz's calculating machine, from Jacob Leupold, *Theatrum arithmetico-geometricum: Das ist: Schau-Platz der Rechen- und Meß-Kunst* (Leipzig, 1727), plate VIII. (Courtesy of Rare Books and Manuscript Library, Columbia University.)

are the material representations of sociotechnological networks; unless working processes were centralized, their realization relied on a whole series of exchanges."[8] Gaining access to such networks and exchanges was difficult; so was bringing and holding a network together.

In 1671, Leibniz announced to the French theologian Antoine Arnauld two machines he had envisioned: one for arithmetic, the other for geometry. The first promised to perform all four operations of arithmetic "mechanically without any work of the mind." The second promised a new way of "determining, using a machine, analytical equations and the proportions and transformations of figures without tables, calculation, or the drawing of lines." This machine would enable the "perfecting of geometry, so far as is necessary for utility in life." The implications of this geometric machine were massive: "If we could transform all figures into thinkable ones, I do not see what could still be desirable for use."[9] The transformations of the envisioned geometrical machine were realized not in any physical machine but in his new symbolic calculus that permitted just such transformations of figures.[10] Leibniz spent decades and a small fortune attempting to realize his arithmetical machine. No prototype machine was functioning adequately when he died in 1716. His legendary failure served as a great inspiration for later inventors, who typically knew only the machine's appearance and promised function (see figure 2.1).

Saving mental labor required Leibniz to become deeply involved in coordinating mental and manual labor. The previous chapter discussed representations of the roles of artisans in building the machines of Pascal and Morland. That artisans were important is clear; how they were, less so. The actual contributions of the artisans remain opaque; so too do the ways in which artisans and philosophical inventors worked together.[11] This chapter shows how Hooke and Leibniz organized and coordinated the skills and knowledge of others. These cases reveal no stable, strict hierarchy between inventor and artisan, creator and implementer, between intellectual and manual labor. Such a stable hierarchy was more ideal than real, more desired than experienced.

Continuing the previous chapter's focus on the role of artisans in creating calculating machines, this chapter adds the dimension of credit—financial as well as intellectual. Michael Sonenscher has argued, "The work of making shoes, or hats, or coaches, involved recognition of rights and obligations of some kind. Work did not simply happen—it was a continuous process of negotiation."[12] Such negotiation involved an ongoing struggle over how to engage and reward the creative and mimetic aspects of artisanal labor. To grant too much importance to artisanal knowledge at any moment was to threaten the exclusivity of the claim of the inventor to intellectual, honorific, and financial credit. Yet motivating skilled artisans to enable philosophic projects depended on financial, honorific, and symbolic recognition of their importance.

The contributions of artisans and the processes of coordination are often hidden from historians; most interactions occurred face-to-face and left few unambiguous traces. Since a number of Leibniz's working papers on his machines still exist, we can investigate his case in unusual detail. When Leibniz moved from Paris to Hanover in 1676, his interactions with his Parisian artisan Ollivier became mediated exclusively through written correspondence, with agents in Paris attempting to supervise the artisan and with Ollivier himself. These letters allow us to see a world of everyday management of technological development and of labor; they let us see what kinds of knowledge and skill Leibniz hoped his artisan would apply toward perfecting the calculating machine; they let us see Ollivier's and Leibniz's ingenuity at work in struggling to design mechanisms given the properties of the materials of the world; finally, they let us see how Leibniz drew on the experience of depending on an artisan in his philosophical work.

Philosophers found it hard to enroll artisans, weights, and springs into their projects. Social coordination and the technical development of the machines came together. To build a machine, a philosopher needed knowledge about materials; to gain any practical form of such knowledge in the early-modern era, he needed to learn about people possessing such knowledge; he then needed to gain access to them and their knowledge.[13] To build a machine, a philosopher-engineer needed competence in coordinating with others in order to allow a cross-pollination of abilities and knowledge. Building a machine was no mechanical process.

Presenting a Multiplying Machine

In February 1672–73, Leibniz presented a model of his calculating machine to the Royal Society of London.[14] He had been working on calculating machines for several years but only began his serious work once he arrived in Paris in 1672.[15] The records of the Society note that Leibniz "shewed them a new arithmetical instrument, contrived, as he said, by himself, to perform mechanically all the operations of arithmetic with certainty and expedition, and particularly, multiplication." His model of an instrument was far from finished: "He gave some proof of what he said, but acknowledged the instrument to be imperfect, which he promised to get perfect, as soon as he should be returned to Paris, where he had appointed a workman for it."[16] Soon after Leibniz returned to Paris, he wrote up the first draft contract for an artisan to produce the machine. The artisan was to declare Leibniz's "design" good and to promise as "an honest man to keep the affair secret, and to speak of it to no one." For a fixed sum, the artisan was to produce the machine in three weeks.[17] It took instead some forty years.

Before turning to Leibniz's two solutions to mechanizing multiplication, consider one way of performing multiplication by hand. If we were to multiply 4567 by 245 by hand, we would probably write something like the following:

$$
\begin{array}{r}
4567 \\
\times \quad 245 \\
\hline
22835 \\
18268 \\
+ \quad 9134 \\
\hline
1118915 \\
\end{array}
$$

We would probably produce the intermediate stages, such as $4567 \times 5 = 22835$, using memorized multiplication tables for single digits: $7 \times 5 = 35$, so we write 5 and carry the 3. We could also find the intermediate stage simply by adding 4567 five times: $4567 \times 5 = 4567 + 4567 + 4567 + 4567 + 4567 = 22835$. The multiplication of multiple-digit numbers can be reduced to a series of repeated additions. Then sums of each series of additions (22835, 18268, 9134) are added. Mechanizing multiplication done in this fashion requires the following:

1. Mechanizing multiple additions of the same number—for example, 4567 by 5, then by 4, then by 2
2. Mechanizing the process of adding each intermediate amount to a higher decimal place—that is, the mechanization of the process of offsetting each intermediate amount one column to the left:

$$
\begin{array}{r}
22835 \\
18268 \\
+ \quad 9134 \\
\hline
\end{array}
$$

Let's start with the first stage in the mechanization: the process of mechanizing multiple additions of the same multiplicand (4567) for each digit of the multiplicator (245). In Pascal's machine, the user has to dial each addition for each place value manually. To multiply, say, 4 by 5, the user would have to dial 4 into the machine five times (like dialing the phone number 44444 on a rotary phone). To multiply 45 by 5, the user would have to dial 5 on the first input and then 4 on the second into the machine five times: 5 then 4, 5 then 4, 5 then 4, 5 then 4, 5 then 4. Obviously, such a process quickly becomes tedious. Machines such as Grillet's and Morland's used forms of Napier's bones to speed up multiplication.

Leibniz sought to have the machine do the entire calculation. To automate the process of performing multiplication via repeated addition, Leibniz needed a way to input the amount to be added for each place value and then to have that amount added to result wheels as many times as necessary. In our example above, Leibniz needed to create a way to set the four digits to be multiplied on the machine to 4 5 6 7, and then have the machine add 4 5 6 7 four times in their respective place value columns, without having to dial each digit manually multiple times. He needed what he called a "multiplying wheel" that could be set to add a given digit multiple times.

During his time in Paris, Leibniz devised two solutions to perform these additions mechanically, each of which was reinvented multiple times in the next two centuries, likely with no knowledge of Leibniz's ideas. In one solution, he envisioned a "wheel with mobile teeth," now called a variable cogwheel (upper left in figure 2.2). In such a wheel, "as many teeth" will protrude from the cogwheel as are needed to equal "the number to be multiplied." To multiply 5 by 4, for example, "you make five teeth" protrude from the cogwheel and make it turn it four times, adding a total of 20 in a given column (and thus causing two carries).[18]

His second solution was the "wheel with unequal teeth," now called a stepped drum, which was central to nearly all his machines (figure 2.3).[19] The stepped drum is a cylinder with nine teeth of increasing length arranged

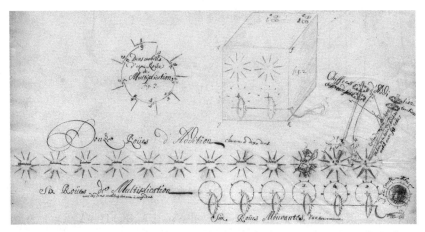

FIGURE 2.2. Manuscript drawing of Leibniz's multiplying machine, including variable cogwheel—"Dens mobiles d'une Roüe de multiplication" (upper left). c. 1673. LH 42,5, f. 29r. Courtesy of Gottfried Wilhelm Leibniz Bibliothek—Niedersächsische Landesbibliothek, Hanover.

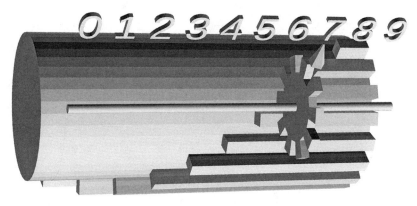

FIGURE 2.3. Schematic diagram of stepped drum. Author's image, OpenSCAD.

radially. A small gear moves along the axis of the drum. Depending on the position of the gear, a different number of teeth will be communicated when the stepped drum rotates 360 degrees. Setting the number 4 into the machine would move the small gear so that each full rotation of the stepped drum would move its corresponding gear four steps. The stepped drum provided the multiplying mechanism in Leibniz's machines from 1674 at the latest until his death.

Whether with variable cogwheels or stepped drums, the machine would, in principle, work similarly. Take our example of multiplying 4567 by 245. This involves setting 4 5 6 7 into the machine and then causing all multiplying wheels to rotate five times by turning a crank (as shown being held by a hand

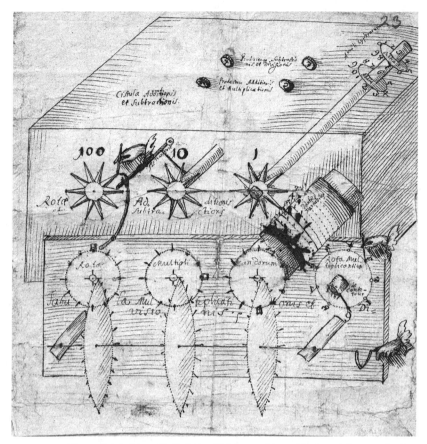

FIGURE 2.4. Manuscript drawing of Leibniz's multiplying machine, with stepped drum mechanism at right., c. 1673. LH 42,5, f. 23r. Courtesy of Gottfried Wilhelm Leibniz Bibliothek—Niedersächsische Landesbibliothek, Hanover.

in figure 2.4). To multiply the next digit, 4, the user moves the entire multiplying mechanism one digit to the left and then turns the crank, causing the multiplying wheels to rotate 4 times, and so forth.

The diagrams of 1673–74 (see figure 2.4) suggest that Leibniz showed the Royal Society the basic idea of using a multiplication carriage (in front and on the bottom) that could freely move left and right along the primary addition apparatus (behind and on top). The diagrams of 1673–74 do not show the adding mechanism and offer little indication of how carries would be performed. Leibniz may have been protecting his current mechanism from prying eyes; more likely, the machine displayed to the Royal Society was incomplete precisely because it could not automatically perform carries. Given that any multiplication would necessarily involve sequences of additions, solutions to the mechanization of addition had to be robust, fast, and accurate. Leibniz realized quite early on that he would need to solve both of the key problems of carrying mechanisms discussed in the previous chapter: the sufficient-force problem (the need to propagate carries across multiple digits) and the keeping-it-digital problem (the requirement that the machine report only counting numbers, not something in between). His working papers show him considering the merits of using springs versus weights in providing sufficient force.[20] He expressly noted that the gears in the machine must advance single digits, "no more, no less."[21] These documents reveal how far Leibniz was in 1673 from a functioning mechanism: they articulate more the general requirements for a machine than an actual design of one. For example, he described the carrying mechanism in open-ended terms: "The preceding cylinder of numbers, in finishing its rotation, and passing beyond, releases or delivers a suspended weight or a tightened spring, which, in falling on or in hitting the teeth of the following ring with enough speed, pushes it beyond, or makes it advance the tenth part beyond the movement that the ring receives from the addition wheel."[22] Bringing the carry mechanism into practice would require that Leibniz and his artisans convert all the vague, qualitative descriptions in his working papers—"a reasonable length," "enough speed"—into physical mechanisms grounded in the properties of available materials. Before Leibniz and his artisan did even that, they would need to decide whether to use springs or weights. Much remained to be done when Leibniz presented his machine in London in the winter of 1673.

A Hooke Look

When Leibniz presented his machine to the Royal Society, the behavior of a prominent member took him aback. To Leibniz's great surprise, Robert Hooke studied the machine in detail: "He removed the back plate that covered it, and absorbed every word I said; and so, such being his familiarity with mechanics and his skills in them, it cannot be said that he did not observe my machine." Hooke did not, however, "distinctly trace out all its wheels, I readily admit."[23]

A month later, in early March 1673, Hooke "produced his [own] arithmetical engine" before the Society "and shewed the manner of its operation which was applauded."[24] Hooke recorded in his diary that he had "Shewd Arithmetick Engine all understood and pleased,"[25] though he omitted to note that the Society "desired" him "to bring in the description of it, that so it might the better appear how it differed from that of Mons. Leibnitz."[26] Hooke appears to have reached more or less the same stage as Leibniz in a matter of weeks rather than years. We know little about Hooke's mechanism, other than he claimed that it performed all four arithmetical operations up to twenty places and made no use of Napier's bones. No models or manuscripts concerning Hooke's machine are known to exist today.[27] Some model for it existed and functioned well enough to gain assent from the Royal Society; Hooke's inventory of instruments testifies to the existence of several arithmetical machines.[28]

Leibniz saw Hooke's creation of a rival machine as a major violation of intellectual decorum and protocols concerning new inventions.[29] In a letter to Henry Oldenburg, Leibniz noted angrily that Hooke "should leave the development and perfection" of the machine up to Leibniz, "as the substance of the invention is mine."[30] Leibniz had done the work of envisioning the fundamental invention, which he referred to with the traditional scholastic term of *substance*, which was capable of taking on different material forms with different accidents. The perspicacious Hooke had glimpsed and instantly grasped this substance while examining Leibniz's model. Whatever work remained to be done, Leibniz suggested to Oldenburg, the essential work of inventing the substance was finished. Only some details—the accidents, as it were—remained to be worked out. Leibniz's manuscripts suggest otherwise.

What enabled Hooke to produce a model so quickly? When Leibniz set out to implement a calculating machine, he had to build up a knowledge of machines, materials, and, above all, people necessary to produce *even* a model of his calculating machine: Leibniz needed others less to "express" a fully formed idea and to construct something based on a well-specified plan than to develop his preliminary work in a practically meaningful way. Hooke had all this at the moment he peered into Leibniz's machine. Hooke's diary reveals the dense interactions that led to his model machine. He rapidly assembled people with diverse skills and knowledge to help him produce a model of a calculating machine sufficiently functional or illustrative to garner the informal approval of the Royal Society. Rob Iliffe notes, "Hooke could have machines made and instruments or watches repaired as fast as expertise then allowed."[31]

Hooke's diary reveals how quickly he involved his friends and usual artisans in producing a machine. On February 2, Hooke noted, "Sir Jonas Mo[o]re and Cox here: compleated arithmeticall engine in the contrivance for a product of 20 places." A well-known practical mathematician, Moore was a producer of popular arithmetic textbooks (mentioned in chapter 1) and an organizer of all sorts of engineering projects, such as drainage.[32] The optical instrument maker Christopher Cox built many of Hooke's microscopes and telescopes.[33] Two days later, more colleagues became involved: "Mr. Haux, Mr. Wise, Mr. Colwall, here, compleated Arithmetick Engine." On February 3, Hooke "Told the Society of Arithmetick engine." The Royal Society journal book notes that Hooke "mentioned, that he intended to have an arithmetical engine made, which should perform all the operations of arithmetic, . . . much more simply than that of Mons. Leibnitz." The Society "encouraged" Hooke "to make good his proposition."[34] Hooke made good on his proposition by using Richard Shortgrave, the operator of the Society, to produce a model of his machine.[35] He saw "Shortgrave about Arithmetical Engine" on February 26. The same day that he demonstrated the machine for the Royal Society, Hooke "prepared Arithmetick Engine by Shortgrave."[36] Shortgrave appears to have continued working on the engine. In the meantime, another artisan had become involved: on March 11, Hooke "bespoke Arithmetick Instrument of Stanton"; nine days later, on March 20, "Mr Stanton shewd me his module of Arithmetick engine." In the next year, Hooke mentioned his machine in his publication attacking Hevelius. Hooke explained to John Pell, "I have a mechanicall way of calculating and performing <u>Arithmetical</u> operations."[37] After that, little more is heard of Hooke's arithmetical engine.[38]

Hooke famously had mechanical knack, vision, and insight.[39] As importantly for our story, he had contacts, relationships of trust, and skill in profiting from organizing others. Like Paris, London had a great concentration of skilled artisans in many domains, though information about artisanal knowledge and skills was massively imperfect and poorly distributed. Hooke knew which people knew what, and he kept records about them. Iliffe argues that Hooke's diary is "a deeply enriched account of locally available techniques and skills which are either put into practice immediately or saved for his use a few months later."[40] Hooke participated in a constant circulation of knowledge, gifts, and favors among mathematicians and artisans in London. The knowledge and techniques of others helped Hooke refine and develop his design rather than simply implement an idea.[41]

Despite creating a now lost calculating machine, Hooke was largely dismissive of the utility of such machines compared to normal ways of calculating using pens or jettons (counters used to perform arithmetic). He mocked Morland's machine as "very silly."[42] He was harshly critical of Leibniz's machine: "It seemed to me soe complicated with wheeles pinnions Cantrights springs screws stops and Truckles that I could not perceive it ever to be of any great use especially common use. . . . [I]t could only be fitt for great persons to purchase and for great force to remove and manage and for great witts to understand and comprehend." The machine would be too big, too expensive, and too prone to break ever to be of any real use. Hooke noted another problem: how could one know whether the machine had functioned correctly? "I saw noe meanes of examining whether the operation had beene truly performed without trying it over again which is intollerable." He found but one thing to praise: "The designe indeed is very good which is the onely thing

I was able to understand of it which is to give the Product and Quotient of a Multiplication or Division which Sr Samuel Morelands Instrument is not all adapted to."[43]

Even if Hooke dismissed the utility of calculating machines in general, he did not criticize Leibniz for presenting only a model of his machine: inventors were supposed to present models early in their projects.[44] Elsewhere, Hooke explained the utility and purpose of models of machines. An inventor does not "exhibit" a model of a machine "for any other use than to show the ground and a reason of thing, and as a sensible object, upon which to reason and discourse, and for the more plain demonstration and explanation of all material doubts, that might arise."[45] Leibniz's model allowed Hooke and the other members of the Royal Society to see the overview of a proposed machine and envision difficulties. Enabling that inventor to claim priority, the model stage allowed others to aid an inventor. Hooke saw plenty of difficulties in Leibniz's model. He was right: the "multitude" of the parts of Leibniz's machine made "it exceeding hard to putt into good order and extraordinary apt to be putt out of it." Putting Leibniz's machine into order required the process of putting into order the knowledge, skill, and diligence of many others.

Finding Artisans, Pinching Their Secrets

When Leibniz arrived in Paris in 1672, he was interested above all "in understanding machines and in inventing them," with special attention to his calculating machine.[46] "Wanting to have my Arithmetical Machine made," Leibniz explained in 1673, he came "to enter into knowledge" of the artisans of Paris.[47] As the "Metropolis of Galanterie," Paris was filled with remarkable artisans and their wares: "They are marvelous in the art of varnishing in the Chinese style; . . . The art of molding or casting every kind of material in sand-forms is in its greatest perfection, as is the casting of medals. . . . They work well in iron, and they have people with the secret of casting iron into forms, and of filing it in ways not commonly possible. . . . In sum, there is an infinity of curiosities, in gold making, enameling, glassmaking, clock making, leather working, pewter pottery,"[48] The rewards from taking the secrets from just one of these trades, Leibniz wrote, would be well worth the cost.

Paris had structural advantages allowing for such a concentration of artisans, skills, and materials. Leibniz remarked, "Manufactures, for the most part, are in the most flourishing state that could be wished for, in part because of the ingenuity of the nation, in part because of the particular care of the king, who had made the best artisans from all around come, and . . . [took] from them their secrets and their inventions."[49] Leibniz greatly admired Colbert's policy of attracting artisans—the great as well as the middling—and then stealing their secrets.[50] He undertook to do it a bit himself. Gaining knowledge about artisanal skill and the trust necessary to gain access to it took a little effort: "It will be important, to poach from workers here the tricks and refinements of their secrets, which one can do sometimes with a bit of shrewdness combined with a little liberality." He praised himself for having done so: "As for me, I had occasion not just to visit a number of good artisans, but also to draw also something from them. Had I wanted to pay out a little, I would have learned yet more."[51] A little cash and a little skill eased the way. From knowledge about artisans, Leibniz could move to gain a bit of their knowledge.

Among Leibniz's manuscripts now in Hanover are many sheets, often un-labeled and mostly lacking dates, with numerous kernels of artisanal knowl-edge Leibniz picked up over the years. These concern materials, techniques, and mechanisms. One charming example begins, "I have found a way of hard-ening ordinary cardboard [*carton*] so much that I make ordinary clocks and pendulum clocks from it."[52] Leibniz's correspondence includes examples of contracts with philosophers and chemists to channel secrets to him in exchange for regular payment.[53] When Ollivier planned to move to Hanover in 1679, he was instructed to collect all information about his craft and trades allied to it. Ollivier "promised to fill his head with all the most beautiful inventions of this country" to reproduce in Hanover.[54] Moving artisans was a crucial part of technological transfer in the early-modern world, and something Leibniz thought particularly necessary for the economic development of Germany.[55]

Early during his Parisian stay, Leibniz explained, "I am reduced to a vol-untary hermitage, to the extent of speaking to almost no one except my two artisans." Working with them on his machine required "a man entirely."[56] Leib-niz chose the Parisian clockmaker Ollivier to construct the calculating ma-chine. Leibniz explained that he had "run all over Paris before picking this one." He found fault with many artisans: "Skilled ones are either too expen-sive or colorful, some debauched, others self-interested . . . I had great trouble in finding a handy person who took pleasure in machines and who worked on them because of inclination, rather than self-interest."[57] An artisan driven only by self-interest would never dedicate himself enough to the project: Leib-niz needed someone with a fascination with machines, not just a job making them. Ollivier was not among the famous mathematical instrument makers of early-modern Paris, most of whom were familiar to Leibniz; but for his work with Leibniz on the calculating machine, he would be unknown.[58] Ollivier did not come cheap. For a steep price, he was willing to dedicate himself entirely to the perfection of Leibniz's calculating machine.[59] Transforming Leibniz's calculating machine from the nonfunctional demonstration model presented to the Royal Society to something functional enough that Leibniz could sell it to the French crown and receive the approval of the learned proved far more difficult than Leibniz, or his artisan Ollivier, ever expected.[60]

When Leibniz denounced Hooke's conduct during the demonstration of his model to Oldenburg, he conceded that Hooke did "not distinctly trace out all its wheels." Someone like Hooke hardly needed to, given his ability to grasp quickly the essentials of the mechanism: "In such cases it is enough for a man who is clever and mechanically-minded to have once perceived a rough idea of the design" and then "to add to that a little of his own, consisting only of some involvement of the wheels that can be effected by different people in different ways."[61] Since Hooke could grasp almost instantly the essence of an invention, he ought not, in this view, even be given credit for a simultaneous invention.

The story of Leibniz's machines underscores just how nontrivial the "involvement of the wheels" proved to be. To give credit only for the overall design was to disregard the work necessary to fill in those complications—work that was often dependent on the skills and knowledge of an "ingenious and mechanical man," such as Ollivier.

A Propagation of Carrying Difficulties

In May of 1677, the clockmaker Ollivier wrote to Leibniz, now in Hanover, to excuse his slow progress, to complain about his pay, to ask for clarification, and to promise to return to the project before long. He noted, "I still have difficulty with the carry, which cannot be made as easily as we had proposed."[62] While Ollivier, when pressed, acknowledged the machine in general to be Leibniz's, the solution to the problem of carry evidently had been, and continued to be, a collective enterprise. Solving the problem was hampered by Leibniz's secrecy and distance: when he left Paris, Leibniz had taken "the imperfect machine" that "would have helped" Ollivier. In a lost letter, Leibniz pressed Ollivier to explain the difficulties clearly. Ollivier responded, "I cannot explain the carry or your own drawings upon which we worked together. The machine would need to be present to let you understand [the difficulty], but since you have decided to send the wheels of the small model with the rolls it will suffice for that."[63] Choosing not to send Ollivier the requested model, Leibniz wrote instead a remarkable "memoire" setting forth an improved design for the carry mechanism of the calculating machine.[64]

When Leibniz moved from Paris to Hanover, much that would have been oral and haptic had to be written down and drawn. Moving from face-to-face to exclusively written and pictorial collaboration was difficult for Leibniz as well as Ollivier. Apparently literate but little used to written communication, Ollivier rarely managed to make his technical difficulties clear. Leibniz was better at communicating solutions in ways meaningful and helpful to Ollivier, who clearly longed to discuss things in the presence of the parts and models in question; Leibniz was poor at directing and informing his agents in Paris who managed Ollivier's work on his behalf. In the letters and the memoire exchanged between Leibniz and Ollivier, as well as between Leibniz and his go-betweens, we can discern in unusual detail the financial and technical negotiations and interactions necessary to produce an adequately functional device. The evidence reveals the scope of creative activity Leibniz expected from Ollivier, even if it is less clear about Ollivier's precise contributions toward the better-working machines of the 1680s and 1690s.

Leibniz's memoire for Ollivier offers an unusual glimpse into what a "philosophic" inventor thought important to include and exclude, what skills he thought artisans could provide, and what expectations he had of the range of artisanal inventiveness and labor. Unlike many of the other draft documents concerning Leibniz's machines, the memoire is a carefully, if quickly, drafted composition for a specific master artisan with whom he had worked closely, shoulder to shoulder, in the previous years.

Leibniz's memoire is a testament to the difficulty of technical communication before the age of standardized written and pictorial communication practices in engineering.[65] Leibniz first described the various parts of the apparatus and their basic mechanical relationships. He then described how the mechanism operates and how a carry works. Finally, he gave a theoretical "proof of concept" by showing that the described mechanism can, in principle, accommodate all the possible states of the machine. Leibniz wrote characteristically in the language of combinations and claimed to demonstrate that the machine so specified would work correctly for all six possible combinations of situations. Whatever the insufficiencies of the description and depiction in the memoire, Leibniz wanted it kept secret, and he wanted it back. The longer Ollivier kept the memoire, the more Leibniz worried it might fall into the hands of other inventors or philosophers. The memoire, after all, contained much greater detail about his mechanism and techniques than he had ever shown or discussed in a public forum.

In the memoire, as in nearly all his writings on the calculating machines, Leibniz provided few metrical specifications of the parts involved: he provided only their relative relationships in space, their proposed motions, and a general sense of the proportions of the pieces to one another. Other than specifying the number of teeth of various gears, he gave no measurements of size or angles for the parts. Considerable recent work by engineers and historians attempting to reconstruct Leibniz's machine demonstrates how carefully the machine must be calibrated or tuned in order to function with any degree of accuracy.[66] So does Leibniz's later correspondence from the 1690s and early 1700s, concerning the later version of the machine that has come down to us. Scholars of Leibniz's work on machines and mechanism have underscored this inattention to detail as a cause of his failures as an engineer.[67] Leibniz left a wide array of crucial details up to Ollivier. After all, he employed Ollivier precisely because he was the sort of inventive artisan capable of resolving such difficulties.

In the previous chapter, we saw two major problems to be solved in carrying mechanisms: the sufficient-force problem and the keeping-it-digital problem. Leibniz's multiplying machine had to solve yet another problem of carry: it needed to be able to accommodate an addition of up to 9 and a reception and/or communication of a carry all at the same time for each digit of the machine. If both operations were simply to be effected on gears mounted upon or driven by the same axis, the axis would get stuck and cause the carry or the addition mechanism to jam, if not both. To solve these problems, Leibniz and Ollivier needed to devise a mechanism to accommodate any of the following situations for a given column (in these examples, the tens column). Leibniz clearly set them out:

1. An addition alone, as in 189 + 10 = 199
2. Receiving a carry alone, as in 189 + 4 = 193
3. An addition and receiving a carry, as in 179 + 12 = 191
4. An addition and communicating a carry, as in 189 + 30 = 229
5. An addition and receiving a carry as well as communicating a carry, as in 189 + 31 = 230
6. Doing nothing or, as he put it: "no action or passion"

The basic setup in the July 1677 carry mechanism involved an axis $2A2B$ with $2G$, a ten-toothed gear, and $1H$, a disk with a hole punched in it (see figure 2.5). A cylinder $lmnopq$ can rotate freely around $2A2B$; it is placed between $2G$ and $1H$. The gear $2G$ is driven by the variable toothed wheels of the adding mechanism. Attached to $lmnopq$ is qp, a roll upon which are written the numbers for displaying the output of the machine for one digit. (10) is a ten-toothed gear fixed on $lmnopq$. RL is a ring attached on one side to $1H$, with a peg LZ extending through the hole on $1H$. A spring θ, not pictured in figure 2.5, presses RL against $1H$. Most of the time, this spring θ keeps the peg LZ meshed in the gear (10), as shown in figure 2.6. So long as the peg LZ is meshed in (10), the axis $2A2B$ and the cylinder $lmnopq$ rotate together, along with their respective mounted gears.

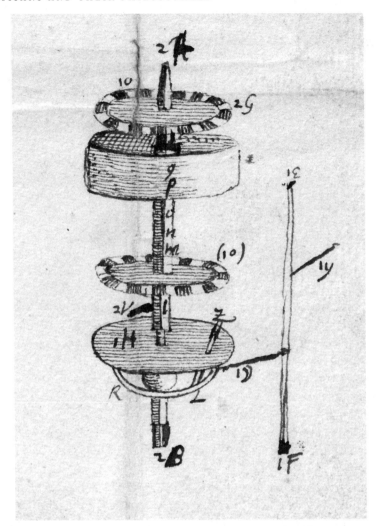

FIGURE 2.5. Carrying mechanism of 7.1677. *LZ* disengaged from *(10)*. Leibniz, "Memoire pour Monsieur Ollivier touchant la machine arithmetique perfectionée," 15.7.1677. LH 42,4, f. 9r. Courtesy of Gottfried Wilhelm Leibniz Bibliothek—Niedersächsische Landesbibliothek, Hanover.

In the case of receiving a carry from the previous digit, *lmnopq* and *2A2B* need to be able to rotate independently of one another, so *LZ* must no longer be engaged in *(10)*. To perform a carry, then, some mechanism must first move *LZ* out of *(10)* and then advance *(10)* and thus advance *lmnopq*. Two pins *1D* and *1y*, both fixed on the axis *1E1F*, perform these two steps. First, *1D* inserts itself between the ring *RL* and *1H*, thus pushing them apart. *1D* thereby briefly pulls the peg *LZ* out of *(10)*. (This is the moment shown in

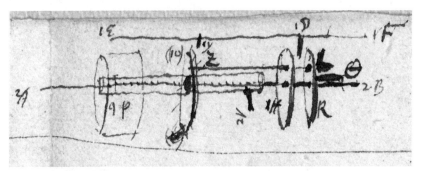

FIGURE 2.6. Carrying mechanism of 7.1677. *LZ* engaged in *(10)*. Leibniz, "Memoire pour Monsieur Ollivier touchant la machine arithmetique perfectionée," 15.7.1677. LH 42,4, f. 9r. Courtesy of Gottfried Wilhelm Leibniz Bibliothek—Niedersächsische Landesbibliothek, Hanover.

figure 2.5.) Just after this, the pin *1y* advances *(10)*. It must do so before the spring *θ* can push *RL* back to *1H* and the peg *LZ* back into *(10)*.

In other words, *1D* very briefly disengages the cylinder *lmnopq* from the axis *2A2B*—just long enough to allow the reception of a carry without jamming the mechanism if an addition is happening at the same time. Leibniz and Ollivier had worked out much of this mechanism before Leibniz left Paris in 1676, particularly the need for a cylinder capable of rotating freely on the axis but usually connected to it. Leibniz's 1677 memoire offered a new approach for temporarily releasing the cylinder from the axis. The delicate mechanism where *1D* briefly separates *RL* and *1H* was new—a replacement for an unknown mechanism in the earlier design, where *RL* was replaced by a simpler spring mechanism.

Putting this ingenious mechanism into practice required a careful calibration of the speed of the machine, the size of parts, and the angular distance between the pegs *1D* and *1y*, as well as the strength of the spring *θ*. Leibniz left the precise tuning and placement to Ollivier to figure out. Leibniz added a marginal note about the shape of the "sword" *1D*: "Note that the piece *1D* will move out of the space between *LR* and *1H* only when a carry occurs; thus *D* must be of a size that the width diminishes toward the sword." Ollivier was to make sure that the piece had this peculiar thickness. The vague terms Leibniz used in setting forth the timing involved with *1D* and *1y* are even more striking:

A piece *1D* of the shaft *1F*, entering a little before like a sword between two [missing word] will cause the collar *LR* to pull back with its pin *LZ*, which will

happen at the moment that the cylinder *lmnopq* receives a carry by means of
the piece y attached to the same shaft with the piece *1D*, but the separation will
be done a little before the carry is done, *which will happen in a moment*; after
this the collar will go back in place and the pin *LZ* will go back between the
teeth of the wheel *(10)*.[68]

It was up to Ollivier to put "a little before" and "in a moment" into some practical quantity and to translate the metaphorical language of a "sword" dividing
two things into a realized mechanism. Leibniz's earlier working papers use
similarly vague and open language. In one of his early sustained discussions of
carry mechanisms from 1673, as we saw above, he called for a "ring" of "reasonable size" and discussed using a weight *or* a spring "falling, or hitting the
teeth of the following ring, with enough speed, to push it before, or make it advance one tenth more than the ring receive from the addition wheel."[69] The size
and even choice of mechanism appear to have been largely left to the artisan.

Criticizing Leibniz here for providing insufficient detail and poor drawings
is anachronistic and misleading: he was providing the detail sufficient for the
kind of working relationship he had with Ollivier, grounded upon his understanding of the distribution of their skills and knowledge. The drawings Leibniz
provided Ollivier were only a rough approximation of what was necessary, as
Leibniz remarked in a marginal note: "I have not had the leisure to make the
design with all the required exactitude. For this reason, there are some pieces
that are neither well placed or proportioned and some of them correspond
rather to the model than the correction I have made in the written description." Ollivier could make up for this: "It will be easy to compensate for these
absences, as much in comparing the various drawings with the same letters . . .
as by the means of the written description that seems clear enough to me."[70]
Ollivier's work using the drawings and description was no standardized task,
his labor was not directed by a set of proscribed actions, and it cannot be characterized by obedience to a fully or well-specified design. Leibniz clearly expected Ollivier to be able to determine the appropriate parameters necessary
for making the carry mechanism function, and he expected Ollivier to be able
to produce all the parts embodying those parameters (or at least to be able to
have his workers produce them).

Historians of technology have stressed that different styles of technical
drawing and description imply and accompany different conceptions and organizations of intellect and labor.[71] More specified drawings aid the engineer
and manager in controlling and regulating production; less specified drawings
suggest the need for the discretion and competence of the producers; more

completely specified drawings widen the gap between inventor and producer; less specified drawings lessen it. The indeterminate nature of Leibniz's description and drawings—the quite deliberate gaps and omissions in Leibniz's memoire—testify to the creativity and the innovative use of outside knowledge and discretion Ollivier was supposed to provide. Ollivier's work was not that of the properly obedient artisans Pascal envisioned or of their modern heirs in factories. His ingenuity and personal initiative did not produce monsters and "abortions," as Pascal complained: his inventiveness was necessary to the devising of the mechanism for a well-functioning machine. His conceptions were not always misconceptions.

Throughout the early development of the arithmetical machine, Leibniz depended on Ollivier for minor tweaks, connections to suppliers, knowledge of metals and lubricants, and the organization of a workforce. Given basic topological parameters and arrangements, Ollivier was responsible for choosing among design options, such as whether to use a spring or a weight in a given place, as well the fine metrical and timing details of the design. In a 1718 book review, Leibniz described the long outstanding need for a work capable "of making all the practice of the art [of clockmaking] known, not only in its principal [purpose]—the measuring of time, but still more in relation to the accidents that comprise a number of pretty inventions in practice among the maker of the art."[72] Ollivier was charged with providing the "jolies inventions" necessary to bridge the gap between a model of a calculating machine and a functioning machine.

Documents that Leibniz drafted for Colbert and the Académie des sciences around 1675 make it clear that Leibniz valued Ollivier as more than just an ordinary craftsman and set out parameters for compensating him accordingly. In those papers, Leibniz relayed an argument he attributed to Ollivier. The clockmaker's compensation needed to factor in the risk involved in producing a new invention rather than simply producing already perfected devices such as clocks and pendulums. Skilled artisans capable of innovating needed incentives to do so: "It is just to pay a skilled master not only for the time he has worked—without speaking of the novelty and the risk of the enterprise—but also for his industry and his skill in distinguishing himself from an utter ignoramus." According to Leibniz, Ollivier "protests that he prefers to make his living easily in the ordinary way, rather than to embark for nothing in an enterprise full of disquiet and risk, and capable of turning off the most patient man in the world."[73] An artisan, in this view, needed appropriate motivation to seek distinction through the use of his inventive skills and to pursue what David Pye calls the "craftsmanship of risk" rather than the repetitive "craftsmanship of certainty."[74] To compensate him for the risk of innovation, the French crown accordingly paid Ollivier 300 livres.[75]

At the end of the brief memoire of 1677, Leibniz remarked, "See everything that I believe is necessary to say in order to understand the pieces and their functions." If Ollivier had any difficulties, he needed to express them clearly, in words and pictures: "If there's something doubtful, it will be good to write me right away and to explain, in words, and using a figure if necessary, for I will attempt to divine what he means."[76] While Ollivier initially claimed that the memoire had resolved his uncertainties, he soon ran into difficulty. In April 1678, he explained, "There is more difficulty in making these carries work than we had envisioned, and more pieces to make; I hope that the three digits will be made to perfection in a month, if money isn't lacking; I have worked too much at my own cost."[77] It took some time, and much coaxing from Leibniz's agent in Paris, before Ollivier attempted to explain more about the difficulties in question: "First, the multiplying wheels, the unequal teeth, that of addition, [and] those of carry are made. The springs for turning them one way or the other—with them comes all the difficulty of making them move. This has been done only through lots of time, always done and undone. As you know, fifteen days have passed without any advance." Progress on the machine had stalled for the moment, and Ollivier was not clear about the technical difficulties: "That's everything I can tell you about the progress of the machine."[78] The artisan seems to have made a machine within Leibniz's specifications, only to find it too difficult to use. Ollivier's impasse prompted Leibniz to rework his design some months later, to confront, in particular, issues of weight and timing. His working sketches of 1679 show that he concluded that he had been too cavalier in thinking about the issues of timing and speed in the carry mechanism, particularly in cases of carries upon carries. "First of all," he began, "the great speed of carries makes the person turning [the crank of the machine] feel the difficulty. Especially if ten or twelve transports must happen all at once." A "very great difficulty" was that the "speed will always be ever increasing by multiplication when there are carries upon carries."[79] More work was needed—so much so that several years later, Leibniz abandoned the approach to transport detailed in the memoire almost completely.

The problems with the machine were perplexing; so too the problems with payment. Ollivier continued: "[I have] received 195 *livres* on your account. That leaves 105 *livres* that you owe me, for that the entirety of the recompense that you promised me. I would rather make the rest for 100 écus than to make the Cause for 800 livres." In trying to decipher the letter, Leibniz wrote "I do not understand" next to "Cause."[80] The slow progress came no less from a fundamental breakdown in trust grounded in the clash between the expectation that Ollivier should act as a skilled and innovative master and the practice of remunerating him as if he were a day laborer.

Ollivier, Payment, and Work-Discipline

In early 1679, Leibniz prepared a detailed contract for Ollivier. We do not know if Ollivier ever signed it or ever saw it; it seems unlikely that Leibniz ever sent it to Paris.[81] The contract offers precious details on the machine Ollivier was to build: a fully functional trial version with only three multiplying wheels and keys for input; the carry mechanism was to act simultaneously with the additions and involve no delay.[82] The contract presents a normative organizational and disciplinary framework for managing Ollivier's work, to be used by Leibniz's agent in Paris, a German diplomatic figure named Friedrich Hansen. Clause fifteen, concerning nonperformance, reveals how difficult and nonproductive the relationship of the philosopher, the artisan, and the go-between Hansen had become: "I will work on it from that time without stopping, so much as is possible for me, and if I fail to achieve it in three months . . . I grant to M. Leibniz or his agent Mons. Hansen . . . [the right] to seize my goods and sell them . . . [and] I will be obliged to work on it under physical constraint, . . . and to decry myself among masters and apprentices and others as the infamy warrants, until I have accomplished what I have promised."[83] The final clause of the contract concerns work-discipline: "I am obliged to demonstrate weekly to Mons. Hansen or another of his choice, some considerable thing that I have accomplished, with a note for M. Leibniz that I will bring every week explaining what has been done, and what is to be done the subsequent week."[84] Such a system of regulated inspection, documentation, and payment fit poorly with Ollivier's position as a master artisan, and indeed with the creative activity he was called upon to do.

Once he moved from Paris to Hanover, Leibniz depended upon his agent in Paris, Hansen, to encourage, regulate, manage, and pay Ollivier. Deeply aggrieved, in debt, and perhaps a bit lazy, Ollivier proved difficult to organize on Leibniz's terms (or rather Hansen's imperfect understanding of Leibniz's terms). Ollivier's work for them waxed and waned in accordance with his satisfaction with his remuneration, as well as an array of unfortunate events in his life. Leibniz's draft contract set to systemize Ollivier's work as well as Hansen's supervision and management of that work. On Leibniz's behalf, Hansen had been attempting to coax, reward, and threaten Ollivier into completing the calculating mechanism and keeping it secret. Managing Ollivier was frustrating work for Hansen—he remarked in 1680 that he had not recovered from the "displeasure M. Ollivier caused me."[85] Again and again, Ollivier failed to complete parts of the machine and to follow up on his promises. He likewise failed to write to Leibniz explaining his difficulties and successes. Ollivier worked on clocks and instruments for other savants; he was often not at home or at his shops, sometimes not even in Paris; he possibly went bankrupt; he and his wife became extremely ill; relatives visited him. Hansen was often reduced to speaking with his children, his neighbors, and his employees to monitor his progress and even to ascertain his whereabouts.[86] Attempting to recover the secret manuscript memoire, Hansen twice sent his "lackey" to exhort, and likely to threaten, Ollivier.[87] Time and again, Hansen sought to force Ollivier to submit to weekly inspections of his progress in making the machine. Ollivier typically resisted simply by being away from his workshop.

Extraordinarily concerned about keeping the details of the machine secret, Leibniz likely chose Hansen just because he could not steal ideas or stake any credible claim about having invented or improved the machine.[88] Leibniz resisted Hansen's pressure to involve more learned savants, such as Edme Mariotte, in the supervision of Ollivier before finally relenting.[89] One morning, Hansen sent his valet to summon Ollivier, only to find him in bed. At this point, Hansen begged Leibniz to explain the nature of Ollivier's contract.[90] Leibniz drafted the contract, with its systematization of the rewards and coercions available to Hansen, a few months later. He probably never sent it and instead left Hansen to continue as he had previously.[91]

Hansen's stewardship began with trust between Leibniz and Ollivier already broken. This breakdown of trust involved amounts of payment and the significance of payment. Ollivier constantly demanded to be paid, and yet he refused at times the money Hansen offered.[92] Ollivier went so far as to threaten that "he would rather melt down all the wheels" into a "mass of copper" rather than be remunerated in a manner ill-fitted to his real contribution and kind of labor.[93]

Just as Leibniz was departing Paris in 1676, he summoned Ollivier to discuss the machine. Leibniz left their meeting thinking he had paid Ollivier adequately; the clockmaker thought otherwise.[94] Ollivier maintained that his pay should be proportional to the total amount promised for the finished machine: based on his advancement toward completion, the pay was not an hourly or weekly wage proportional to time.[95] Lacking guidance from Leibniz, Hansen initially agreed to pay Ollivier in some proportion to the final amount. He balked, though, at the high figure of 1,000 livres Ollivier claimed Leibniz had promised for the machine.[96] Hansen sought to pay Ollivier "at the end of each week in proportion as he had advanced," but at nowhere near the level Ollivier deemed suitable. At just this point, Ollivier threatened to melt down the work rather than suffer being treated in such a shabby manner.[97] Ollivier did not want to be paid as unskilled labor capable merely of reproduction and imitation, or more precisely, as a noninventive, noncreative, simply mechanical laborer. To accept small payments for daily work, and to be subject to certain forms of work-discipline, was tacitly to accept something like Pascal's normative picture of artisanal practice as the mimetic production of machines devised by philosophers and engineers rather than a vision of the productive interaction of philosophers, mathematicians, and the higher sort of artisans, all free and creative members of liberal métiers drawing together their complementary competencies.

We know more than Hansen did about the financial agreement. Documents Leibniz wrote for Colbert and the French Académie des sciences around 1675 reveal Ollivier's care in setting forth his demands for remuneration. Developing the machine was a risky endeavor that would monopolize the time of his workmen as well as his own. Ollivier therefore insisted on a subsistence to be paid for him and one apprentice biweekly or monthly, so long as "the work advances."[98] Not a wage for work done apportioned by the hour, this subsistence payment was to support him during the work.[99] Beyond this, Ollivier was to receive a sum when "he could produce at least some part of the machine . . . perfectly well made."[100] The proportion to the total was not specified. Ollivier maintained that he could not earn his keep and abandon his other work for anything less than 200 Louis d'or "for an accomplished machine of 12 digits with the keys" to be paid in installments as the work progressed with a final payment for the perfected machine.[101] This figure is approximately the 1,000 livres in question in 1678. Such terms were hard to stomach, Leibniz admitted, but they were the steep price necessary to monopolize the time and skills of a sufficiently innovative craftsman who could improvise given Leibniz's overall design.[102]

In his letters, Ollivier stressed that he considered at least some major parts of the machine to be a collectively inventive project—a mutual enterprise of Leibniz and Ollivier, drawing upon the skills and competencies of both. Some parts were Ollivier's: "You have seen the movement of these two little mills, which is mine."[103] He noted that the transport mechanism "cannot be made as easily as *we* proposed."[104] Ollivier feared the alienation of his creative labor once he had brought the machine to perfection.[105] He worried that Leibniz would simply take all his solutions to the problems of the machine and give the perfected, well-adjusted design to less skilled and cheaper German craftsmen, thought to be capable only of reproducing machines and incapable of innovating themselves.[106] (Leibniz elsewhere characterized German artisans in just such terms.[107]) The draft contract provided Ollivier no guarantee of exclusivity: in it, Leibniz "promised me that he would have me make other machines of the same kind so that I could earn something reasonable."[108]

According to the draft contract, Ollivier was to be subjected to work-discipline involving a weekly supervision, while all the time remaining the risk-taking, innovative artisan Leibniz needed to move his machine from model to viable machine. Ollivier was to be obedient in his habits and rhythm of work while constantly being creative in that work. He was to be paid more or less as a noncreative, ordinary artisan while doing the work of an extraordinary one. Embodying these contradictions, the contract restated the very tension that had led to the breakdown of trust in the first place.

At stake here is more than an academic question of discerning the contributions of philosophers in contrast to those of their artisans: this story involves disputes over compensation and recognition based on the value of different kinds of work toward technical development. Advancing the process of conceiving and building the machine required temporary solutions to these disputes; as noted in the introduction, such negotiations were crucial to the process of work itself. At times, Leibniz worked to assign essentially all real credit to himself; he recognized that a particularly skilled artisan, one who would become more skilled in the process, had a legitimate right to profit based on his skill and ingenuity, including the right to keep producing machines whose "complication" he had perfected based on Leibniz's general design.

So frustrating did the situation with Ollivier become that Leibniz began to inquire among his far-flung correspondents for information on suitable replacement artisans and the salaries they would demand to come and work in Hanover.[109] With his fortunes in France fading quickly, in the meanwhile, Ollivier had resolved to move to Hanover. Much to the annoyance of Hansen and Leibniz, numerous events delayed his journey; Hansen made him send a signed promise to come.[110] While no hard evidence demonstrates that Ollivier traveled to Hanover, various traces suggest he did so.[111] Ollivier abruptly disappears from Leibniz's correspondence: this suggests that they had returned to face-to-face interaction, with Ollivier again becoming an invisible technician, his struggles and innovations leaving no trace in the archive.[112] Whatever the case, some invisible artisan in Hanover appears to have continued to work on the machine. In the mid-1690s, the clockmaker Georg Heinrich Kölbing worked on the machine alongside Leibniz's butler; a junior clockmaker, Adam Sherp, worked on the machine until 1700.[113] While all apparently worked to develop Ollivier's machine, they probably did not produce Leibniz's surviving calculating machine, which was rediscovered and partially reconstructed in 1876 and currently resides in a vault in the Leibniz Library in Hanover.[114] This "newer" machine was built and "debugged" over many years, alongside Ollivier's "older" machine.

Failure to Coordinate: Hanover to Helmstedt to Zeitz

The effort to "perfect" the calculating machine continued on and off until Leibniz's death.[115] Work on the machine continued sporadically in Hanover until 1700, when Leibniz transferred supervisory responsibility to a client of his, one Rudolf Christian Wagner, "a young learned scholar and mathematician much practiced in mechanical things,"[116] with interests in architecture and fortification. Leibniz helped Wagner become a professor of mathematics at the University of Helmstedt; Wagner paid Leibniz back in part by dedicating years of service to supervising the activity of a variety of artisans on the calculating machine.[117] Perfecting the carry mechanism was the central task. Wagner proved a better manager than Hansen, as he could relay the technical quandaries and speak with some confidence about the results of the artisans. Yet the progress of the work remained extremely slow. Wagner's frequent illnesses ground work to a halt, the artisans became preoccupied with other work, and Leibniz was slow to respond to queries about potential changes to the design.[118]

In 1711, Leibniz transferred the working pieces of the machine again, this time to officials within the court of Moritz-Wilhelm of Zeitz.[119] Responsibility for the machine was divided among a court official named Buchta, given overall supervision of the work; a court deacon named Gottfried Teuber, given technical supervision; and artisans, charged with implementation. Finding artisans willing to be "diligent and docile" continued to be a challenge, and Leibniz worked to bring a skilled clockmaker, Haas from Augsburg, to Zeitz. Like Ollivier, Haas protested the schemes for payment from Leibniz and his agents.[120] Real work began only in 1714.

In Zeitz, carry remained the primary concern.[121] Whereas Hansen supervised but had no grasp of the design or of machinery in general, Teuber, like Wagner, understood the basic principles of the machine and many of the particulars and was to innovate as seemed necessary.[122] Leibniz granted Teuber broad authority to change the design as needed: "These things are left to the judgment of M. Teuber, and I ask him to act as if it were his own business."[123] When working with an artisan, Teuber was charged to "direct him and animate him to diligence," and an entirely new machine was begun.[124] Despite considerable effort, neither the old nor the new model of the machine was perfected at the moment of Leibniz's death in 1716.[125]

To this day, scholars studying Leibniz's machine disagree on how well its different iterations functioned—or could have functioned.[126] As we will see in chapter 4, the failure of the "universal genius" Leibniz to reduce the machine to working practice became an eighteenth-century trope testifying to the challenge of mechanizing calculation and of productively collaborating with artisans.

Against Pascal's Cartoon: Leibniz's Account of Artisanal Knowledge

Famous for his calls for making knowledge as deductive as possible, Leibniz greatly valued artisanal skills. Such skills were a problem for philosophical analysis and for concrete statecraft. His epistemological and practical views on artisans and their skills informed his theology and political economy alike. Late in the 1690s, Leibniz complained about the dangers of producing only raw materials and exporting them: "We work properly for foreigners, more or less like the journeymen who help artisans, who make less money although they do the hardest part of the work, since they work using force rather than cleverness, like the animals who serve us and whom we direct."[127] According to Leibniz, those states whose industries produce finished cloth deserve greater recompense: they act less through force than through skill and intelligence. A state's economy ought to be organized around its trained artisans, not its "journeymen."

The hierarchy of artisans we've seen in this chapter informed Leibniz's famous denunciation of the Newtonians' narrow focus on God's power to the detriment of His wisdom and goodness. The Newtonians misunderstood what characterized the finest machines and those that made them: "The true ground that makes us praise a machine, is taken from its effect, rather than its cause. We take account less of the power of the Machinist than of his artifice." Leibniz criticized the reason given for praising "the Machine of God"—that He alone could make the entire machine Himself, "whereas an artisan needs to seek out his matter." Such an evaluation, Leibniz complained, comes "only from power" rather than from his wisdom. God's artistry is great because of His wisdom in crafting His machine: "His machine lasts longer, and works better, than any other." A person buying a watch does not care "if the artisan makes it all himself, or if he had others make the pieces and he simply adjusted them, so long as it goes as it should." In evaluating any artisan, the fact of subcontracting is neither here nor there: even if God had granted an artisan power even to create the matter of the wheels, "we would not be content" with that artisan "if he had not also received the gift of adjusting them well."[128] A master artisan is characterized by his ability to design, configure, and adjust all the parts of a mechanism such that it works as a unified whole. Even though artisan-helpers may do some of the work, God gets all the intellectual credit for the invention, precisely because He can design things in the smallest detail. What matters in understanding credit for invention is not who makes the thing but who designs all its details ad infinitum.

In Leibniz's account, God's mechanisms are perfectly fitted and unified machines, all the way to the smallest parts.[129] Leibniz reworked the famous Aristotelian distinction between the products of nature and art. God's intentions are stamped onto every part of machine, no matter how far one should subdivide it. God never has to make do with the material properties of parts on hand; within the constraints of this best of possible worlds, He can make all the pieces, in all their material qualities, to spec. Every part is distinct, intentionally crafted to work perfectly in the mechanism as a whole. Such machines are "organic"—organic wholes. Taking apart the parts of a machine made by human beings eventually results in a natural part, one not perfectly fitted to the goal of the machine: "For example, a tooth of a wheel made of brass has parts and fragments, which are no longer something artificial to us, and have nothing that mark the machine in regard to the use for which the wheel was destined." When we break down our machines, we always find something produced by nature, not machined by us. In contrast, "the Machines of Nature, i.e. living bodies, are still machines in their least parts, to infinity."[130] No human artisan can ever hope to make his or her parts fit their functions with the greatest possible perfection. Within the limits of a possible world, God makes all His parts and all His materials; human beings are stuck making parts out of materials whose properties they do not know, much less fully understand. Leibniz's account of organic machines became central to late Enlightenment accounts of artistic unity and of intellectual property.[131]

Leibniz's theological invocation of artisans and the habitual modes of acting that define them rests on a picture of artisanal skill and knowledge more layered and less dismissive than that of Pascal's polemical account discussed in chapter 1. Balancing a sophisticated account of how to mind the hand and manualize the mind, Leibniz recognized the necessity for artisans characterized by their industry and their real skill in inventing and adjusting machines. At the same time, Leibniz saw such practical artisanal capacities as decidedly inferior to theoretical knowledge as a way of knowing, however much artisanal skill surpassed theoretical knowledge in understanding the particulars of nature and the means to manipulate it. In a vein akin to the deskilling effort of the Encyclopedists and the projects of Francis Bacon, he called for artisanal skills and knowledge to be explicitly stated and conceptually clarified.

Leibniz regularly denigrated lower forms of skilled labor as animallike. He claimed that, in general, most skilled labor is capable only of responding in precisely defined ways to stimulation; it lacks the conceptualization and generalization that allows one to respond to any situation: "Human beings act like animals insofar as the consecutions of their perceptions are produced only from the principle of their memory." He conceded that "we are only Empirical in three quarters of our actions," even though such ways of acting do not stem from judgments of the reason and do not thus involve what defined us as human.[132] Like animals, simple empirics act as if things will always happen as they have before; they cannot register and contend with exceptions, whether in human affairs, in the human body, or in the technological world. Only a theoretically informed practice can recognize and contend with exceptions in a general way.[133] Alas, Leibniz conceded, the method of using reason is not yet perfected: the crude current state of fundamental propositional knowledge precludes contending with all problems—technical and political.

In Leibniz's account, God has perfect theoretical knowledge and thus perfect practical knowledge. He has simultaneous recall of all knowledge and can discern what is necessary to do in all situations in an instant. Simple human beings are not so lucky: "We cannot take enough precautions in practical enterprises, and since the method of reasoning has not yet achieved the perfection it is capable of, and that, moreover, our passions and distractions often prevent us from profiting from our own lights, I maintain that is it necessary to distrust reason by itself: we must have experience or consult those who have it." Throughout his life, Leibniz worked hard to improve techniques for focusing the attention and quickly making judgments about experience and deductions alike, in theoretical as well as practical matters.[134] The longer such techniques remained underdeveloped, the longer natural philosophers, engineers, and rulers would continue to depend on bearers of habitual skills.

Like Pascal, Leibniz distrusted inventors anchored only in theory, just as Leibniz dismissed artisans lacking any theory as mere empirics. Unlike Pascal, he recognized that skilled artisans frequently had just the blend of the practical and the theoretical necessary, given the current state of theoretical knowledge: "For an artisan who knows nothing of Latin or Euclid, when he is a skilled man, and knows the reasons for what he does, he truly has the theory of his art, and is capable of finding expedients for all sorts of events." An artisan who possessed the theory of his art did not have to act like an animal or a simple empiric—that is, someone only able to act like a machine, following procedures and rules kept in memory. In searching for an artisan to produce his calculating machine, Leibniz had explicitly sought such a higher sort of craftsperson. Better artisans could contend with the unexpected and were capable of innovation using their theory. Using terms likely borrowed from Pascal, Leibniz mocked philosophers and projectors who lacked the knowledge gained from practical experience with the actual qualities of the world: "A demi-savant puffed up with an imaginary knowledge projects machines and buildings that could not succeed, for he lacks all the theory necessary. He understands perhaps the vulgar rules of movement . . . , but he does not understand that part of Mechanics I call the science of resistance or of hardness."[135] A good Baconian, Leibniz insisted at many points throughout his writings that theoretical progress should attempt to uncover artisanal knowledge and give it a formal expression allowing articulation, replication, and distribution.[136] This knowledge is particular but essential in the transition from ideas to the making of things: "We find this through experience in passing from theory to practice when we want to make something. It is not at all that this practice cannot be written, since it is at base only another theory, one more composite and particular than the ordinary one." Neither workmen nor savants are likely to work to write it out: workmen "do not explicate intelligibly through writing and our [philosophical] authors jump above these particularities, which, however essential, they take for trifles."[137] For example, the practical how-to knowledge concerning the hardness and resistance of materials learned through practical interaction and experience will, with time and work, become "distinct" knowledge. As this knowledge becomes more distinct, the elasticity and springiness of various materials will be cataloged, made recognizable, and given theoretical explanation. Such distinct knowledge can be defined in terms of "certain marks," such as the stages of the work of an assayer that provide certain marks for judging whether something is gold. Problems of hardness and elasticity, a major natural philosophical interest of Leibniz, were at the center of the material difficulties that made it so challenging to perfect a well-functioning calculating machine.[138] Few natural philosophers had investigated these "particular" phenomena, so inventors had to resort to the knowledge of artisans.

Leibniz insisted that most unexpressed artisanal knowledge could, would, and should be given an explicit conceptual and written formulation in terms of fundamental physical and metaphysical principles. Once the art of reasoning and its attendant tools were sufficiently developed to allow human beings to cognize theoretical knowledge quickly, a philosopher-engineer would always be able to devise mechanisms far more sufficiently and far more completely than any skilled artisan. Lacking sufficient theoretical knowledge and tools for abridging reasoning using theory, however, inventors continued to need skilled artisans to help them foresee, recognize, and solve problems and difficulties of all kinds. Until theory and the arts necessary to use it expeditiously were further developed, artisans were needed to do far more than simply build according to specified designs. In all this, Leibniz echoed his experience of attempting to realize his calculating machines.

In appreciating artisanal skill, Leibniz recognized the importance of knowledge that was neither cognized nor distinctly expressed. He furthermore praised the power and necessity of habitual modes of perceiving and acting, even as he underscored their limits and called for their replacement through techniques meant to perfect reasoning insofar as this was possible for fallen human beings. In his *Meditations on Knowledge, Truth and Ideas* of 1684, he described confused knowledge as "when I cannot enumerate one by one the marks that are sufficient to distinguish the thing from others." As examples, he gave our knowledge of colors, odors, and flavors. Our judgments concerning these things are often correct, even though we do not have consciously cognized and expressible bases for making them. He noted, "We sometimes see painters and other artists correctly judge what has been done well or done badly." Despite being correct, they cannot explain the grounds for their judgments; they can only say that "the work that displeases them lacks *je ne sais quoi*."[139] Recall Leibniz's concerns about the mechanically trained, ingenious eye of Robert Hooke, capable of quickly grasping the fundamental idea of a mechanism.[140] While praising the superior knowledge of mathematicians, Leibniz years later conceded that the skill of a "worker or engineer . . . could have this advantage above a great mathematician that he could discern among them [polygons] without measuring them," like colporteurs who can eye the amount of weight they can carry—"in which they surpass the greatest experts in statics in the world." He conceded, "This empirical knowledge, acquired through long exercise, is of great use for acting quickly, as engineers often have need of."[141] Such "confused" knowledge was still inferior to what Leibniz called "distinct" knowledge.[142] Elsewhere he gave the examples of Galileo and Archimedes, whose knowledge of statics gave them distinct ideas and thus superior knowledge. Yet in practice, in times of need, at the moment, speed trumped distinctness.

Domains other than engineering also required such unarticulated knowledge to be made into habits, which are useful, productive, and fast. Leibniz described a wide range of artistic activities pursued by polite people in a machinelike fashion:

> There are things, above all those depending on the senses, where we will succeed rather and better in letting ourselves act machine-like, as in imitation and as in practice, than in remaining in the dryness of [theoretical] precepts. Just as in playing the clavicle requires a habitude that the fingers must take, so

in imagining a *bel air*, in making a beautiful poem, for creating promptly the
ornaments of architecture, or drawing a tablet of invention, our imagination
must develop a habit, after which one can give [the imagination] the freedom
to take flight, without consulting reason, in a sort of Enthusiasm.

Once this freedom has taken flight, reason should reflect on what has been
produced and consciously apply the rules of art. Lest anyone mistake what he
thought was the proper domain of theory, he explained, "There are things that
depend rather on a game of the imagination or a machine-like impression than
on the reason, and where one needs habit, as in the exercising of the body, and
even in some exercising of the mind. In such cases, we need a person of practice
to succeed."[143] With the development of techniques for reasoning, many more
things could be reduced to logical judgment based on the principles of reason
and inferences from experience, using "an exact method of the true logic or the
art of inventing." Theory may, at some future time, become a practice capable of
providing quick resolutions. In the meantime, Leibniz argued, finding "a good
solution to a muddled difficulty," still required "that we have the force of extraordinary genius or that we have a long practice that makes the answer come
into our mind machine-like and by habit, when that answer would need to be
sought through reason."[144]

Future developments in techniques for abridging the process of reasoning
would eliminate the dependence on genius and habit. Leibniz underscored
again the dangers of such habits: even theoreticians without any knowledge
of practice will surpass practitioners when they are forced to deal with "some
difficulty very different, from those they have practiced." Someone with theory can "find exceptions and remedies." Equipped with some understanding
of practice and the properties of nature known to practitioners, in addition
to their formal theoretical training, they can react creatively to difficulties.
At this point, Leibniz presents what can only be seen as an autobiographical
encomium: "We see every day that people of good sense, who have need of
some artisans, after having understood the matter and reasons of practice,
know how to provide openings in extraordinary cases, which the craftspeople
do not perceive at all."[145] This ability to provide openings in extraordinary
cases alone could lead human beings to their highest potential, by inventing
techniques and machines capable of overcoming the difficulties of mind and
body alike. Archimedes, Leibniz wrote, had a talent lacking in Descartes and
Galileo: "He had a marvelous mind for inventing machines useful for life."[146]

Leibniz's work on the calculating machine testified to his skill in finding such openings. Or at least it was supposed to. To convey the importance
of his innovations in calculating machines and the integral calculus, Leibniz

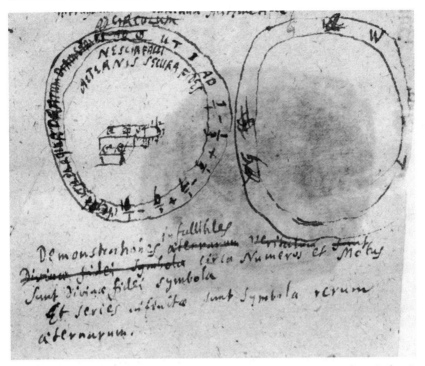

FIGURE 2.7. Leibniz, "Numisma Memoriale meorum Inventorum Tetragonismi et Machinae Arithmeti-
cae," c. 1676. LH 35,2,1, f. 2. Courtesy of Gottfried Wilhelm Leibniz Bibliothek—Niedersächsische Landesbibliothek, Hanover.

sketched a commemorative medal around 1676 (see figure 2.7).[147] Just before
describing his machines to his future patron Johann Friedrich, he offered
"to bring new inventions and curiosities from extraordinary cabinets. . . .
In these matters, I have had enough lucky opportunities . . . having learned
from workmen themselves many things, some curious, others important
for commerce and manufacture."[148] In including his calculating machine in
the list of achievements he presented to princes, he used his new technology as evidence that he could be a new kind of state counselor, essential in a
cameralist age—a counselor capable of seeing the openings necessary to improve technology and polity at once. Leibniz presented himself as a superior
form of projector, schooled in theory, grounded in practical knowledge, and
capable of producing the improvements necessary for social order, health,
economy, and morals.[149] He could convert the imperfectly understood and
ill-distributed knowledge of artisans into practices essential to develop the
state and its people.

11 Second Carry

Babbage Gets Funded

In granting Charles Babbage £1,500 toward perfecting his Difference Engine, the chancellor of the exchequer emphasized that the government generally did not financially "encourage" inventions: "If they really possessed the merit claimed for them the sale of the article produced would always be the best reward to the inventor." Babbage and his allies had succeeded in convincing the government that the Difference Engine warranted violating this general principle: "The construction of such a machine could not be undertaken with a view to pecuniary profit arising from the sale of its produce and that as the tables it was intended to produce were peculiarly valuable for nautical purposes it was deemed a fit object of encouragement by the Government."[1] Babbage had to convince the government he was not a "projector," setting forth a dubious scheme to defraud government and the people alike. Memories of the South Sea Bubble and of schemes for finding longitude and producing perpetual motion died hard.[2] Babbage apparently addressed skepticism toward projectors head on, remarking to the chancellor of the exchequer that he "must be applied to almost daily by a multitude of quack speculators wanting rewards for their inventions." In response, the chancellor "observed that the opinions of the most competent persons in the country had been taken" and that he was "satisfied from those opinions that [Babbage's] was not of the kind . . . alluded to." The chancellor likewise had decided that Babbage was not in it for himself.[3]

Circumspection about inventors, manufacturers, and projectors has long been essential for government officials. Identifying the most competent persons to adjudicate between projects and schemes was hardly trivial. The early nineteenth-century British state, like most early-modern polities, rarely supported inventive activity directly, especially before the processes were thoroughly developed and reduced to manufacture. One friend of Babbage noted, "You will find it hard to Persuade Parliament to address the Crown for a manufacture of a machine." He could think of only two parallel examples of direct government support.[4]

Through a publicity campaign and the support of scientific bodies, Babbage obtained unusual direct funding for his project in 1823. Before the project collapsed, the state continued its support of the project, though not without persisting qualms and objections both inside and outside of government. Ultimately, the Treasury contributed no less than £15,288 ls. 4d. for engineering work and £2,190 13s. 6d. for constructing a new workshop for the machine and its drawings.[5] The legal conditions of this support were surprisingly murky throughout, a fact that eventually doomed the effort.

Publicity and Reports

By the middle of 1822, Babbage demonstrated a model of his engine—called Difference Engine 0 by historians—to select audiences:[6] "I have done nothing to the engine but it begins to be talked about. I hope to find on my return sufficient reasons to induce me to go on with it."[7] By this, he meant cash—he was not (yet) an independently wealthy man. Babbage made his call for external support in a privately published and widely circulated letter to Sir Humphry Davy, president of the Royal Society: "Whether I shall construct a larger engine of this kind, and bring to perfection the others I have described, will in a great measure depend on the nature of the encouragement I may receive." Babbage stressed that the usual financial incentives for invention would accrue too slowly, as success in building the machine would "be attained at a very considerable expense, which would not probably be replaced, by the works it might produce, for a long period of time." The delay between profit and research investment was too great. Babbage personally needed encouragement to focus so exclusively on the mechanical: bringing the machine to perfection is "an undertaking I should feel unwilling to commence, as altogether foreign to my habits and pursuits."[8] Privately, Babbage commented to a friend that it "would be unpleasant to myself and I think disgraceful to the country that I should be compelled to repay myself by printing tables and consequently keep secret the machinery which is capable of producing so much."[9] The standard route for a practical inventor, even if it were financially possible, would preclude Babbage from acting as a disinterested natural philosopher; he would be forced to act as a secretive tradesman and not disclose his invention

to the cosmopolitan world of natural philosophers. Babbage needed inducements to become disinterested; his self-interest had to be fulfilled so that it
would make him free of interest, thus a philosopher-inventor, not a grasping
projector.[10]

In his appeal to Davy, Babbage avowed that his proposal would prompt visions of Jonathan Swift's Laputian philosophers.[11] Home Secretary Robert Peel
complained to a friend in the Admiralty that a member of Parliament, Davies
Gilbert, "has produced another man who seems to be able to vouch at least for
Laputa."[12] A member of the Royal Society and friend of Babbage, Gilbert pressured Peel to turn to the Society in order to "induce the House of Commons
to construct at the public charge a scientific automaton."[13] Whatever his skepticism in this case, Peel conceived the sciences as integral to British power.[14]
He needed some assurance that Babbage was not a projector caught in delu-
sions of scientific grandeur before bringing before "a thin house of country
gentlemen"—that is, Parliament—"a large vote for the creation of a wooden
man to calculate tables from the formula $x^2 + x + 41$."[15] His correspondent
doubted the engine "would be the most useful application of public money,"
but he encouraged Peel to refer the matter to the Royal Society.[16] Four days
later, the Treasury formally asked the Royal Society for "their opinion on the
merits and utility of this Invention."[17] The Royal Society seemed a good venue
for Babbage: before the committee met, one ally encouraged Babbage, "You
should (entre nous) draw up the heads of such a report as would be borne
out on examination: leaving the detail to be filled up by some friend at the
Com[mitte]e."[18] A committee of the Society, comprising primarily Babbage's
friends and allies, concluded quickly: "Mr. Babbage has displayed great talents and ingenuity in the construction of his machine for computation, which
the Committee think fully adequate to the attainment of the objects proposed
by the Inventor, and that they consider Mr. Babbage as highly deserving of
public encouragement in the prosecution of his arduous undertaking."[19] However friendly to Babbage, the Royal Society was not the obvious forum for
examining the practicality of his project. Since its inception in the seventeenth
century, members of the Royal Society had profound ambivalence about the
organization's role in the validation of inventions. Tainted by the efforts of
disreputable projectors early in the eighteenth century, the Royal Society had
distanced itself from its earlier goals of improving and policing the arts and
trades. At the same time, other claimants to technical expertise had vied for
this function, notably the Society of Arts that formed the template for "Societies of Emulation" across Europe and America.[20] The Royal Society did
not have—and never had or received—control over validating inventions on

behalf of the state. In the second quarter of the nineteenth century, a community envisioning itself as "scientific" coalesced in part around the articulation and defense of scientific expertise as crucial for advising state and private actors alike on issues of industry.[21] Integral to this conception of expertise was a vision of invention led by science and great geniuses; a parallel, radical critique rejected such accounts in celebrating invention through cumulative accretion of skill and technique.[22]

The report of the Royal Society, however convincing and celebratory, proved insufficient. Even after the Treasury appeared to have accepted the Royal Society's expertise and its report, the chancellor of the exchequer decided to quash the project. Davies Gilbert wrote on May 28, "The administration have turned short round, and declared an extreme unwillingness to consent to encourage or assist any invention whatsoever." Overcoming this return to form required that Babbage "exert [himself] and get several persons to interpose with the Government."[23] A group of members of the Society, led by Gilbert, sent a follow-up letter commending the scientific utility of the machine: a "Grant of Fifteen Hundred Pounds, does not exceed the amount, which will be requisite, for enabling Mr Babbage to proceed with the construction of an Apparatus quite new in all its parts, and demanding the labour of the most skilful workmen."[24] Fortunately for Babbage, his friends had nominated him for the medal of the Astronomical Society, which he received. In his address presenting the Gold Medal to Babbage for his invention, the president of the Astronomical Society emphasized the great potential of the Difference Engine for numerical analysis. The encomium was calibrated to stress the philosophical and practical qualities of the project alike. The medal was awarded not for an idea of a machine but rather "for the machine in the finished form of a calculating instrument." The president was emphatic that the model had been sufficiently reduced to practice: "I speak of it as complete with reference to a model which satisfactorily exhibited the machine's performance." He tempered expectations about the facility of producing the "more finished engine," which "may not yet for some time be in a forward state to be put in activity and receive its practical application."[25] Babbage subsequently met with the chancellor personally, who expressed the government's standard refusal to fund inventions while nevertheless agreeing to grant Babbage £1,500 "to perfect the engine."[26] More than just a personal victory, the support Babbage garnered appeared a victory for a political economy that connected elite natural philosophers to improved technology and British economic advance. This sociology of technical production notoriously transferred poorly to the workshop.

Babbage under Fire

On August 8, 1828, an anonymous article attacked Babbage for his apparent lack of progress on the machine: "Although it did not appear probable that this invention would ever arrive at that high degree of perfection" promised, "it was to be expected that *something* would arise out of it to warrant the high encomiums of the distinguished members of the Royal Society . . . , and of the liberal patronage it received. As yet, we have heard nothing more of the invention." The article intimated—not unjustly—that Babbage had secured government support through his friends: "The inventor and his friends owe to the public, to state how far they have succeeded in the accomplishment of this national invention." Given the silence, any progress must be minor or nonexistent. "If, like many others, it has ended in disappointment," the article insisted on "some account of the manner in which the money awarded by the Government for its completion has been disposed of." The implication that Babbage had profited from the public purse shadowed him for decades. The anonymous attack ended snidely: "Surely philosophers of all other men, should be above the vanity of refusing to acknowledge that they have failed in any undertaking."[27] Herschel replied on behalf of Babbage, still out of the country, in a letter to the *Times*.

What account did Babbage owe? The £1,500 given in 1823 was to "enable [Babbage] to complete and bring to perfection a Machine invented by him for the Construction of numerical Tables."[28] Was it a one-off award? A commitment to fund the machine to completion?[29] Babbage thought the latter; the various governments and treasury officials in the years to come thought not. Babbage apparently believed the government had promised in 1823 that "whatever might be the labor and difficulties of the undertaking, he should . . . not suffer any pecuniary loss from it."[30] Babbage took the view that the government had urged him to set aside his other affairs and complete the machine for the nation on its behalf; in contrast, members of the various governments over the time period tended to view themselves as having rewarded a well-connected man of science who possessed a potentially important contrivance.

Throughout the process, Babbage and his allies had to overcome continuing opposition to the very notion that the state might do more than reward such a project afterward. In 1829, for example, the chancellor of the exchequer offered an explanation in a private note that infuriated Babbage: "The view of the Government was to assist an able and ingenious man of science whose zeal had induced him to exceed the limits of prudence in the construction of a work which would if successful redound to his honor and be of great

public advantage." The precedent of agreeing that the government sponsor-
ship entailed any commitment to fund the development until completion was
dangerous.[31] With the aid of a number of supporters, the Royal Society, and
the surprisingly enthusiastic support of the Duke of Wellington, Babbage was
able to overcome this skepticism up through 1833, despite changes of government and slow progress.

Integral to the continuing support of the Difference Engine project was
showing that the engine had proved to be more than just an unpractical philosopher's dream after five years of generous governmental support. On December 24, 1828, the Treasury turned once again to the Royal Society to consider
whether to advance additional funds for the construction of the engine, in
sum to answer "whether the progress made by Mr. Babbage in the construction
of His Machine confirms them in their former opinion that it will ultimately
prove adequate to the important object which it was intended to attain."[32]

A carefully chosen committee of the Royal Society soon produced another celebratory report that answered positively the lord treasurer's question
of the "adequacy" of the project. Chaired by Babbage's friend Herschel, the
committee this time included numerous senior engineers, mostly friendly
to Babbage.[33] "Adequacy" was a standard legal term used in considering the
sufficiency of disclosure in patents—and thus served as a proxy for determining whether a machine had already been constructed or could practically be
constructed in some short amount of time. The committee put "adequacy" at
the center of its considerations: a list of "Objects to be attended to" in the working memorandum included first, the "Actual state of the machine and Works
Drawings," and second, the "Adequacy of the \contrivances/ and Workmanship."[34] Although earlier drafts of the report included more detail on the
theoretical principles of the machine and Babbage's innovative mechanical
notation, the committee stressed instead precisely those elements of the
machine that indicated its practical precision—less Babbage's theoretical and
philosophical acumen than his striking ingenuity in devising mechanisms
that produced perfect calculations, despite the troublesome quality of actual
metalwork, however well done. Praising him as "fully aware . . . of every
circumstance that might possibly produce error," the report presented Babbage
not as a philosophical projector but rather as someone working through
the arduous but worthy course of reducing the machine to practice.[35] The
committee report gushed that "far from being surprised at the time it has
occupied to bring it to its present state, . . . they feel more disposed to wonder
it has been possible to accomplish so much," given that plans needed to be
drawn, tools needed to be made, and the long process of moving from plans
to realization through careful experimentation had to be undertaken.[36] The

report's authors strove to dispel the notion that little effort had been undertaken to move from ideas to implementation.

In a late addition, as the most powerful evidence that the nitty-gritty of actual reduction to practice was well advanced, the report adduced the "contrivances" for keeping the machine digital: "In all those parts of the Machine where the nicest precision is required, the wheelwork only brings them by a first approximation (though a very nice one) to their destined places; they are then settled into accurate adjustment by peculiar contrivances, which admit of no shake or latitude of any kind."[37] The draft report reveals that the parts in question are those "on which the calculations are registered" (precisely the security mechanisms discussed in the First Carry).[38] Herschel's draft notes for the report included additional details about the concrete implementation of the machine for the committee to consider, including the following:

> 1st by loading the axes as little as possible—the calculating wheels do not touch the axes.
> 2nd by always employing two springs on opposite sides.
> 3rd any spring may be replaced without taking to pieces the machine.[39]

The work—and ideas—of Clement and his staff were crucial to making the case that the machine was progressing toward a reduction to practice. The final version of the report noted, "The movements are combined with all the skill and system which the most experienced workmanship could suggest." An earlier draft stressed that Babbage was using one of the "the ablest . . . Engineers \Draughtsman/ in the Metropolis (Mr. Clement) . . . for that express purpose." Clement and his team had in fact made the machine far more than a philosopher's idea—but the point was canceled and marked, "Not to be."[40]

Although the report emerged from the Royal Society, the recommendation only had force because the committee was *practical*, not too philosophical. In approving the glowing report, the council of the Royal Society stressed that the committee comprised "among its members several of the first practical engineers and mechanicians in the country."[41] In their internal memo, the treasury lords likewise underscored that the report rested on the expertise of "Practical Mechanics"—not natural philosophers—in judging the work.[42] The report had force because it wasn't the usual Royal Society at work. While officially intended for the government, the report and the council's approval were soon published to dispel public doubt.[43] Judged not to be a projecting philosopher, Babbage obtained another £1,500.

Babbage signed no formal contract with the government in 1823. The drawn-out series of meetings, negotiations, and referrals to the Royal Society in the early 1830s ended with an informal agreement that the government would continue to support the construction of the project; eventually, the government agreed to ownership of the drawings and made pieces. On February 24, 1830, Babbage was informed, again informally, with no written contract:

> Government would not pledge themselves to complete the machine.
> Govt were willing to declare the machine their property.
> Govt were willing to advance 3000 more than that already granted.
> At the end when it is completed they were most willing to attend to my claim for remuneration.[44]

Babbage's allies had found a middle course in which the government would support the project and perhaps pay costs after its success but would not guarantee the continuation of support. Considerably more financial support proved forthcoming, especially once the government began to pay Clement's bills directly. In part because of Babbage's disputes with Clement, the project to complete Difference Engine 1 stalled in 1834; Babbage attempted, on and off, to obtain government support for completing it until 1842.

Reward Deferred

Production of the drawings and parts halted in 1833 amid disputes' with Clement over the cost of moving to Babbage's new workshop near his residence, said to be a more convenient and fireproof space. Once more, the Royal Society agreed, and the Treasury paid for the renovations. Clement demanded £600 a year to pay for him to move workshops. Babbage and the Treasury rejected the claim. When Clement asked for some advance money, Babbage refused. Clement soon dismissed his workmen; work on making the engine never resumed.

Clement was not the only one to find his hopes for reward denied. The final efforts to fund the completion of the first Difference Engine failed in 1842. The new prime minister, Robert Peel, had long been skeptical of the engine: "What shall we do to get rid of Babbage's calculating machine . . . worthless to science in my view."[45] He was dubious about the advice the Royal Society had provided previous governments, and particularly of the influence of Babbage's friends and allies. The Astronomer Royal Airy, long an opponent of Babbage's, proclaimed the machine "worthless." When pushed, even Babbage's ally Herschel was privately far more equivocal than he had been a decade earlier.[46]

After the government finally decided to abandon the funding of the Difference Engine and surrender it to Babbage, he obtained an interview with an "excessively angry and annoyed" Prime Minister Peel. In preparing for the interview, which went very badly, Babbage articulated the grounds of his desert:

Grounds of Claim

1. 12 years of hard work
2. Gift of invention, made in previous years
3. 8 years of delay preventing all other arrangements
4. Large expense: all the time loan of money to government [*sic*]. Banker, in Life Ass[urance with likely salary of £]1500
5. Public will assert that I have profited by the money—friends urge to make statement.
6. They decide to give it up, it is not my fault.
7. Other men of science rewarded—[Babbage being the] exception.[47]

Babbage's laundry list—one of many prepared before the meeting—combined numerous forms of reward. No longer was the logic of desert narrowly based on a machine of some potential importance to the state; rather, Babbage turned to a logic of opportunity cost, in monetary terms, coupled to a logic of reward due to scientific eminence. Babbage expressed his deep sense of injustice about his treatment relative to his peers.[48] Typically for him, he named names, allies and enemies alike. Peel averred that these were mostly "professional rewards." Babbage rejected this: "It was perfectly well known that they were <u>not</u> given for professional services for that though they were eminent in science they <u>had</u> not any of them <u>ever done</u> anything to distinguish them <u>professionally</u>."[49] They received rewards for scientific glory, not their labor as professionals. Babbage was probably right about the failures of the course he had chosen. The logic of government preferment fit more comfortably in the early-modern state than did direct subsidies for research and development.

Improvement for Profit: Calculating Machines and the Prehistory of Intellectual Property

In early 1675, Gottfried W. Leibniz drew up terms for the French Académie des sciences concerning his "reasonable" compensation from the French crown upon delivery of well-functioning calculating machines. He would first receive a privilege, limited neither by number of years nor by any "other reserve"—a bold demand. By virtue of this privilege, no one would get "complete or partial machines except from me or my designees." Setting the price of the machine, he wrote, involved two major considerations: first, his past and future expenses, and second, the "reasonable advantage that he could expect from an invention as considerable and difficult" as the calculating machine. While quantifying expenses proved relatively straightforward, quantifying "reasonable advantage" was trickier. Inventing the machine, Leibniz explained, "occupied and will continue to occupy me almost entirely for some time"; it will thereby "prevent me from profiting from other opportunities," as he could show via "written offers from Princes and Ministers." The crown needed to pay Leibniz's opportunity cost. Using the language of early-modern contract law and a dash of emerging probability theory, Leibniz set out the just compensation given the risks. Lest the reader forget that reasonable advantage was a legal concept, Leibniz wrote, justice demands recognizing "the risk inventors expose themselves to, in advancing costs at their own expense, and in putting their reputation in jeopardy." Reasonable advantage also had to price in novelty: "For embellishments and curiosities for cabinets, novelty and rarity are paid for, as is seen everyday with the examples of pictures, prints, drawings and medals, [all of] which are but dead beauties lacking action and effect." The difficulty of such innovation likewise needed consideration: when first introduced, pendulums cost thirty, even fifty Louis d'or in Paris, even though "their beauty consisted entirely in the invention without the need for extraordinary work by artisans."[1] Perfecting and constructing a calculating machine would monopolize an innovative artisan and his atelier, just as it would monopolize its inventor's time and energy. Leibniz seemed set to straddle the profitable world of contracting for the state alongside the honorable world of the institution of the Académie des sciences. On top of the costs associated with someone of his skills and abilities abstaining from his other opportunities, Leibniz sought to monetize the risk to his reputation—to quantify risk in the nonfinancial sphere of international and local honor or glory. Leibniz's concerns about the danger to his reputation were well founded, for the calculating machine project quickly tarnished it, in France and England alike. The insufficiencies of his machine brought Leibniz little symbolic or financial credit in Paris and London.

In making his case, Leibniz defended the novelty and distinctiveness of his calculators compared to the famous machines of Blaise Pascal, built in the 1640s, and those of Samuel Morland, built in the 1660s. In this chapter, I study the calculating machines of these three philosopher-engineers within early-modern systems for protecting and encouraging manufactures and, indirectly, invention. The calculating machines emerged within an early-modern proto-capitalism joined to the subcontracting world that constituted much governance in early-modern polities.[2] Pascal, Morland, and Leibniz all sought to make the most advanced natural philosophical and artisanal knowledge of the day pay off in practical applications for state and market alike. All three were philosophical entrepreneurs who sought to be subcontractors and princely sanctioned monopoly vendors of machines and processes.[3] Unlike many elite practitioners in the sciences before and after their time, Pascal, Morland, and Leibniz cast the quest for monetary gain as complementary to natural philosophical and technical achievement and capable of spurring it. Leibniz explained the possibility of unifying personal gain and charity in a letter to his patron, Johann Friedrich of Hanover: "He who is happy enough to establish his fortune by advancing the public utility, can unite charity with prudence."[4] A servant of the state or an entrepreneur could become wealthy while doing public good.

Early-modern calculating machines were initially designed to aid calculation in early-modern governance, financial accounting, and astronomy. The production of these "philosophical" machines was parasitic on artisanal skill and knowledge; the legal protections afforded "philosophical" machines were likewise parasitic on legal devices devised to support artisanal, not philosophical, activities. These protections drew upon legal devices that primarily enabled industrial espionage by rewarding the movement of artisans, their techniques, and their organization of work into new jurisdictions. Before inventors got patents, artisans did. Before speculative designs for machines became intellectual property, the actual procedures of production of machines were protected through royal and princely monopolies as well as numerous other forms of preferment.

Creating and legally protecting the modern Romantic author meant effacing the craft dimensions of writing in favor of privileging the inspired mind and protecting its written products.[5] In the case of machines, creating and legally protecting the philosophical inventor meant effacing the craft dimensions and the managerial practices traditionally protected by privileges in favor of protecting the ideational designs of philosopher-inventors. "Intellectual" property in the form of patents came by the wayside, a side effect of extending the temporary monopolies protecting manufactures to ever more "philosophical" instruments and their makers. Denying the value of artisanal insight and labor to the conception of a machine—to its essence—was tantamount to denying the artisans' contribution to that conception and thus their ownership in it; it was to confect a legal and philosophical divide between mere manufacture and creative invention belied by actual processes of innovative making.[6] Countering this denial, however, often entailed an implicit concession that a machine and the process of producing it could be understood as having a mentalistic essence independent of that entire production process. When artisans and other inventors defended their propriety in machines and processes in terms of "authorship," they contributed to an understanding of invention that ultimately excluded the full scope of their work and promoted an understanding of property that excluded many of their competencies. Codifying the productions of philosophical inventors within the system of privileges helped make invention more intellectualized; so too did defending against such codification within an intellectualist concept of authorship. The proliferation of new written and visual techniques from the legal and bureaucratic sphere reified this codification.[7]

Calculating machines were not important commercial commodities— like pins, stockings, china, or watches—until the late nineteenth century. The machines of Pascal, Morland, and Leibniz are best understood as candidates for commodities produced with explicit and tacit state support to serve the state alongside individual consumers. Due to the fame of their philosophical inventors, an unusual amount of documentation exists about them. With this substantial existing documentation about efforts to monopolize the machines, we can see the process of gaining privileges and patents in unusual detail generally lost from the historical record of more successful commodities. The philosophical preoccupations of their makers illuminate how it became possible to envision machines in mentalistic terms and then eventually to create legal regimes of property protecting such an intellectualist understanding. The clash of interests and jurisdictions, of regimes of glory and of money within absolutism, offered a matrix for the contingent production of mentalistic conceptions of machines and legal techniques for protecting them. The history of calculating machines and like-minded projects lets us glimpse absolutist governance and its limits—its dependence on skilled people coupled with its pretensions to near omnipotence and independence.

Pascal: A Misleading Example of Seventeenth-Century Intellectual Property

In drawing up his terms for the Académie des sciences and Colbert in 1675, Leibniz referred to language in Pascal's 1645 pamphlet requesting royal protection for his calculating machine. In his request addressed to Chancellor Séguier, Pascal called for an unusual privilege that would "suffocate, before their birth, all these illegitimate abortions that could be engendered otherwise than by the legitimate and necessary alliance of theory and art."[8] The privilege, granted in 1649, gave him a monopoly on the production of calculating machines in all the realms controlled by the king of France for an unlimited length of time: any artisan producing machines without Pascal's authorization or approval would find his machines subject to seizure and himself subject to fines of up to 3,000 livres: one-third to be given to the king, one-third to the Hôtel-Dieu hospital, and a final third to Pascal.[9] More unusually, the privilege offered support for continued development of the machine independent of any demand that the machine be perfected in short order and brought into regular manufacture. Unlike the vast number of privileges of the time, Pascal's seems to have offered him something like a patent protecting an idea. More precisely, the privilege covered all possible machines with any mechanism or material that performed arithmetic with automatic carry.

The awarded privilege remarks that Pascal had made "more than fifty models" with various mechanisms, sorts of motions, and materials.[10] In all the "different manners the principal invention and essential movement consists in that each wheel or rod of a numerical order when it makes a movement of ten arithmetical digits, makes the next one move one digit only."[11] At the heart of the privilege, the invention underlying all the distinct mechanisms, is Pascal's isolation of the key problems of carrying tens (the sufficient-force and keeping-it-digital problems).[12] More precisely, the privilege appears to cover the *goal* of automatically performing carries, *and not any particular mechanism for doing so*; rather than covering the idea of any particular carrying mechanism, the privilege protects the goal and all possible solutions to it.[13]

Commentators have seen in Pascal's privilege early glimmers of a necessarily unfolding patent system offering mentalistic or intellectual property (IP) to those willing to specify their inventions to the public: "Here we have the vigorous beginnings of specification-writing. The object of the machine is fully suggested; next comes an outline of the variants of the basic model; and most importantly, the recitals end with a clear, generic definition of the invention. This definition appears as a forerunner to modern claims. It was drawn to point out the gist of the invention."[14] In fact, the privilege Pascal requested and received was profoundly *atypical*. To see more precisely how Pascal's privilege was, as he said, "far from ordinary" requires an understanding of the difference between modern patent regimes and the system of protecting inventions with privileges. We need to explain the conditions that made the unusual, "modern" qualities of Pascal's privilege *possible* in order to see how, in part, modern patent systems and conceptions of invention that developed alongside them came to be; we need an account of the development of a conception of an invention as having a gist, a substance, or an essence capable of being isolated and protected independent of any particular materialization.

Early-modern polities did not have patent systems of the sort known since the nineteenth century. Their closest equivalent—the system of privilege— was not about protecting intellectual property in the modern sense.[15] Privileges covered the introduction of trade or art with government protection; they could involve some major technological innovation but did not have to. In contrast, patents provide temporary monopoly rights over principles of a technological invention, as embodied in a specification and/or model disclosing those principles; they *can* lead to an introduction of trade or art but do not have to. In Italy, France, and England, the system of privileges emerged in the late middle ages and Renaissance out of "measures for recognizing and rewarding craftsmen's skills" and in particular for encouraging the transmission of technical "know-how" from one territory to another. Venice first institutionalized such a model in 1474; France and England copied many of the features of such a system.[16] Such privileges involved no innovative legal doctrines about "intellectual property": they were the stuff of royal and princely governance involving "gifts" of economic and political concessions of all sorts to all kinds of people and corporations. Essential to the toolkit of late medieval and early-modern governance, such privileges were then applied to craftspeople, their knowledge, their organizational procedures, and their skills.

Early-modern privileges for inventions generally served to aid technology transfer into a territory or to serve as gifts for favored courtiers and bureaucrats. Modern patents cover some key component mechanism of an invention, always embodied in some way (not quite an idea, though the law, particularly in the United States, has been moving ever closer to making ideas and natural laws patentable). Early-modern privileges protected not ideas or abstract designs but processes of manufacture and provided for regulation of labor, religious exceptions, payments, and naturalizations. Privileges often included freedom from guild or religious restrictions and sometimes pensions and housing.[17]

Novelty requirements for privileges concerned novelty within a given territory, not absolute, global innovation.[18] Written descriptions and public disclosure of the "essence" of an invention or process were generally not required, whereas a demonstrable, and quick, "reduction to practice" of an invention or process was crucial. Failing to produce a working device or process in a short period of time—typically as part of an entire process of production—invalidated or nullified most privileges.[19]

In his 1645 pamphlet requesting royal protection, Pascal maintained that his machine had been reduced to practice and that it was both robust and accurate.[20] The royal privilege subsequently granted to Pascal denied his claims: the machine had not been reduced to practice in a meaningful way: "And since the aforementioned instrument is now at an excessive price . . . [and is] therefore useless to the public . . . and so that it might come into regular use, all of which he intends to do through the invention of a simpler mechanism, . . . he works continually in search of such a mechanism, and in training little by little workers still too little habituated to it, which things depend on a time that cannot be limited."[21] Here, the failure of reduction to practice did not mean that the individual exemplars of the machine did not function as claimed; rather, it meant that there was not yet a form of the machine that could be produced with ease at a price level that would permit it to be more generally useful to the public. In other words, the privilege suggests that Pascal could make machines as one-off luxury items priced for collectors, but he could not yet manufacture them in a standardized way, as commodities at a price, fit for more general use. As a rule, privileges did not protect speculative ideas of projects to be worked out and then realized at some future time; they protected the manufacture of particular objects or processes already reduced, or soon to be reduced, to practice. Royal privileges often recognized the exceptions and emoluments necessary to obtain and regulate a diversely skilled workforce; they often called for the continuing perfection of new manufacturing processes, but they were rarely issued, if at all, for processes not yet even devised. Pascal had neither the secret of a simpler version of the machine nor an organized work-process necessary to produce machines without difficulty.

Pascal's awarded privilege explicitly states that the king's provision of incentives for Pascal to bring the machine to a practical form of perfection is a gift aimed at "exciting him to communicate more and more the fruits of [his capacities] to our subjects." The gift serves further to encourage Pascal to continue to innovate and to share the benefits to be accrued by Louis and his subjects from his mathematical and natural philosophical skills. The privilege is a private economic gift to support Pascal in his role as a philosopher-engineer working for glory and profit alike; the expectation is that Pascal in time will publicize his discoveries and innovations for the public good. A logic of theoretical discovery and publicity—a logic central to modern IP—intrudes into the logic of the privilege. Given all this,

> We permitted and permit . . . to the said sieur Pascal and those having rights from him, from now and for ever, to have constructed and made by whichever artisans, in whatever materials, and in whatever form that he judges to be good [in the future], and in all places subject to us, the said instrument invented by him to count, calculate, to perform all additions, subtractions, multiplications, divisions, and other rules of arithmetic, without pen or jettons, and we very expressly forbid everyone, artisans and others, whatever their quality or condition, to make them, or have them made, or sold in any part of our realm without the consent of the aforementioned Sieur Pascal.[22]

Pascal could have his machine made in any form, in any part of France, by any sort of artisan of any guild (or not), for an unlimited length of time.

Pascal's privilege appears to grant a monopoly of unlimited duration. Thirty years was rare—an unlimited grant was extremely unusual for manufactures.[23] No evidence has been found that the Parlement of Paris or of other localities registered the privilege: the privilege may not, in fact, have had legal binding force.[24] The duration of Pascal's privilege may have been more limited than is first apparent. The crown systematically used a seal of green wax for perpetual privileges and one of yellow for limited ones—and Pascal's was yellow. Only one known privilege for an invention had a green seal—and the parlement duly limited its duration to thirty years.[25] The crown likely was granting Pascal a monopoly of unlimited duration as a major symbolic gift, one according with protocols of glory, without actually ever having to worry about enforcing it for an unlimited period; if the machine were to come into regular production, the Parlements of Paris and Rouen would have likely restricted its terms, per their customary practice.

The royal gift to Pascal went even further. The logic of the privilege involved no "intellectual property," only the protection of manufactures.[26] Pascal's privilege endorsed a fundamental and perhaps incommensurable injection of a concept of invention and the inventor's mind into the logic of the privilege. Protecting manufacturers against counterfeiting was a central function of French privileges for inventions; such patents often only covered a limited jurisdiction within France: one could own the exclusive rights to make a given kind of clock in Paris but not in Lyons.[27] Rather than simply stealing Pascal's property, counterfeiters hindered Pascal from developing the machine into something more than an expensive curiosity. Counterfeiters undermined the credibility of Pascal's ability to create forms of his machine that could be sufficiently reduced to practice to be manufactured for everyday calculation:

> As the aforementioned instrument can be easily counterfeited by various workers, and that it is nevertheless impossible that they would come to execute it in the exactness and perfection necessary to use it fruitfully, if they have not been directed expressly by the said Sieur Pascal, or by someone with *a total comprehension* [entière intelligence] *of the artifice of its movement*, the fear would be that, were it permitted to anyone to attempt to build similar machines, the faults created by artisans that would be recognized in them would render this invention as useless as it should be profitable, being well executed.[28]

Even as it chides him for his failure to bring the machines into regular manufacture, the granted privilege accepts Pascal's polemical account of invention as essentially philosophical and mental, something requiring "a total comprehension." Making the machine practical required, the privilege argues, an unusual grant to maintain the distinct roles of intellectual invention and artisanal implementation.[29] Whereas artisans can only produce a "monster that lacks its principal limbs," savants ensure that a unified ideational essence undergirds and makes possible a mechanical unification; they ensure that the matter could possibly, in the right conditions of production, embody some unifying form.[30] This account divided form and matter, inventor and implementer, inspiration and implementation in ways foreign to actual early-modern manufacture and the legal systems organized around it. The account tears apart the amalgam of form and matter of early-modern making and thus allows an ideational conception of machine independent of any particular instantiation of it. Pascal depicted a normative hierarchy of invention and production that did not in fact exist as part of his attempt to secure something like that hierarchy in practice.[31]

In his important survey of the development of intellectual property in early-modern Europe, Carlo Marco Belfanti argues that the privilege was "a tried and tested instrument" taken from the "institutional 'kit'" available to early-modern polities. Privileges were "solely intended to reach a concrete economic policy objective," not to provide "an explicit safeguard for intellectual property." And yet, as Pamela Long has stressed, privileges suggested that craft knowledge—know-how—was a form of intangible property; accordingly, privileges were remodeled over time to protect artisanal inventors.[32] In accepting Pascal's account of labor, knowledge, and skill, the privilege for the calculating machine draws upon a concession of intangible property in their skills to artisans. This concession, however, abstracts their skill from the realm of actual production and tacit know-how in order to protect a more mentalistic account of inventive activity. Not quite expressing the concept of a truly intangible idea that is owned, the privilege grants Pascal ownership of all possible machines—all possible expressions—incarnating his essential breakthrough. Pascal's privilege, highly unusual for 1649, allows a glimpse into the process by which the monopoly protection of actual processes of production and the know-how involved in that protection could be transformed, under certain conditions, into the protection of abstract (but not necessarily functional) designs produced by an intellective author.

How did a logic of an intellectual inventor and noncreative artisans, of a machine conceptualized mentalistically, of support for a project far from reduction to practice, come to figure in a legal document produced within a privilege system in which ideas of machines in inventors' minds had no place? Pascal's privilege declares itself to be a royal gift to a favored client or, rather, a gift to the son of a favored client.[33] Whereas modern patents are rights of citizens, something the government is obliged to issue and protect for those meeting the appropriate criteria, early-modern privileges were legally gifts freely presented to preferred subjects.[34] More generous privileges with long durations or peculiar clauses, such as Pascal's, tended to go to favored clients like Pascal and his family.[35] However atypical Pascal's privilege, his route to it followed a path early-modern clients trod often. The Pascal family knew the protocols and informal procedures for obtaining support and custom from the French crown through its powerful ministers. Pascal's privilege for the calculating machine came largely thanks to his father's patron, Chancellor Séguier, to whom Pascal dedicated his machine.[36] By dedicating his machine to Séguier, Pascal offered a gift that helped repay the outstanding debt of honor his father and his family owed Séguier.[37] In a classic example of a continuing gift exchange within a network of clients and patrons, Séguier honored the son and his machine with a privilege, one that restricted the rights of the guilds of Normandy, among others. The unusually expansive terms of Pascal's privilege—its infinite duration and support for something far from perfection—likely stemmed less from the technical promise of the machine, however great, than from the importance of the connections, and the complicated set of reciprocal debts, between the Pascal family and Séguier.

Plum privileges came primarily from such relationships. Artisans and entrepreneurs seeking privileges often allied themselves with noblemen with connections to the court. Such an alliance resulted in the lengthy thirty-year privilege given for establishing a manufactory of porcelain and faience in Normandy in 1673.[38] Pascal pursued such a strategy in establishing the first bus system in Paris in the early 1660s.[39] The privilege for the bus system was granted to Pascal's patron and friend, duc de Roannez, and to several other higher-status noblemen, not to Pascal with his relatively low status.[40] When the issue was as important to the crown and commerce as the practical measurement of longitude or the import substitution of porcelain, the crown went to great lengths to draw upon expertise to test claims and procedures. Whatever its potential utility, Pascal's calculating machine underwent no such detailed examination. A calculating machine probably did not matter enough.

Together with the relative unimportance of the calculating machine as a commodity, the connections and reciprocal debts with Séguier permitted Pascal to secure a privilege for his calculating machine that imposed a mentalistic account of a machine onto a form of regulation foreign to such an account of invention and production. The privilege consequently appears to protect something more like intellectual property than almost all other privileges of the period. Séguier was at once a patron of natural philosophers and alchemists, the protector of the *Académie française* after Richelieu's death, and the one responsible for processing requests for privileges to print books as well as many privileges for manufacture.[41] A key node at the intersection of royal economic and symbolic patronage, Séguier could assent to Pascal's request in projecting the more intellectualized understanding of invention from the domain of glory into the nonintellectual monopoly protection of the privilege.[42] By recapitulating Pascal's polemical account of the separated roles of intellect and artisanal skill in the process of invention, the crown sanctioned a view of invention and manufacture grounded in a highly intellectualized inventor.

The privilege given to Pascal figured into a larger political program. In giving a royal privilege covering all of France to Pascal, Chancellor Séguier was challenging the extant rights of more local groups, be they the regional parlements or local guilds. He was infamous for such challenges to local prerogatives.[43] Giving a privilege to Pascal involved abrogating the rights of the clockmakers and goldsmiths of Normandy and circumventing the parlement that policed and protected those groups. The Parlement of Normandy long guarded its own power to issue privileges, even maintaining its own system of privileges for publishing books well into the seventeenth century.[44]

The political edge of Pascal's picture of artisans upsetting the epistemological and social order of production was likely not lost on Séguier. Pascal invented his calculating machine initially to help his father, a royal official, in a time of crisis.[45] Cardinal Richelieu and Chancellor Séguier had dispatched Pascal's father to Normandy in the early 1640s to help quell a tax rebellion there.[46] This tax rebellion, known as the revolt of the *Nu-Pieds*, was among the greatest challenges to the internal stability of the seventeenth-century French state and to any pretense that the monarchy was absolute.[47] The revolt in Normandy underscored that the king of France and his ministers no more had absolute power over his subjects than Pascal did over his artisans. Pascal's father was part of a new form of centralized bureaucracy, its power stemming from the king.[48] Theorists and propagandists disseminated images of the correct hierarchy of monarchs and subjects; these images were constantly belied by the traditional powers of French elites to check and challenge the king.

A few years later, Pascal articulated the analogy between sovereigns and inventors and their parallel troubles with subjects and artisans. In 1652, Pascal sent Queen Christina of Sweden a gift of a calculating machine. In his accompanying letter, Pascal compared sovereigns and savants. Good sovereigns were like good savants: they mixed knowledge (theory) and power (practice): "We must avow that each of these is great in itself; but, Madame, if Your Majesty will let me say . . . *one without the other seems defective.* However powerful a monarch might be, his glory will be lacking, if he has not the preeminence of mind; and however illuminated a subject be, his condition is always lowered by his dependence."[49] Christina has achieved what "men," implicitly including the kings of France, had long sought. In her alone, "power is dispensed by the light of knowledge; and knowledge raised by the brilliance of authority."[50] Alone among monarchs, Christina combined knowledge and power: "the union found in her sacred person of the two things that overwhelm me with admiration and respect equally, namely, sovereign authority and solid knowledge."[51] The proper relationship of savants and their listeners is much like that of sovereigns and their subjects: one of instruction and obedience: "The power of kings over subjects is, it seems to me, only an image of the power of minds over minds inferior to them."[52] The success of governance and science alike required the proper order of sovereigns and subjects. The claim was true in neither governance nor technology.

Specifying and Contesting Absolutism: A Hypothesis

Enforcing privileges often required the coercive powers of the state, especially when privileges infringed on the traditional activities and rights of others. As royal gifts, grants of privilege and patents often provoked protest and ire— sometimes leading to lèse-majesté and violence, sometimes held to legitimize

rebellion against the abuse of royal authority.[53] In February 1664, "Simon Urlin, at a meeting of wire drawers summoned by him to oppose Mr. Garill's patent, said in passion that the last King lost his head by granting such" patents.[54]
Such views of the causes of the English revolution testified to the dangers of
the untrammeled use of royal prerogative. The wire drawers were ultimately
successful in blocking the patent. In France, local judicial bodies such as the
Parlement of Paris had to approve privileges; they nearly always modified their
terms. On occasion, they rejected them outright. These judicial bodies often
also restricted the granting of privileges to court favorites who had no real innovation or locally new process.[55] In 1621, for example, the master baker Denis
Mequignon received a royal privilege lasting ten years for a new sort of mill;
in registering and approving the grant, the parlement reduced the duration
to "five years only" and limited the price to fifty sols within its jurisdiction.[56]
The Parlement of Paris approved the proposal for the Paris bus system with
two provisos: first, that there be only a single price, and not one prorated by
distance traveled, and second, that soldiers and liveried servants be excluded.[57]
In its customary way, the parlement checked the power of the crown to grant
privileges by stressing that extant privileges and liberties were not to be infringed. Such changes and reductions served as a daily reminder of the real
limits of royal authority. The crown was likewise prone not to upset existing
rights. When Christiaan Huygens sought a privilege for his pendulum clock
in 1658, Pascal's patron, Chancellor Séguier, refused three times, because he
"did not want all the master clockmakers of Paris crying after him." No matter
how grounded "in reason" Huygens's appeal was, "these difficulties and obstacles" precluded the chancellor from exercising his grace and freely granting
the privilege.[58] At the time, Huygens wasn't worth the trouble.

The continual contest over privileges was part and parcel of the quotidian
jostling by the crown, representative bodies, and various evaluative bodies
to retain and to gain control over aspects of governance. In France and England alike, the crown's claimed prerogative to issue patents, privileges, and
monopolies was constantly challenged. Early-modern princes with pretensions to absolute power liked to present themselves as offering "gifts" freely,
unconstrained by obligation, just as their propagandists presented them as
able to rule, in principle at least, unconstrained by legal traditions. Early-
modern monarchs were caught in worlds of traditional obligations and legal
constraints, which they could modify only with some difficulty, usually with
considerable resistance, and sometimes only through violent coercion. Royal
privileges extended and ratified royal authority; resistance to them checked
that authority in the name of traditional prerogatives, local sovereignties, and
more rarely, rights.

Scholars writing histories of patenting have readily found evidence for a teleological narrative; they have discovered examples of global novelty, specification, immateriality, and intellective invention. Revisionist historiography rightly downplays these Whiggish examples, but we ought to be able to explain their sizeable production. The evidence used in the teleological accounts reveals neither a hidden inherent "substance" of a modern patent "regime" in the process of unfolding nor mere accidents within an early-modern privilege "regime" to be disregarded. Rather, the dispersed evidence of teleological accounts was systematically generated by the clash of interests constitutive of actual governance in early-modern sovereign states and from the weakness of crowns invested in portraying their states as strongly centralized and unified around a single power. The clash of legal regimes within sovereign states produced resources that the modern patent regimes of the rights-oriented states of the late eighteenth century drew upon and then reinterpreted within a doctrine of rights.

At the center of the intellectualization of modern patents systems is the specification—a written document that, in principle, should allow those with "ordinary skill in the art" to replicate a disclosed invention. Such specifications played no part in the early-modern privilege system. The requirements for writing specifications did much to put the "intellectual" into intellectual property.[59]

Before the late eighteenth century, written specifications were not required by law or by bureaucracies as a standard procedure to receive a privilege or patent. New forms of specifications emerged out of attempts to combine the different sets of logics at play in practically lived absolutism. Two are apparent in the history of calculating machines: first, the interplay of economies of glory and of lucre, and second, the interplay of the crown's prerogative and the defense of privileges and rights already granted.

First, the attempts at synthesizing the international "glory" (symbolic credit) of philosophers with financial credit required a rejection of local, territorial novelty in favor of temporally defined, global novelty. Isolating this global novelty could take a variety of forms: among the most obvious was the written articulation of some mentalistic form held to be essential to the invention. More important than the ability to capture accurately such an essence, through new forms of technical description or drawing, is the belief in such an essence independent of any particular material instantiation. Pascal's privilege isolates an essential ideational core to be protected just as it denies that that the machine actually has been reduced to practice. As we will see, Leibniz had to demonstrate his novelty through written or ostensive specification.

Second, ad hoc forms of specification became more central as a means for protecting the liberties and rights of others already granted privileges (or understood to have them by custom).[60] Such specification happened largely under duress, as the sovereign's prerogative was checked by extant privileges and their associated rights and liberties, either by representative and judicial bodies or by groups agitating in public or complaining to authorities. Such contestation was as present in more absolutist France or in German principalities as in parliamentary Britain.[61] Samuel Morland, to whom we turn next, was forced to specify publically in a quest for a royal monopoly.

The rights-based patent regime drew upon atavistic products and practices of the tensions within absolutism and centralizing states. The "intellectualization" of patents and privilege stemmed from many sources, which were only later crystallized into formal bureaucratic and legal features.

Morland: Protected and Unprotected Inventions

Demonstrating the novelty of his calculating machine led Leibniz to articulate the distinct innovations of Blaise Pascal and of Samuel Morland. Morland chose not to attempt to perform carries automatically, in order to avoid the difficulties of using "toothed wheels."[62] Morland chose likewise not to attempt to procure a privilege, patent, or monopoly for his calculating machines, though he did pursue such grants for other inventions and projects. The mathematician, inventor, spy, and turncoat Morland drew upon the full range of available techniques for protecting and promoting his inventions and for garnering support for their improvement. Morland's career illustrates the varied strategies for drawing upon formal and informal governmental patronage and protection for invention and manufacture.

King Charles II gave Morland a prize of £1,000 "for providing severall mathematical Instruments for our own use according to Our particular Direction."[63] As importantly, he was given the precious and rare right to advertise. At the end of the April 16, 1668, edition of the London *Gazette*, readers were promised a reward for a lost greyhound. Immediately following, readers found an advertisement for some new machines: "Sir Samuel Morland, having for divers late years, by His Majesties special Command and Encouragement, closely applied himself to the painful study of Numbers, and having at last . . . though with the expense of considerable sums, found out two very useful Instruments; The one serving for Addition and Subtraction . . . ; The other, for the ready performance of Multiplication and division."[64]

The king had supported Morland's efforts; to publish this brief advertisement was a sign of favor that cost the crown little. As an official paper, the

London *Gazette* initially forbade advertisements "as not properly the business of a Paper of intelligence,"[65] excepting, as one historian notes, "official announcements and the notices of courtiers for the return of their lost falcons and greyhounds."[66] Rather than granting him a privilege or monopoly for his machines, the king permitted Morland greater publicity, something likely of greater value when the goals were selling machines by informing the public about where to purchase them and, above all, advertising Morland and his remarkable range of services.

Morland might have been able to secure a patent on his calculating machines, but no evidence suggests that he attempted to do so. Getting patents was expensive enough that they were only worthwhile for high-priced or high-volume items or for those seeking to cement a contract with the government. Inventors who stuck it out to obtain a patent often did so to secure a business relationship between themselves and the crown or to circumvent guild restrictions.[67] Morland made a living from many kinds of mathematical and engineering labors, as well as secret intelligence work. For wealthy clients, Morland designed and sold instruments such as custom-designed and custom-installed way-wisers (early odometers) as well as calculating machines and pumps.[68] Owners of estates called on him to provide pumps and to advise them on improving their water systems.[69] For a wider public, Morland marketed numerous books offering basic arithmetic and the calculation of interest, as well as a perpetual almanac that sold well into the eighteenth century. He likewise sold tables for converting measurements of different jurisdictions. Parties interested in his machines could come to his showplace at Vauxhall and see any number of his curious inventions, including machines for making breakfast. Morland offered a larger number of items, as well as his expertise, and sought to become a contractor to the crown, particularly the navy, in a highly competitive market. Selling pumps was potentially lucrative, as was consulting about them. In the case of calculating machines, Morland was likely working to develop a clientele interested in his services rather than attempting to profit from the sales themselves. In the years after the Restoration, Morland regularly sought to use royal preferment to aid him in selling this variety of services and devices. Morland's career was marked by moving constantly between high and low disciplines and practitioners: from selling mathematical tables to designing the water system at Windsor, from selling simple calculation aids to attempting to sell his calculation machines in luxurious as well as simple versions.[70] The king aided him in doing so with advertisements, with patents, and perhaps by paying printing costs.

Morland became an inventor more by necessity than choice; his career as a mechanic was a structural by-product of the inefficiencies in governance and the internal politics within the circle of King Charles II. Finding "myself disappointed of all preferment and of any real estate," he wrote, "I betook myself to the Mathematicks, and Experiments such as I found pleased the King's Fancy." The king was Morland's greatest patron, but an all-too-typical early-modern one. To his great dismay, the Restoration failed to provide Morland with regularly paying employment as a state official or courtier, and certainly nothing as elevated as the position of secretary of state he thought his due. Morland claimed he spent most of his money on devices for the king.[71] Such was the life of courtiers across Europe.

Morland's financial position was probably more the norm than the exception. Morland provides a classic case of an early-modern client whom a patron cannot really afford to pay. The king and his councilors sought means to help Morland without having to pay him in a regular way or give him a comfortable and stable position in government, or even a sinecure. Giving him a knighthood and a baronetcy was one way. Patronizing Morland's calculating machine by allowing it to be advertised was another informal way to help, as was granting Morland an unusually strong monopoly for an innovative new pump. As so often happened, a Stuart king overplayed his hand by trying to play an absolutist monarch; in so doing, he inadvertently forced Morland to specify in print.

Before the process of specification was given a secure legal footing around 1778 in Britain, specification emerged from the limited nature of a sovereign's prerogative in granting privileges.[72] In the early 1670s, Morland began to market and sell a new form of water pump involving a leather collar.[73] In March 1674, he secured a patent for "severall engines for raising great quantities of water with farr less [proportion] of strength than can now be p[er]formed either by chaine pump or other engine now in use."[74] As usual, the patent included no description of the substance of his innovation, much less a specification.[75] He demonstrated his pumps before various audiences, including the king. In the spring of 1673 and then again in 1677, he sought a bill from Parliament offering something stronger—a monopoly of rare scope.[76] His contemporaries complained in bitter and striking terms that the absence of a clear description of his invention made Morland's bill a platform for any number of abuses. A printed "REASONS Offered against the Passing of Sir Samuel Morland's Bill" first described the long-standing law against monopolies from the reign of James I. So vague was the bill that Morland could easily claim all new innovations as his. More than just the liberty of other inventors is challenged, however, for Morland's bill "takes away the liberty of every Owner of Houses and Mines, as to their Pumps." Finally, the bill subverts the ordinary judicial protection, "which yet is never omitted in Letter-Patents of this kinde, and, rarely, if at all, in Acts of Parliament."[77]

An Italian traveler to England once noted of Morland, "In truth, his talent for politics is not wonderful."[78] Morland had chosen a poor time to attempt to subvert an important check on monarchic power. The 1670s saw growing suspicion that Charles II and his advisors sought arbitrary power. Patents had long been key sites of dispute over royal power and its limits: after the Restoration, Charles II had been told that his father had been beheaded because of disputes about patents.[79] Morland's opponents drew on a rich language of defending traditional liberties and procedures against monarchic abuses and encroaching absolutism: the bill would take away the liberties of makers and consumers alike, and it would destroy due process. The opponents turned resisting Morland's quest for a monopoly into an act of resisting the movement toward Stuart absolutism. As was usual in early-modern practice, his patent protected the secret behind his pump; no disclosure was required. Once the possibility of the monopoly became public, the political process of securing a stronger form of property in the pump forced him to specify his "principal parts." In attempting to draw upon an unchecked monarchical authority to impose a monopoly, Morland found himself forced to reveal his secrets to a public capable of influencing representative institutions.

In his printed response, Morland described the "principal parts" of his new pump in unusual detail for the time:

1. *The Playing of a Braß Bucket or Forcer into a Chamber of Water through a narrow Neck of Braß, in which is placed a small Fillet of Leather curiously prepared,*
2. *The reducing the Circular Motion of a Crank into a Perpendicular Motion, within a Frame of Iron, between two Rollers of Braß.*[80]

He claimed he had only "petitioned for an Act for the sole Use of those Engins, and of no other Engins whatever." He insisted upon the global novelty of his invention.[81] As best we know, the bill never passed. By 1681, Morland had lost what exclusivity he had with even his own partner.[82]

The debate over the bill reveals a moment of the crown attempting, as a favor to a preferred subject, to insist on a kind of monopoly that Parliament had long since abolished. It illustrates the tensions constitutive of absolutist governance: the constant inability to pay subjects and the inability to pursue policy and preferment without external checks. These tensions pressured Morland into specifying features of his invention in a public forum and thereby giving it a mentalistic essence. Artisans and other innovating parties resisting the granting of patents and privileges helped produce a practice of written justification that favored an understanding of invention and the process of making that was ever more mentalistic and less based in the entire array of capacities necessary for creating. For all their lack of standardization, the push toward kinds of specification in the conflicts around monarchical prerogative helped make a mentalistic conception of an invention foreign to the privilege system ever more possible—indeed, thinkable.

Like Morland, the young Leibniz had hopes for high state office and schemes for using machines to promote the public good while enriching himself. Absolutist incentives motivated both Morland and Leibniz; the process of inventing under monarchies left both dissatisfied with their professional and pecuniary status.

Leibniz: Protocols of Glory, Protocols of Financial Credit

On the back of a undated, autographed document titled "Things to be fixed in the [arithmetical] machines," concerned mostly with carrying mechanisms, Leibniz made a list of things necessary in order for "the machine to be put into use"—to be brought into practice. His list included what we might call his business model. Like savvy cosmopolitan artisans of his day who possessed valuable techniques, he set out a plan to get privileges from "many republics" across Europe before securing them from the Holy Roman emperor and the king of France. He outlined the various groups that would buy such a machine: universities and academies, merchants, and collectors of curiosities. Leibniz ends his to-do list by noting that "The King"—clearly Louis XIV—"can make it become fashionable."[83]

As so often in early-modern governance, getting a privilege involved the arduous personal grooming of powerful patrons. Leibniz had to cultivate personal connections to powerful people around kings and princes who alone could grant privileges. Leibniz appears to have focused on France, rather than the smaller states, even before his trip to Paris. He pursued at least two paths to Colbert over the next few years. A correspondent suggested Leibniz contact the mathematician and librarian Pierre de Carcavy, "all powerful around Mgr. Colbert in everything concerning letters."[84] Leibniz sent Carcavy news of his calculating machine and several printed works concerning natural philosophy and other instruments he claimed to have invented. Neither Carcavy nor the members of the Académie des sciences were much impressed—Leibniz had previously committed the sin of wasting their time and goodwill on insufficiently developed ideas and projects.

In December 1671, Leibniz received a letter from Carcavy that was at once admonishing and encouraging. He advised Leibniz not to send so many unclear and half-baked schemes and proposals to the Académie. Colbert "is satisfied only with what is real and solid," so Carcavy would present something to him only once Leibniz "had begun to send something effective to present to" the minister. Despite his reservations, Carcavy remained interested in Leibniz's plan for a new calculating machine and explained what he knew about Pascal's. Carcavy explained the protocols for evaluating and possibly rewarding Leibniz. If Leibniz "wish[ed] to send [Carcavy] something worthy of being seen," Leibniz could be "assured about three things" involving the protocols of considering and rewarding invention and new techniques. The suggestions offer a glimpse of the unwritten informal protocols around invention in Colbert's and Louis XIV's France.[85] First, Carcavy promised that Leibniz need not fear his invention would be stolen: "No one here will usurp what another has done"; he also "pledged to conserve all the glory to whom it is owed." Carcavy carefully avoided granting that Leibniz had something truly new; he simply maintained that credit would be fairly apportioned and that Leibniz would receive his due, if in fact he deserved any. Second, Leibniz could set the conditions for the use and dissemination of any machines or descriptions of them he imparted to Carcavy: "I will absolutely use whatever you send me only as you proscribe." Third, appropriate financial credit would be granted to those worthy: "I will procure from it, for you and for those deserving, the reasonable advantage necessary."[86] Though reason and justice need not have constrained the crown, Carcavy assured Leibniz they would. Finally, Carcavy intoned about the dangers of the "self-love" (*amour-propre*) of "authors" who overestimate the novelty of their inventions. However skeptical Carcavy may have been about claims to novelty, he explicitly treated inventions as things produced by authors, not as products emerging from an entire manufacture— that is, those produced by skilled artisans.

The clarity of these terms shows Carcavy's gatekeeping function on Colbert's behalf.[87] As Leibniz often stressed in his economic and political writings, Colbert had made the granting of privileges to foreign artisans and their workshops a centerpiece of his policy.[88] Much as Leibniz collected secrets from artisans, he collected precious secrets of Colbert's policies in practice: "I will find some way with M. Colbert's people to learn the details of this great plan for commerce and police [of industry]."[89] Previously charged with bringing Christiaan Huygens to the Académie, Carcavy worked to protect Colbert from mere projectors with empty schemes and also to assure inventors, artisans, and savants that their projects, glory, and economic interests would be protected. He needed to insulate Colbert from scams while recognizing and encouraging useful inventions and techniques. Credulity was dangerous; so was too much suspicion. Before any privileges, money, or other gifts were awarded, Carcavy offered a set of guarantees to inventors with potentially important projects. According to its own self-representation and the logic of the privilege, the crown had no requirement to follow abstract rules of justice in rewarding inventors, but Carcavy assured them that the crown would do so. The protocols Carcavy set forth probably were improvised adaptations of standard procedures for finding innovative craftspeople and luring them to France as a matter of economic policy. Adapting these procedures to philosophers and engineers working in realms of glory—international reputation— loosened the locality and materiality central to the privilege system. Leibniz trusted Carcavy to keep his secrets and to apportion glory and money along these lines, and he hoped to be rewarded as Huygens had been.[90]

Since Leibniz's earlier proposed machines to the Académie were seen as vague projections of possible machines, Carcavy pushed Leibniz to disclose more about his devices or to send them so that Carcavy and the Académie could judge whether they could be useful and whether they were, in fact, innovative. For Carcavy, this meant devising some means for comparing Pascal's and Leibniz's machines. Carcavy explained that Pascal "had not provided a particular description of his numerical machine," but he offered to have a detailed specification of the device written up. Alternatively, Leibniz could send an exemplar of his machine: "If you want to send me yours with the manner of working it, I will tell you what is the same and what different."[91] Carcavy rightly surmised that Leibniz's machine existed more as an aspiration than as any actual device, or even as a concrete design.

Global novelty was not a necessity for most early-modern privileges. Carcavy nevertheless explained that Leibniz needed to provide something globally new and useful to the crown. Carcavy was working in two different economies of credit: that of philosophical reputation ("glory") and that of money ("reasonable advantage"). Had Leibniz been concerned exclusively with monopoly protection and the creation of a manufactory, Carcavy probably would have only required local novelty; but because Leibniz was looking for the international glory that would follow a strong approbation from the Académie des sciences, he was subject to a stricter requirement of global novelty.

Proving global novelty in a written description promoted a mentalistic conception of inventions as possessing some essence. Among Leibniz's manuscripts from the Parisian period is an autograph assessing his machine, written in the third person. It is likely a fragment of a report Leibniz prepared on someone else's behalf—likely Carcavy's or Colbert's—to justify granting him a payment, a pension, a privilege, or some other preferment. In comparing his machine with those of Pascal and Morland, he carefully assigned credit while suggesting the faults of the competing machines.

The beauty and ingenuity of Pascal's machine cannot be dismissed, Leibniz explained. Like almost all serious critics of Pascal's machine, Leibniz noted that it could only be used right to left, so subtracting could not be done directly. Leibniz likely saw Morland's machines at Whitehall during his visit to London and possessed a copy of Morland's book concerning the machines.[92] Although Morland's machine failed to perform carries, Leibniz noted, it could be used in either direction, unlike Pascal's machine. Morland's machine was therefore distinct enough that there could be no question of the independence of his invention: "Mr. Morland is the inventor of his machine without owing to M. Pascal the idea and still less the execution."[93] Leibniz claimed that Morland had made clear that he would readily accede glory to Leibniz for his kind of calculating machine if what he had heard about it were true.[94] Morland was famous across Europe, above all for his speaking tube; not only was he a credible witness to Leibniz's innovations, but he could also easily afford to grant Leibniz credit. Leibniz sketched out the ideational essence of the different machines and then used that sketch to partition credit. These distinctions in the realm of glory served to prove global novelty and to justify his demands for a privilege. He would claim for years afterward that both the Royal Society and the Académie des sciences accepted the "infinite difference" between his machine and others, even if both institutions demanded a reduction to practice.[95]

Soon after he arrived in Paris in 1672, Leibniz began working on producing a machine he could claim to be effective—or at least effective enough to gain the support of Carcavy and, through him, Colbert. Along the way, Leibniz cultivated a parallel set of personal connections to Colbert, this time through Colbert's son-in-law Charles-Honoré d'Albert de Luynes, duc de Chevreuse.[96] Leibniz cultivated relationships with Chevreuse's client, Jean Gallois, secretary of Colbert and editor of the *Journal des Sçavans*, as well as one Jérôme Dalencé. A shadowy figure interested in barometers and magnetism, Dalencé served as a sort of privilege broker for inventors of various types on Colbert's behalf—someone less concerned with the philosopher's quest for global novelty than the state's concern for securing artisans with useful manufacturing processes.[97] With the help of this clientele network, Leibniz sought full membership with pension in the Académie as well as a privilege, financial support, and an advance contract for his calculating machine.

Carcavy, Dalencé, and Gallois encouraged Leibniz, advised him on the sorts of things to present to the duc de Chevreuse and then to Colbert, and admonished him when he made outrageous claims or did not produce in time. Leibniz failed time and again to produce a sufficiently working machine, much to his patrons' dismay.

Much as he was never offered a full, pensioned position in the Académie, Leibniz never received a privilege for his machine. Pascal managed to get a privilege for work that still needed to be perfected in order to be put into practice. Leibniz did not, despite this precedent. Based on his models and drawings, Leibniz received preliminary orders from Colbert for machines for the crown, for the observatory, and for Colbert himself, contingent on bringing the machine to practice. Colbert also offered support in the form of payment for the artisan Ollivier to bring the machine into practice. In the documents Leibniz wrote up for Colbert and the Académie, he outlined the package of incentives necessary to motivate a skilled clockmaker to abandon his trade to concentrate exclusively on perfecting and then building the machines. As we saw in chapter 2, skilled artisans who were willing to innovate needed incentives to give up their profitable, accustomed ways: "It is just to pay a skilled master not only for the time he has worked—without speaking of the novelty and the risk of the enterprise—but also his industry and his skill in discerning himself from an ignorant." The payment of artisans must register the higher creative abilities of the superior sort of artisan.[98] As discussed in chapter 2, Leibniz insisted on the importance of such superior artisans for the development of technique and of economy.[99]

Leibniz's brief evidently worked—at least temporarily. The accounts of the "Bâtiments du Roi" record that on December 15, 1674, a "Sieur Ollivier, clockmaker" received 300 livres "in consideration for a numerical machine he has made."[100] Other evidence suggests that this money was dispersed to Ollivier not as a lump sum but as a sequence of payments to encourage his work, with a final payment to motivate and reward completion. According to a letter of April 1675, Ollivier was to receive twenty pistoles a month for five months to support him, and then "he will be paid nothing more, until he has made a piece and can furnish it all ready."[101]

In these various briefs, Leibniz translated arguments about motivating artisans into arguments about philosopher-inventors such as himself. Like Ollivier and later Babbage, Leibniz claimed he needed incentives to focus exclusively on machines. The first version of his briefs set out Ollivier's arguments about his just compensation; subsequent versions transmuted many of these arguments into the claims about Leibniz's just compensation with which this chapter began. Leibniz's manuscripts illustrate how arguments about rewarding the industry of inventive artisans were adapted to justify the rewards for philosophical invention—an improvisational reworking of ideas of proper recompense and incentives.

In his narrative, Leibniz explained that his project was nearing fruition and that he was moving from active invention to simply directing "the execution of this work."[102] In this claim—soon to be falsified—the ideational work of invention was complete; all that remained was materialization. He claimed to have already reduced his proof-of-concept model to practice—a claim that appears often in his correspondence. Early-modern inventors were given only small windows, often six months, to provide working versions of their model instruments. Leibniz attempted to secure more time—and money—in his contracts for reducing his model to practice. Even though it was generous by early-modern standards, the extra time Colbert and the Académie gave him proved to be not nearly enough. At no stage was the model as perfected as Leibniz boasted. His patrons and supporters in France quickly grew irritated, as did others, such as Henry Oldenburg and Robert Hooke.

In late October 1675, Leibniz received a summons from the privilege broker Dalencé to come to the house of the duc de Chevreuse in Saint Germain the next day: "Given that you have taken the trouble to tell me that the *machine* is *all ready*," he asked Leibniz "to *bring*" the machine "as it was before one began this fourth wheel; I beseech you *not to fail* to come to my place tomorrow at one hour exactly after noon and *to bring the machine*" to take to the duc's house.[103] The emphasized words, in the original, strongly suggest appointments missed and promises not kept. Leibniz failed to show. A few days later, he lamely wrote through an intermediary that an "indisposition" had prevented him from making the appointment, and he dared not write the duc directly.[104] Leibniz may have presented a model of the machine to Colbert at Saint Germain in late 1675.[105] Despite the support of the duc de Chevreuse and others, Leibniz did not receive a pensioned position in the Académie des sciences and could not remain in France. The failure of a timely reduction to practice likely contributed to undermining Leibniz's candidacy for a rare permanent—and pensioned—position in the Académie.[106]

Given these failures, Leibniz's overconfident language of the just recompense due someone who contributes to the glory of the crown and the common good disappeared. By the time he wrote to Colbert in January 1676, Leibniz no longer drew on a language of just compensation; he wrote in a tone of some desperation, as a submissive and unworthy client begging for any recognition at all. Addressing Colbert, Leibniz explained, "is a sort of recognition [*reconnoissance*]: we owe you the presentation, but you owe nothing in exchange, and the liberty of choice remains entirely yours."[107] With no timely delivery of the machines, Leibniz had moved from demanding his reasonable due to begging for favor based on the novelty, promise, and interest of his models. In so doing, he moved from the idioms and practices of the workaday legal, financial, and commercial world of early-modern France, with its ever-checked sovereignty, to the idioms and practices of the self-representations of absolutist France, from an independent contractor to a self-effacing courtier. He moved from representing himself as the sort of person upon whom the success of the crown depended, and thus deserving of credit, to seeking favor out of the unconstrained and undeserved goodwill of a patron.

Leibniz did not long remain abashed. Having failed to attain a privilege for a calculating machine and a position in the French Académie des sciences, he sought to revolutionize mining in lower Saxony. When Leibniz arrived in Hanover as a minor court official, he attempted to apply his newly acquired knowledge of the laws of motion, the minimization of friction, and the optimal design of gear-teeth to the problem of draining the mines of the Harz mountains using a new form of windmill. The story of Leibniz's windmills is one of technical failure and constant dispute with mining experts and mining labor.[108] When Leibniz first envisioned his new means for powering pumps to remove water from mines, he asked his patron, Duke Johann Friedrich, for a due recompense in terms much like he posed to Colbert: "Two things are necessary, namely the costs of the enterprise . . . and some advantage for the entrepreneur or inventor, which is just."[109]

Leibniz's framework for defending the privilege that he requested from Duke Johann Friedrich reveals the tensions between an absolutist account of untrammeled power and traditional obligations of justice. If the obligations the duke owed to the inventor Leibniz were to be regulated by appeal to the "justice" due to inventors, the traditional obligations the duke owed to other subjects must be overturned in the name of the public good. The logics of obligation here are contradictory: if "owing" is subject only to the duke's will to serve the public good, he hardly need have any "just" obligation to provide the Leibniz with anything. He might have had an instrumental reason to do so based on Leibniz's contributions to the public good, but he had no particular obligation as an absolute monarch to recognize those contributions as deserving of anything.

Leibniz's own account of charity suggests that he ought simply to reveal any inventions to the public or to a prince on behalf of the public. "Nevertheless" he argued, "reason requires" that he "stipulate some advantages, before coming to reveal the essence [fonds] of the invention." Such reason— here meant as a natural law of property—need not constrain the prince, even if it ought to: "Reason demands it and Your Exalted Serenity being as just and generous as he is, does not disapprove it." Using his unchecked sovereign power, the sovereign should *freely* choose to do what natural justice demands—much as Leibniz's God chooses freely to create that which is best and most just. In 1678, Leibniz noted, "The best means that an individual can find for doing good for the public is to have the approbation of a prince who is enlightened [enough] to recognize and powerful [enough] to execute what is useful."[110] Leibniz was prophetic here: doing what reason recognized to be both just and in the public good would require sovereign authority and coercive power, be it from monarchs or representative bodies.

In some written notes on Colbert's industrial policy, Leibniz remarked on the legal maneuvers to fight off "contestations" against new privileges.[111] He applied this knowledge when he requested a privilege for his windmill for draining mines. Leibniz rightly predicted that the interested parties would object and would frustrate the project, and he devised a set of measures to ward off such contestation. Above all, his authorship had to be hidden. He urged his duke to portray the entire project as a direct and personal application of his sovereignty to serve the public good. In order to succeed legally and practically, the entire project had to be represented as emanating personally from the duke's will. Under an absolutist account of ducal authority, the duke possessed the right to set aside the ordinary laws governing mines and the laws of privileges and contracts by which he would usually have been bound. Even before the name of the inventor and the plans are announced, Leibniz suggested, the duke should send a state counselor to "fix up the situation on the ground, and create a certain and just order." Only after having established such a properly pacified and properly regulated zone, "will it be time for Your High Serenity to declare what will follow": the building and testing of Leibniz's windmills with Leibniz in a position of authority.[112] The duke chose not to comply—he needed his miners more than he needed Leibniz, and he owed them more than he owed Leibniz. Even if he had been willing to overturn the prerogatives of the miners, Leibniz's sovereign just wasn't strong enough.

Creating an intellectualist patent regime required an intensification of sovereign power, to overcome traditional valuations of labor, skill, and intelligence. Establishing rights to patents demanded a regime capable of enforcing them against traditional prerogatives and one capable of transforming work and the ownership in it against artisanal and traditional practices and understandings of property. The transformation of the patent bargain from an individual gift of a sovereign to a subject into a generalized contract between the public and an inventor was predicated not on the liberation from sovereign power but on the (ever-imperfect) actualization of that sovereign power.

111   Third Carry

Babbage Claims His Property

In 1834, Charles Babbage wrote to the Duke of Wellington, "My right to dispose, as I will, of such inventions" as the Difference Engine "cannot be contested; it is more sacred in its nature than any hereditary or acquired property, for they are the absolute creations of my own mind."[1] Brave talk of natural rights in intangible property notwithstanding, inventors still had no such rights by statute or judicial decision in Britain.[2] A review of one of Babbage's books by David Brewster complained that authors have their "property secured to them by statute," as do engravers, draftsmen, geographers, and hydrographers. He "who has invented a new steam-engine cannot, like the author of a new romance, dispose of it forthwith. He must devote himself night and day to the practical application of his principle: he must construct models and perform experiments."[3] Brewster complained that inventors had to do far more than offer an idea for a mechanism. Britain retained its privilege system, even if it included a judicial requirement for specification from 1778. All patents formally remained monarchical gifts with high fees attached.[4] Brewster minced no words: the patent laws of Britain comprised "a system of vicious and fraudulent legislation, which, while it creates a factitious privilege of little value, deprives its possessor of his natural right to the fruit of his genius."[5]

Babbage's overconfident and modern-sounding claims about his natural property rights appear in the middle of a far more traditional warning to his government. Like artisans seeking privileges and preferment, Babbage threatened to move his manufacture abroad. Babbage could "collect together all that is most excellent in our own Workshops—those <u>Methods</u> and <u>Processes</u> which are equally essential to the Perfection of Machinery, but which are far less easily transmitted from Country to Country," and that "would be at once brought into successful practice under the Eyes and by the Hands of Foreign

Workmen." Creating a new corpus of engineers would give "a lasting Impulse to the whole of the Manufactures of that Country, and that the secondary Consequences of the Acquisition of that Calculating Engine might become far more valuable, than the primary object for which it was sought."[6] Bluster about rights in ideas aside, Babbage recognized that transferring manufacturing practices and the people embodying them—not the ideas behind the Difference Engine—was the real risk he, as a philosopher-entrepreneur, could pose to the state.

Babbage's confident articulation of natural rights in the products of genius masked more immediate anxieties about ownership around machines and invention. Babbage was so uncertain of his rights that he asked the government at one point, "Suppose Mr. Babbage should decline resuming the machine, to whom do the drawings and parts already made belong?"[7] The question should have been resolved long before. Just as he had been insouciant about the difficulty of carry when he began the Difference Engine, Babbage had been nonchalant about the legal and financial coordination for the project. After informing his friend Herschel that he was likely to receive £1,500 from the government in 1823, he noted, "As I have liberal people to deal with I shall not be annoyed about pence and the particular mode in which I may think it right to distribute them and I shall I hope to be able to bring the thing to perfection or at least to a good practicable working state and that in a few years we shall have new (but not patent) stereotype logarithmic tables as cheap as potatoes."[8]

By the late 1820s, he greatly regretted this disregard and his assumptions about the "liberality" of his engineer. Clement worked in an alternate moral economy. In 1828, as his doubts began to rise, he noted, "I trusted to his [Clement] being an exception and thought him a sincerely honest man. Perhaps I was a fool . . . and I will profit by my experience which in this instance will be rather costly."[9] During his Italian trip, Babbage was having premonitions about disputes over ownership: "Clement is spending much time in making tools." This, Babbage said, "is to a certain extent necessary and requires considerable supplies of money but I should wish you incidentally if possible to find out whether it is not Clement's intention to make me pay for the construction of these tools and then to keep them as his own property—from the multitude he is making it looks so."[10] Though he held no privilege or patent, Babbage pushed his mechanic, Joseph Clement, to agree, "It would be manifestly a great injustice for the contriver of such a machine whose sole risk it was made that any other should be made by the same workman with the same tools."[11]

At issue was less Clement's honesty than a clash of conceptions of ownership and the organization of work.[12] Much as Babbage never had a clear contract with the government, Clement and Babbage had no clear understanding about ownership of tools, of drawings, and indeed of the right to reproduce the machine. In a long series of inquiries into the propriety of Clement's outstanding bills, Babbage engaged the expertise of two noted engineers "to examine Mr Clement's charges and report to me their opinion of their propriety observing to them that such labors attended with such loss of health as he had experienced deserved to be well remunerated."[13] For all his frustration, Babbage never gainsaid Clement's skill and ingenuity. After a long delay, Clement was paid and work resumed in the spring of 1830.

In addition to auditing Clement's bills, the two engineers took on the question of ownership more generally. They reported Clement's understanding:

> Question 1st: To whom do the Tools belong.
> Answer: To Mr Clement. Mr C says that all the Lathes have been made at
>   his own expence.
> Question 2nd: To whom do the patterns belong.
> Answer: To Mr Babbage.
> Question 3rd: To whom do the Drawings belong.
> Answer: To Mr Babbage.
> Question 4th: At whose risk is the Machine Tools and Drawings.
> Answer: The Machine, Drawings, and Patterns are at Mr Babbage's risk,
>   the Tools at Mr Clement's.

Ownership, in Clement's account, was of tangible things, not intangible property: Babbage owned the machine, to be sure, insofar as he owned the metal pieces of the machine, the drawings, and the paper and wood patterns. But Clement did not accept that Babbage owned intellectual property in the machine or an exclusive right to authorize its creation:

> Question 8th: Ought not Mr Clement to engage, not to make another
>   Machine without Mr Babbage's written permission?
> Answer: Mr Clement declines doing so.[14]

Why should he? By tradition, his work gave him ownership in the things produced—claims of exclusive ownership thanks to philosophical ideas and Romantic authorship be damned.

Babbage knew this. The technical, organizational, and material obstacles to bringing his Difference Engine into practice led Babbage to recognize the dispersion of creative skills necessary to produce machines in actual practice. In discussing the vagaries of his own invention, he stressed time and again the importance of a creative engineer such as Clement. The "first necessity" for the Difference Engine was "to preserve the life of Mr Clement. . . . [I]t would be extremely difficult if not impossible to find any other person of equal talent both as a draftsman and as a mechanician."[15] Like Leibniz before him, Babbage left major design decisions up to his engineer.[16] Even as he fantasized about a future division of labor giving all initiative in design to intellective inventors and thus eliminating the need for engineers such as Clement, Babbage warned of the difficulties of transforming ideas into working machines, and he retained reduction to practice as the standard of success: "When the drawings of a machine have been properly made, and the parts have been well executed, and even when the work it produces possesses all the qualities which were anticipated, still the invention may fail; that is, *it may fail of being brought into general practice.*"[17]

In his 1832 *Economy of Machinery and Manufactures*, Babbage offered the programmatic dream of a mechanical reproduction of parts as fully specified by theory: "Nothing is more remarkable, and yet less unexpected, than the perfect identity of things manufactured by the same tool."[18] Such a system of manufacture was a goal of the reorganization of labor and technique, not something yet achieved; the difficulties in producing calculating machines served as an emblem for needed reforms of work and as a major spur for the development of new machining techniques and organization of labor.[19] Once standardized manufacture with reproduction of parts had been achieved, a conception of the invention of a mentalistic "form" independent of a process of the actual production became far more plausible. Manufacture considered in this way severed form and matter and justified dividing machines into intangible ideas and concrete instantiations.[20] Such manufacture, should it come to pass, would eliminate the sagacious and creative artisan and engineer. Only then could his actions be seen as merely repetitive, machine-like, and thus undeserving of ownership.[21] Recent disputes about traditional knowledge in the global south underscore the continuing political potency of denying and recognizing novelty and innovation.[22] In reflecting upon the novelty of the Difference Engine, Babbage and his allies collected the "prior art" of the eighteenth century—a rich tapestry of machines built in ignorance of earlier efforts, imitations, and improvements.

Reinventing the Wheel: Emulation in
the European Enlightenment

Every improvement . . . has arisen from our imitation of foreigners; and we ought so far
to esteem it happy, that they had previously made advances in arts and ingenuity. . . .
[H]ad they not first instructed us, we should have been at present barbarians; and did
they not still continue their instructions, the arts must fall into a state of languor, and
lose that emulation and novelty, which contribute so much to their advancement.[1]
DAVID HUME, "Of the Jealousy of Trade"

In 1784, a major literary and philosophical periodical of the German Enlightenment, the *Teutsche Merkur*, published an article celebrating the novelty of a
calculating machine created by Johann Müller, an engineer from Darmstadt
(figure 4.1). The author of the announcement, the literary critic Johann Heinrich Merck, reflected upon the honor and glory due to multiple inventors of
some "thing."[2] Judging the merit of a second inventor required determining
the extent of knowledge of previous efforts: "One can clearly be the second
discoverer of an already invented thing in two ways: either with or without
the assistance of the initial inventor." Even when secondary inventors had
some knowledge of the means for "perfection" of the thing in question, much
merit and glory could rightfully be theirs: "The more the proper ingenuity
of the second discoverer was necessary in this problem, and as the work that
he created is more or less perfected, the more that he should be rewarded for
how he departed from the already known details of the prior invention."[3]

The Pietist minister and clockmaker Philipp Matthäus Hahn had published a description of his calculating machine in the same journal in 1779.[4]
The cylindrical machines of Hahn and Müller seemed at first glance much the
same. Merck argued that Hahn provided "no instruction about how a similar
machine might be prepared."[5] While Hahn had explained nothing about the
internal mechanism, his limited disclosure was productive: "The very process
of making [the machine] known provoked a similarly inventive mind to seek
out a similar work; so it appears that the second Prometheus worked with his
fire, without having stolen it from the first."[6] In delineating what Müller did
not know about the details of Hahn's machines, the article acknowledged that
the experience of reading about Hahn's machine was an emotional and intellectual catalyst for Müller. Contemporaries had a category, tracing back to

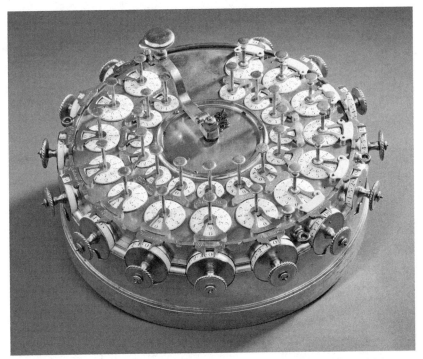

FIGURE 4.1. Johann Helfrich Müller, Rechenmaschine, Inv.-Nr. Ph. C. 60/90, c. 1784. (Hessisches Landesmuseum Darmstadt. Photo: Wolfgang Fuhrmannek.)

Aristotle, for understanding how just such affective reactions could provoke authors and inventors to attempt to transcend inspiring models. The emotive and moral state of desiring to outdo another was called "emulation," defined as "a noble and generous passion, which, while admiring the merit, the beautiful things, and the actions of another, attempts to imitate them, or even to surpass them, while working to do so with courage through honorable and virtuous principles."[7] An honest competition without mere "aping" or imitation of the work of another, emulation inspired and challenged creators to produce yet better things. Emulation allowed—indeed, encouraged—one Prometheus to be inspired by another.[8]

Taken by the description of Hahn's machine but worried about the machine's propensity to err, Müller transformed the basic outline of the design through the addition of a warning mechanism—a bell—that sounded when certain errors occurred. The author of the article judged this bell favorably: "Reliability is the primary feature of Müller's machine, on account of its [warning] bell."[9] He concluded, "The merit of invention belongs to him just as to the minister Hahn," and he noted that such merit "would equally belong to a third or a fourth" who, after having learned about this "second invention in the *Teutsche Merkur,* achieved a similarly new way and brought it to practice."[10]

Imitation was a major wellspring of technological innovation in the eighteenth century and was explicitly recognized as such in political economy and practical statecraft alike.[11] Writing amid the enthusiasm around the new notion of genius, Merck, the author of the defense of Müller, struggled to fit a model of cumulative and simultaneous invention into a newer account of creation focused on identifying single authors and inventors in the public sphere. Merck had to shoehorn the rather conventional story of the collective and cumulative inventions of Müller and Hahn into a model of unitary invention crucial to the incipient conceptions of genius and intellectual property so characteristic of the late eighteenth century. Yet Merck and his subjects remained too close to real collective invention in the years before serial manufacture to downplay concrete technical solutions developed over time through iterative design, accretion, and trial and error—through craftsmanship.

Merck closed by pointing out how the refusal to acknowledge multiple inventors created a competitive economic disadvantage: "How proud France would be, to have two men of such an inventive genius as Hahn and Müller at the same time. How . . . to pull it out from the darkness, to honor, to reward, to encourage the greatest possible use of its talent!—And Germany?"[12] Too narrow a vision of the merits of inventive activity could only undermine the efforts of the various German states to compete with the might of France or England. German competitiveness required acknowledgment of the productivity of inventive, imitative competition.

Eighteenth-century mechanisms of reward and economic transformation encouraged the "imitative" creation of calculating machines, usually in the absence of detailed technological knowledge of earlier efforts. Such imitation was certainly creative: no fewer than five major ways of organizing the multiplying mechanism for calculating machines were invented and reinvented, some with knowledge of predecessors, others independently. The productivity of this imitative activity cannot be understood exclusively as the history of industrial practices within a narrowly economic history. At one level, the imitation took part in a broader Eurasian process of import substitution and industrial espionage.[13] Particularly inventive forms of imitative processes were valorized and then actualized not just within supposedly universal market incentives but also within a rich understanding of "emulation" as a competitive imitative practice that improved at once the inventor and the technology in question.[14] Polities seeking economic transformation were attracted to the potential of such emulation for "naturalizing" foreign manufacture as an initial step toward producing novel and superior manufactures of their own.

Our story takes us to Lunéville, Darmstadt, Stuttgart, and Venice. Ministers and monarchs in the panoply of small principalities of post-Westphalian Central Europe drew upon an emulative ideal to improve their fortunes and to augment their magnificence. Eighteenth-century calculating machines were largely efforts of artisans, engineers, and natural philosophers attempting to provide tangible proof that they were imitative inventors who could produce import substitutions, naturalize foreign manufacture, and become ornaments to the court capable of enhancing glory and garnering lucre alike. Most of the machines were efforts to make their inventors' competence legible to monarchs and ministers uncertain whom to trust in seeking economic and social improvement through technology. The machines were concrete demonstrations of how their makers could unite their ingenuity with the ability to bring something into practice in drawing upon and coordinating the skilled labor within a polity.

The vision of an emulative invention combined productively with the lack of information about calculating machines and the people who might make them. Together, imitation and an absence of details motivated the creation of a remarkable number of calculating machines designed around strikingly different principles. The history of calculating machines in the eighteenth century is a story of copying—and improving—mechanisms and processes, often known with only the sketchiest of detail.

Philosophical Emulation: The Productivity of Ignorance

An early nineteenth-century history of calculating machines relayed an oft-told tale about failed genius: "Herr von Leibniz invented a calculating machine, which, however, was never perfected and reduced to practice, even though he spent more than 20,000 Thaler on its implementation."[15] Perhaps the most salient fact for the proliferation of calculating machines in the eighteenth century was Leibniz's failure to bring his calculating machine to practice. Although Leibniz belatedly published an account of his machine in 1710, he did not disclose the mechanism in any meaningful way; contemporaries bemoaned this lack of disclosure (see figure 2.1). An article in 1769 noted, "Neither this description nor this figure suffices in any way for forming an idea of the mechanism of this curious machine."[16]

Leibniz's struggle was well known by the early decades of the eighteenth century. That the "universal genius" Leibniz failed to bring a machine to practice made the pursuit of a better-realized machine a tempting goal for professors, philosophers, and clockmakers alike; to produce, to repair, or even to design a potentially more perfected machine could become a tangible display

of knowledge, skill, and ability in coordinating skilled labor. The story of
Leibniz's failure and difficulties became a standard trope for those pursuing
a reinvention of a multiplying calculating machine. Hahn recounted how
the standard narrative of Leibniz's twenty years of work and 20,000 Gulden
spent helped him surmount his doubts once difficulties set in: "So I was not
deterred."[17] A lack of knowledge, far from hindering the creative energies
of inventors, prompted them—and permitted them—to reinvent Leibniz's
wheel in pursuit of glory and profit.[18] Within the economy of glory in the international republic of letters, Leibniz's failure to complete his machine and
his failure to disclose its principles legitimated efforts at reinvention.[19]

Nescience about earlier efforts, for example, legitimized the Venetian engineer Giovanni Poleni's inventive activity, which he pursued in hopes of a
position serving the state. Unable to find the description of the artifice of the
Pascal and Leibniz machines or to make them out from outward descriptions,
Poleni explained, "I desired greatly, either to divine by thinking and reflection
their construction, or to construct a new one, that could produce the same
effects."[20] Poleni designed a machine organized around a new principle for
the multiplying mechanism: the variable cogwheel. A large, wooden structure, Poleni's machine looked like a clock (figure 4.2) He published in 1709 a
detailed, illustrated description of the internal mechanism, the first detailed
printed description of a mechanical calculating machine.

The Viennese "court optician and mathematical mechanic" Anton Braun,
along with his son and his journeymen, invented several varieties of calculating machines for the Holy Roman emperor. At the time of his death in 1727,
Braun left several incomplete calculating machines and sketches of calculating machines. On the basis of his calculating machines, he received a "considerable" payment and a yearly pension of "2000 fl. to prepare mathematical instruments." Invention, not reduction to practice, was his strong suit. A report
just after his death noted that Braun "was in invention more lucky than in
implementation, because in the improvement . . . of machines, he always had
new ideas . . . Someone brought the Emperor a drawing of the machine he
left unfinished. His Majesty honored him with this praise: 'A Braun we won't
come by again.'"[21] Two machines credited as "invented" by Braun exist: both
were reworked and brought into practice after his death. These two models for machines were esteemed and rewarded for their inventiveness *locally*
without any apparent regard to the international realm of glory. Elite artisans
had long been rewarded for the introduction of processes and manufactures
regardless of whether those processes existed elsewhere. Local, not global,
novelty mattered in the Viennese court.

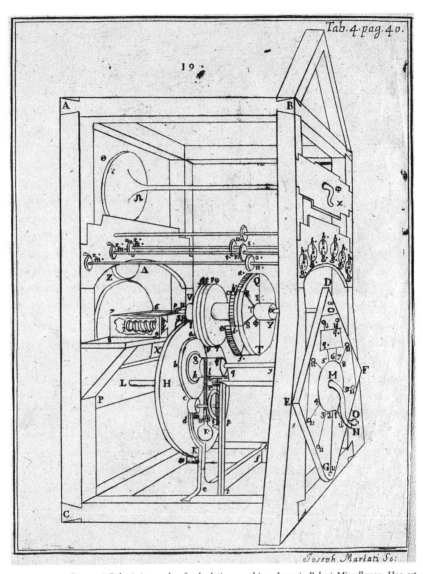

While Braun worked within a local courtly system of reward in which he could largely ignore creations elsewhere, scholars like Poleni who sought glory in the international republic of letters and beyond had to attend to creations elsewhere, not simply locally. A posthumous *éloge* relayed that Poleni protected his reputation with dramatic flair: "While this machine was very simple and easy to use," as soon as Poleni heard of Braun's machine, "he broke his and wanted never to rebuild it." The *éloge* explained, "It was without comparison more glorious for him, especially at his age, to have broken it than to have invented it."[22] Even if apocryphal, the account reveals how inventors within the economy of glory needed to insulate themselves against the possibility of plagiarism.

A second example reveals the care taken in negotiating the transition from a local courtly setting to an international one. In 1735, the Giessen professor of mathematics Christian-Ludovicus Gersten carefully navigated the possibility that he *could* be seen as merely copying: "I took the Hint of mine from that of Mr. de *Leibnitz*." As he could not "hit upon the original Ideas of that Great Man, an exact Enquiry into the nature of Arithmetical Operations furnished me at last with others, which I expressed in a rough Model of Wood, and showed to some Patrons and Friends, who encouraged me to have another made of Brass."[23] As with Poleni, ignorance about the details of Leibniz's machine motivated Gersten's inventiveness. Gersten worried about publishing about his invention lest he discover that Leibniz had in fact reduced his machine to practice. However willing to demonstrate his machine to the monarch and court, he hesitated to publicize it to the broader republic of letters.[24] The approval and support of the monarch, he explained, "would have been Inducements for me to have publish'd at that time an Account of my Machine; but I was checked by the Uncertainty I was under, whether possibly Mr. *Leibnitz's* machine had not been brought to its perfection." He eventually became "certain that none of Mr. *Leibnitz's* Invention has yet appeared in such a State of Perfection, as to have answer'd the Effect proposed, and that these of mine differ from all those mentioned above . . . I make no Scruple to present to the *Royal Society* this invention."[25]

Claims of their ignorance were central to Poleni's and Gersten's justifications of their efforts to make and to publicize their work beyond local spheres. Their protestations about their lack of knowledge were more than rhetorical: their nescience encouraged their distinctive approaches to machine calculation. Each devised novel and otherwise then-unknown mechanisms for making machines automatically perform multiple additions. Unlike Leibniz, they published the details of their inventions in search of local preferment and international esteem once local authorities had judged their devices favorably.

Subsequently, others took the inability of Gersten and Poleni to discern the mechanism of Leibniz's machine as motivation and legitimation in a continuing quest to perfect it. On April 26, 1765, in St. Petersburg, the academician Christian Kratzenstein presented "a completely achieved arithmetical machine to our Academy of Sciences, one more perfect than Leibniz's." Kratzenstein had grasped what others had not: "While Gersten, the mathematician from Giessen, admitted . . . that he never could discern the construction of the Leibnizian machine, I believe that I have discerned it." All existing disclosed machines, including Leibniz's, "are prone to err whenever it is necessary to make a number of 9999 move to 10000."[26] Kratzenstein claimed to have perfected the carry.[27] Nothing more is known of this machine.

We are accustomed to a vision of technical progress attendant upon the greater availability of information—a vision inherited from ideals of the republic of letters and the high Enlightenment.[28] In contrast to a vision of shared information producing change, in the case of early-modern calculating machines, the very lack of disclosure provided authorization to seek glory and lucre in the imitation of a device that was not fully known and was never brought to perfection. Within the incentive systems and structures of the eighteenth century, ignorance about the failed productions of a "universal genius" served to legitimize inventive activity by suggesting the intrinsic value as well as the great difficulty of building machines—both contributed to the estimation of merit of anyone seeking to devise and construct a machine. Ignorance provoked innovative production, just as it had with famous Asian commodities such as hard porcelain.[29] Maxine Berg argues, "Admiration for Asian craftsmanship was followed, however, not by a direct process of copying, but a more subtle process of imitation. The key response to these commodities in Europe was a process of product innovation and invention through imitation. . . . The European mimesis, in turn, was not to produce a direct import substitute, a lessor or perhaps more expensive version of the original, but to turn that imitation into product innovation."[30]

The twin, often intertwined, economies of imitation in eighteenth-century Europe had high-minded as well as hardheaded approaches to imitation and reinvention. Outright industrial espionage, import substitution, and reverse engineering were central to the economic development and the movement of skill, technique, and knowledge around Europe and between Europe and its trading partners and colonies.[31] Entire industries, such as the Genevan clockmaking manufacture, developed a rich skill base, skilled workforce, and practiced industrial organization as centers of counterfeiting before becoming primary sites of innovation. English papermaking overcame earlier Dutch supremacy in a similar manner: "British papermaking entered the craft's transnational mainstream, unsteadily to be sure, by naturalising the workaday as well as the inventive aspects of Continental papermaking's technique, fashion and instruments."[32] Calculating machines partook in a similarly dynamic process of reinvention; this creative imitation of goods without knowledge of the processes used to produce them was central to the upsurge in manufactured goods during the eighteenth century. If the aim was initially reverse engineering, the net result was a proliferation of new techniques, new technical solutions, new management of labor and machines, and new goods. Ceramics offered the prime example: in their attempts to reproduce true Chinese (hardpaste) porcelain, artisans, philosophers, and entrepreneurs of all kinds devised new sorts of ceramics, many of which became sought-after goods associated with the place of manufacture, such as Delftware or Norman faience. Princely privileges protected these efforts at import-substitution through creative reverse engineering. Even after hard paste porcelain was reproduced and sources for the key ingredient kaolin were discovered, the bevy of substitutes occupied their own market niches; as importantly, the knowledge and skills in natural and organizational matters created through imitation allowed any number of further product innovations, large and small. The decades-long efforts in polities across Europe to reverse engineer porcelain led to the creation of a diverse array of new kinds of ceramic wares. The decades-long effort to reverse engineer Leibniz's calculating machine yielded an array of new calculating machines and new principles for their operation.

Attempting to straddle the worlds of cosmopolitan glory and state contracting, Poleni, Gersten, and Kratzenstein sought to demonstrate their talents and potential importance for their patrons by emulating and perfecting Leibniz's machine. The machines offered a concrete way for technically nonsavvy monarchs and their advisors to recognize people worth supporting. The production of the machines made their competencies into a tangible manifestation of their connection to the circulation of machines and skills across Europe and of their ability to police artisans; the machines indicated that their makers were the sort of people one needed to produce import substitutions, to "naturalize" goods, and to organize skilled craftspeople.

Solving a narrow problem of ignorance—about Leibniz's calculating machines—they sought to demonstrate themselves to be capable of ameliorating far more significant absences of knowledge in late mercantilism: ignorance about what knowledge was necessary for a reformed state, economy, and princely magnificence and uncertainty about to whom to turn in economic and technical matters. The numerous jurisdictions of the Holy Roman Empire encouraged the imitation of any number of schemes and devices for reforming state, economy, and people.[33] The proliferation of machines springs in no small part from these many centers of mercantilist policy. Mercantilist competition encouraged imitation, including the blossoming of different sorts of calculating machines.

This philosophical imitative practice, insulated through the careful description of ignorance and defenses of the propriety of rediscovery, drew parasitically from the imitative and emulative practices of innovative eighteenth-century artisanal culture. In his *Conjectures on Original Composition*, Edward Young encouraged writers to follow the model of skilled artisans and move from imitation to emulation: In the mechanical arts, "men are ever endeavoring to go beyond their predecessors," whereas in the literary arts, they endeavor "to follow them." He explained, "since copies surpass not their *Originals*, as streams rise not higher than their spring, rarely so high; hence, while arts mechanic are in perpetual progress, and increase, the liberal are in retrogradation, and decay."[34] If often isolated from the cosmopolitan republic of letters, innovative artisanal culture drew from a parallel, rich cosmopolitan network of skill and knowledge. The makers producing the most successful calculating machines were precisely those able to unite these parallel learned and artisanal cosmopolitanisms in practical workplace environments that combined invention and implementation, skill and natural philosophy.[35]

Artisanal Emulation: Philippe Vayringe, "Apprentice Locksmith . . . Become a Philosopher"

A cylindrical calculating machine, now in the Deutsches Museum in Munich, is inscribed "Braun invenit Vayringe fecit": Braun invented it; Vayringe made it. The ornately decorated machine was likely made around 1730.[36] Its maker, Philippe Vayringe (1684–1746), rose from humble roots to become the pensioned court mechanician of the Duke of Lorraine and eventually "Professor of experimental physics of the Académie of Lunéville."[37] During a visit to the court, Voltaire described "an excellent physicist named M. de Vayringe, who raised himself from an apprentice locksmith to an esteemed philosopher thanks to his nature and the support he received from the late duc de Lorraine."[38] Vayringe's ability to transform himself from a locksmith to a maker of calculating machines and then to a philosopher rested upon a rich European network of craft knowledge, of philosophical knowledge, and of machines in circulation. Vayringe exemplifies the eighteenth-century skilled artisan who, through princely support and for mercantilist ends, was enabled to pursue an educational path combining formal philosophical training, industrial espionage, and the emulation of astronomical machines and clocks—as well as calculating machines.

A ducal desire provided the proximate cause for the Braun-Vayringe machine, according to Vayringe's autobiography: after succeeding to the duchy of Lorraine, Duke Francis Stephan "informed me that the emperor possesses an instrument, by means of which almost all arithmetical propositions could be worked out, and that nobody had as yet made one like it. I undertook to do so, if someone would show me its form and dimensions, and I was therefore sent to Vienna."[39] The prince sent him to the Viennese court to imitate a courtly machine. In Vienna, Vayringe seized a greater opportunity: "On arriving there I was told the instrument was out of order, and the maker dead, so there was no way to see its operations; I said I would undertake to set it to rights if they would let me." It was so commanded: "Without leaving the room, I mended it in six hours, and I made it perform the four rules of arithmetic in the presence of a nobleman of the court, who went to report what I had accomplished to his Imperial Majesty." The speed of this success so impressed the emperor that he honored Vayringe with "a gold chain and a medal weighing one hundred and fifty ducats."[40] Vayringe garnered an impressive reward for fixing Braun's machine—for bringing it into practice.

He reaped greater rewards when he reinvented the machine: he explained, in the language of emulation, that working on the broken machine had inspired him to exceed it: "As soon as I returned to Lorraine I made an instrument a great deal simpler than this, which produced the same results." As a reward, the prince gave him far more than money: "He honored me by choosing me to give a course in experimental physics."[41] The end of the War of Polish Succession forced the duke to exchange Lorraine for Tuscany. Despite tempting offers from France and the new ruler of Lorraine, Vayringe made his way to Florence, charged with a vast array of instruments, only to find it a sorry reflection of its cinquecento glories.

Copying and then emulating machines was the fundamental dynamic of Vayringe's career. He gained the competencies that allowed him to rebuild and improve the calculating machine through the major pathways of the eighteenth-century knowledge economy: artisanal travel with industrial espionage, the mercantilist ambitions of small principalities, import substitution, and emulation. Vayringe embodied the structures of the new knowledge economy, where the "establishment of networks among a wide range of working people, artisans, and entrepreneurs and the acceleration of diffusion made Europe a pool of materials and skills in which producers could discover devices, equipment, and processes."[42] Vayringe strategically dove into that pool: much of his life was dedicated to discerning the construction of machines and instruments, reconstructing them, and then becoming inspired to create improved versions. He proved as interested in copying the tools for manufacture

as in the goods themselves. Vayringe traveled, for example, to Paris with a letter of introduction to a clockmaker: "I observed that his wife was cutting the
teeth of the watch-wheels with an instrument that was unknown to me. I drew
near, that I might observe the mechanism more closely, and instantly compre-
hended it." This classic bit of artisanal industrial espionage occasioned a speedy
movement of technical knowledge. His "first care" as soon as he returned
home was to reproduce the machine "for cutting and notching the wheels"—
"the most useful invention in the whole clockmaking art." While the Parisian
version could cut only "ordinary wheels," he "improved upon it, making one
that could notch from fifteen teeth to a hundred and thirty thousand, and in
which all the numbers, both odd and even, were to be found, for the construction of mathematical instruments."[43] Vayringe's international world had little
or no printed public disclosure but saw the constant circulation of technique,
practices, and ideas—the cosmopolitan culture Liliane Hilaire-Pérez charac-
terizes with the notion of "open technique."[44] Given the new techniques and
tools, and his improvements upon them, his clockmaking business boomed
and he set his sights higher.

Vayringe soon had instruments of sufficient quality to display to the Duke
of Lorraine in Lunéville. The duke displayed them to some English visitors.
Though initially convinced of English superiority in instrument making, they
proclaimed Vayringe's instruments better in quality and simplicity. With this
external validation, Vayringe was appointed a court clockmaker and mechanician. These appointments signal that he had been judged capable of helping
Lorraine compete in princely magnificence and the production of high-quality
goods. Mercantilist political economy required the creation of such goods for
markets local and foreign. The traveler Johann Georg Keyssler used a precise
vocabulary of copying and perfection to capture the nature of Vayringe's
improving imitation of a Copernican planisphere: "At Vayringe's I saw an ingenious imitation (completed with fewer wheels) of the machine made up by
the English Mechanic Rowley for Prince Eugene that puts the Copernican
system before the eyes"; the traveler had seen John Rowley's own machine in
Vienna the previous year.[45] Vayringe traveled to Vienna with a planisphere
of his own as a gift from the Duke of Lorraine to the Holy Roman emperor.[46]

Working for the court in Lorraine soon propelled Vayringe to seek out
more skill and knowledge abroad. After acquiring numerous mathematical
instruments from England at great cost, the Duke of Lorraine decided to send
Vayringe to England to observe its instrument trade firsthand. The great lec-
turer J. T. Desaguliers welcomed him. Vayringe soon gained something even
more enriching: "What benefited me most was, that this eminent professor
ordered his workmen to construct, under my direction, a series of instruments

corresponding with his own." As ever, he improved things along the way: "I managed to simplify them, rendering them at the same time more efficient."[47] Before long, he had built an entire lecture course's worth of demonstration instruments and a short text to accompany them.[48] Voltaire described the "admirable establishment for the sciences" in Lunéville that resulted: "The great room is all furnished for new experiments in physics, and particularly everything that confirms the Newtonian system. A simple locksmith turned philosopher, sent to England by the late duc Leopold, made the majority of these machines by hand, and demonstrated them with considerable finesse. Nothing in France compares with this establishment."[49]

The likely exaggerations of Vayringe's autobiography point to the structural elements he deemed most important for the rapid progress of technical knowledge and production. When he arrived in Vienna to observe the calculating machine and found it broken, Vayringe brought his formal philosophical training, a powerful ability to grasp the functioning of machines, improved tools for making machines, and the social knowledge of a master artisan, as well as secrets collected and poached from masters across northern and central Europe. Vayringe's autobiography celebrates the taking of practices and instruments from other artisans and makers without apology or qualm. He worked in an artisanal world with no need for the mental gymnastics required to prove originality or demonstrate his ignorance of previous efforts. Imitation and improvement of the work of others was not vile plagiary but standard practice—and state policy.

Vayringe was a known quantity in Vienna when he arrived to inspect the calculating machine. The emperor had rewarded him on a previous trip for his elaborate Copernican planisphere. His patron sent him to draw inspiration from a calculating machine and reproduce it in his own court; the Viennese court obliged him and more. "Braun invented it; Vayringe made it": the inscription on the machine carefully divides credit between invention and implementation. Vayringe fixed one machine in Vienna and made one on his own in Lorraine.

Vayringe was not the last to be rewarded for reducing a Braun machine to practice. In 1766, Anton Braun the younger received a hefty award for reworking his father's machine that had been envisioned, constructed, and dedicated to the emperor some forty years previously—perhaps the machine Vayringe worked on in Vienna before building his own. The machine offers the first solid version of a variable cogwheel mechanism, envisioned by Poleni years before. The reconstructed and reworked cylindrical machine was carefully examined and then certified by one Father Hell as "the most perfected [*Vollkommenste*], either than he had previously seen or that he had read about." The Father testified likewise to the quality of the workmanship: "The entire machine—the gears as well as the springs, is built truly strong, well, and to last a long time. Thus, the Artist merits every praise and a worthy award."[50] The finely finished machine was the first using a variable cogwheel to be reduced to practice in contemporaries' eyes. Recent historians esteem it greatly.[51] The machine remains in Vienna, long credited entirely to the elder Braun and dated, quite deceptively, to 1727.[52] Modern and contemporaneous observers have judged the machines of Vayringe and Braun the younger as perhaps the first well-working calculating machines. Emerging from elite artisanal milieux, Vayringe and the younger Braun had the range of coordinative capacities necessary to reduce calculating machines to practice.

Theories of Imitation and Emulation

In his sixth discourse on art of 1774, Joshua Reynolds defended the propriety of borrowing ideas from other artists and weaving them seamlessly into one's own work. An artist "should enter in to a competition with his original, and endeavor to improve what he is appropriating to his own work. Such imitation is so far from having any thing of the servility of plagiarism, that it is a perpetual exercise of the mind, a continual invention."[53] Such hierarchies of more or less valuable forms of imitation mattered in eighteenth-century political economy, aesthetics, and ethical theorizing. The young Kant offered a fairly standard account of the hierarchy: "Imitation is different from copying, and these differ from aping." He continued, "There is no progress of spirit, no invention, without knowing how to imitate what one already knows in new connections."[54] The best forms of imitation were not servile. The major German aesthetic theoretician Johann Georg Sulzer described the nature of the judgment and choices of the "free" imitator: "Whoever thinks and judges at all times, freely imitates. He sees in a work . . . certain things that do not conduce toward their purposes; these he does not put into his work; rather, he chooses another [thing] in its place after his own purpose." For Sulzer, the free imitator never allowed the judgment of another to serve as an indicator of what was the best made. Lower kinds of imitators, in contrast, presume the work of others to be best made and make no effort to do better.[55]

The best sort of imitation had a powerful moral component that separated it from mere aping, as seen in the entry on "emulation" in the *Encyclopédie* of Diderot and d'Alembert quoted above: "a noble and generous passion, which, while admiring the merit, the beautiful things, and the actions of another, attempts to imitate them, or even to surpass them." Beautiful works inspired the emulative soul to higher things: emulation became a theory of the collective improvement of arts and sciences through a lively, moral competition. In his *Conjectures on Original Composition*, Young attempted to distinguish low forms of imitation from higher forms of "emulation": "Imitation is inferiority confessed; emulation is superiority contested, or denied; imitation is servile, emulation generous; that fetters, this fires; that may give a name; this, a name immortal: This made *Athens* to succeeding ages the rule of taste, and the standard of perfection. Her men of genius struck fire against each other; and kindled, by conflict, into glories, which no time shall extinguish."[56] Not just individual artists but entire polities benefited from the competition of fruitful and free imitation of earlier products.

"Emulation" offered a fundamental theoretical means for understanding and defending the process of imitation. The notion breaks easy dichotomies between mere copying and pure innovation. The category of "emulation" provided a key normative concept for legitimating and structuring imitation by distinguishing it morally from envy and jealousy.[57] Emulation offered a vision of aesthetic and technical progress made possible by pursuing the right forms of imitation and eschewing servile copying. Its account of individual progress was closely tied to national progress, as Young's invocation of the development of Athens suggests. The category of emulation helped legitimate commerce as an ennobling form of competition and offered a structure for understanding the emotional spurs to productivity provided by such healthy competition. Commerce was understood to be more than the action of self-interested actors; commerce happened best through efforts at moral and technical improvement to outdo one's national and international competitors. The concept served as ideological cover for international industrial espionage, nascent capitalism, and imperial maneuverings, to be sure; but it also offered a rich account of knowledge production as the novel drawing together of existing knowledge and processes.[58] More than just ideology, emulation structured new forms of economic patriotism across Europe. In his account of the importance of the notion, Istvan Hont explains, "Emulation was a patriotic duty, motivated by the love of country and serving national honor. It was the flagship policy of the huge assortment of patriotic agricultural and improvement societies that spring up everywhere in Europe in the second half of the eighteenth century. These societies organized regular prize competitions and devised systems of meritocratic honors in order to foster assiduous participation in economic improvement."[59] The economic and moral goals of the Genevan Society of Arts, founded in the mid-1770s, exemplified the phenomenon: "At issue is exciting, by means of emulation, the minds of Artists to observations and discoveries; to call forth all the new forms of industry to naturalize themselves among us; to favor useful establishments; and, above all, to support and to accredit our already established manufactures, and to join, for the glory and good of the Arts, insights from outside to those produced from our own inquiries."[60] Unlike many similar societies in Europe and North America, the Genevan Society of Arts was a great success, as it was instrumental in securing nineteenth-century Genevan horological supremacy, built in equal parts upon new technical developments in horology and skills accumulated through decades of large-scale counterfeiting of English watches and watch parts. Geneva's horological ascendance was the political economy of emulation in action.

Within these new economic societies, emulation meant a drive to connect theoreticians and practitioners. Practitioners needed more theory to perfect their skills and to raise their estimative skills in judging something well made. Theoreticians needed to focus on making theory capable of improving practice—and on learning from practitioners. At issue was not a deskilling of artisans in the name of theoreticians but the parallel transformation of artisans and theoreticians alike, the better to collaborate. Transforming artisans and theoreticians demanded the creation of new techniques of intercommunication and collective practice permitting a cross-pollination of activity necessary for emulative progress.

The rise of the cult of genius and of new concepts of intellectual property privileging individualistic invention quickly eroded the centrality of emulation as a way of thinking about cultural production, even as the practice of emulation intensified in many practical sectors. In a famous letter to Kant, Johann Georg Hamann wrote, "He who trusts another man's reason more than his own ceases to be a man and stands in the front ranks of the herd of mimicking cattle."[61] Abandoning his hierarchy of kinds of imitation, Kant wrote, "Everyone agrees that genius is entirely opposed to the spirit of imitation. . . . [E]ven if one thinks or writes for himself, . . . indeed even if he invents a great deal for art and science, this is still not a proper reason for calling such a great mind . . . a genius." Even Newton was no genius: "Everything that Newton expounded in his immortal work on the principles of natural philosophy, no matter how great a mind it took to discover it, can still be learned; but one cannot learn to write inspired poetry. . . . The reason is that Newton could make all the steps he had to take, from the first elements of geometry to his great and profound discoveries, entirely intuitive [*ganz anschaulich*] not only to himself but also to everyone else."[62] Only inventions completely outside production via rule-based behavior were creations of genius.

Even as he divided genius from imitation, Kant retained fundamental aspects of the emulative model: copying was necessary, and so was "academic" instruction. The products of genius, he explains, "must at the same time be models, i.e. **exemplary**, hence, while not themselves the result of imitation, they must yet serve others in that way, i.e. as a standard or a rule for judging."[63] Given that processes of genius cannot ever be articulated, "the rule must be abstracted from the deed, i.e. from the product, against which others may test their own talent, letting it serve them as a model not for **copying** [*Nachmachung*] but for **imitation** [*Nachahmung*]."[64] The challenge alone can force the improvement of the talent. Yet without formal training in the mechanical, nothing can be made.[65] Genius requires a coordination of talent and rules for procedures, as well as sheer skill to produce things of beauty.

Throughout the eighteenth century, incomplete information about Leibniz's calculating machine served as the basis for productive imitation. Few, if any, of the early eighteenth-century machines were reduced to practice so as to allow regular production. Their creators lacked the coordination of practitioners and competencies necessary to bring machines to practice. In the 1770s and 1780s, three makers emerged—Hahn, Müller, and Stanhope—who were capable of integrating design, invention, and practical work necessary to produce well-perfected machines.

Emulation and Circular Machines

Let's return to the twin cylindrical machines of Hahn and Müller and the merit due to secondary inventors. The literary critic Merck explained that merit was apportioned differently in the fine and mechanical arts: "In many cases, especially in works of genius and the fine arts, the merit of the first inventor is far inferior to the merit of one who has brought the thing to a higher degree of perfection; in other cases, however, especially in mechanical discoveries, the one, who first had the thought and happily overcame the initial difficulties, has an advantage over someone who simply reworked it and, instructed by a foreign experience, is able to beat someone to it along an already paved path, if perhaps with less effort."[66] Merck sought to defend the merit due a mechanical inventor for bringing something to a higher degree of perfection. Some account of secondary invention was necessary to understand the development of calculating machines, as Philipp Hahn and Johann Müller were both secondary, if not tertiary, inventors. As both knew, the principles of a circular machine had been published in 1727, and all the Braun machines embodied a circular design.

The first history of calculating machines, the *Theatrum arithmetico-geometricum: das ist: Schau-Platz der Rechen- und Meß-Kunst* of 1727, tells of many failures, and particularly the failure to explain well.[67] Having detailed the whole array of tools for aiding calculation, from the Chinese abacus to

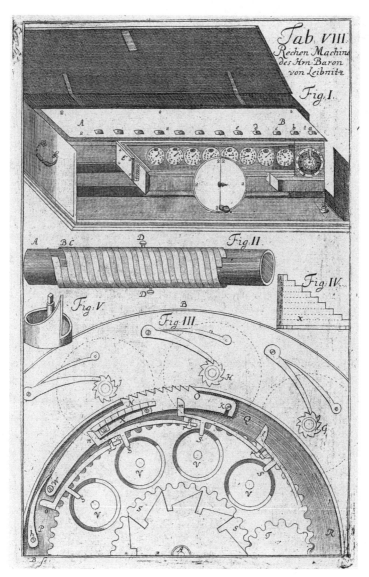

FIGURE 4.3. *Top*, exterior of Leibniz's calculating machine; *bottom*, Leupold's calculating machine, with details of adding mechanism. Jacob Leupold, *Theatrum arithmetico-geometricum: Das ist: Schau-Platz der Rechen- und Meß-Kunst* (Leipzig, 1727), plate VIII. (Courtesy of Rare Books and Manuscript Library, Columbia University.)

Napier's bones, Jacob Leupold (1674–1727) translated and included Poleni's description and diagrams; he complained of Leibniz's failure to describe his machine adequately and reproduced Leibniz's 1710 diagram with added letters to clarify its functioning (see top of figure 4.3). Leupold translated Leibniz's description of the outward appearance and instructions for using the machine. He neither knew the details of Leibniz's stepped drum nor hinted at them.

"Having read for more than twenty years that calculating machines had been found," Leupold continued in the affective vocabulary of emulation, "I developed the appetite, not just to see one, but to invent one myself." Leupold described his invention as a "curious and entirely new calculating machine"; it was a cylindrical design driven by a central crank. He included some detailed drawings with accompanying explanations but apologized for being unable to provide a fully adequate description of his machine. He "would have liked to have given a fully perfected description," but he was unable to. Instead, he wanted "to reveal the entire fundamental principle clearly" with the majority "represented in natural size" so that "a mechanical mind could truly easily have a complete idea" of the machine.[68] Perhaps to underscore the gap between his openness and Leibniz's secrecy, Leupold printed his diagram, disclosing his central driving mechanism, immediately beneath his reproduction of Leibniz's diagram that betrayed nothing of the mechanism. Like Poleni and Gersten, Leupold imitated without merely copying or aping. Leupold did not seek merely to reverse engineer the hidden components of Leibniz's machine. Rather, he replaced the rectangular configuration entirely with a cylindrical one soon widely copied, much as new ceramics were created in the attempt to reproduce Chinese porcelain. He devised a novel alternative to the stepped drum: the selectable ratchet (depicted in the bottom half of figure 4.3). Leupold insisted that his disclosure required a trained eye to grasp, much less to appreciate, its technical ingenuity. At least one of Vayringe's calculating machines was a realization of Leupold's proposal. Leupold's circular design, with its specified interior, served as a major source of emulative inspiration, including to the Hahn and Müller machines, each based on a rediscovery of some form of the stepped drum.

A Machine "More Perfected and Useful than Leibniz's": Hahn and Invention as Process[69]

Karl Eugen, Duke of Württemberg, had a problem. One of his preachers, Philipp Matthäus Hahn, appointed to a pastorate in Kornwestheim, was almost certainly a heretical pietist and advocate of Enlightened ideas; insistent on preaching, the man kept refusing to be appointed a professor of mechanics in Tübingen, where he might mislead a few students but not affect vulnerable parishioners. Alas, nothing was to be done. Philipp Matthäus Hahn was simply too important a maker of machines for the duke. And Hahn had offers from other princes.[70]

On August 9, 1774, the noted physiogonomist Johann Kaspar Lavater visited Hahn and saw his calculating machine: "Leibniz's decades-long effort at an impossible machine is . . . ten times more perfected." Hahn had completed what Leibniz and all those who followed could not. Later that evening, Lavater and Hahn spoke of publicizing the machine and "the difficulty of transport in this calculating machine."[71] Hahn knew well the primary difficulties of bringing machines to practice; he admitted these in his letters to Lavater.[72]

In his description of his machine, published in the *Teutsche Merkur* in 1779, Hahn did more than describe the outward form and use; he chronologically laid out his difficulties and successes as he worked through the machine. He offered a picture of invention as an improvisational and iterative process, necessarily involving far more than rendering an imagined design or form into metal and wood. Finished form does not precede materialization; materialization did not happen independent of the envisioning of design. His private diary entries of the period, however telegraphic, largely confirm his self-presentation. He consulted all the known published material on machines—above all, those of Leupold and Leibniz. Having started work, he soon perceived "how restricted my understanding was; how I had not thought of things, whose necessity appeared in the course of inquiry, how dark and chaotic the picture of the machine remained in me."[73] Not a design preexisting as an image or form in the mind's eye, an invention emerges only gradually through the effort to implement. Invention and implementation required much dialectical work before the picture would become clear as the machine came into being (and not as the cause of the machine coming into being).[74] The greatest darkness was, of course, the carry mechanism that he noted "in the end prevented [Leibniz's] machine from a complete implementation."[75] To focus only on the stepped drum is to miss the real work of perfecting the machine. The creation of the machine involved a long process of continued form-making, not a bifurcated stage of design followed by implementation.

Years passed. Hahn abandoned a rectangular machine in favor of a cylindrical one, until finally he achieved what he called a "simple perfection" of the machine. The desire to perfect did not abate: "An invented thing easily can be much improved; so someone with a simple perfection is not satisfied, when he sees a way, to improve it yet more, so he reflects upon how to fix these imperfections . . . ; for it is impossible to see them at the beginning. Finding improvements requires a certain amount of time, some space and the resting of the mind. One is happy at the beginning, only to have overcome the first difficulties."[76] Productively obscured at first, the first mental picture of an invention merely needed to solve initial problems; only a long period of attempting to improve the machine would reveal the difficulties to be overcome. Ignorance about finer details would allow earlier stages to proceed.

After years of effort, a "well satisfied" Hahn ordered his machine gilded for presentation to monarchs: his duke, the visiting Holy Roman emperor, and other princes.[77] He was urged to announce his discovery to princely academies and to publish descriptions in learned journals.[78] The machine had been reduced to practice enough for machines to be made for presentation purposes and to be a discovery worth publishing. On July 26, 1780, Hahn's wife, "a courteous little woman," demonstrated a calculating machine to one Heinrich Sander. She did not show "the inside," which is "full of gears"; one turned a crank "just like a coffee grinder." She demonstrated "all five arithmetical functions."[79]

After he tried making a rectangular machine, Hahn picked up the basic idea of a cylindrical machine from Leupold, as had, in all likelihood, Braun, Vayringe, and—with certainty—Müller. A circular or cylindrical design allows a turn of a crank both to perform the series of additions and to provide the force necessary to permit sequential carries. Many implementations were possible within the basic framework of a circular design. Hahn's mature design is a circular machine with a main crank for performing a single addition and a circular array of upright stepped drums. In Hahn's machine, a curved arc of teeth completes a full revolution when the main crank is turned counterclockwise. As it does so, it engages with a gear mounted on a shaft just below each stepped drum and causes each to complete a full revolution. The user sets the digits to be added in each place value by moving a series of graduated rods along the diameter of the top plate up and down. These rods carry the stepped drum up and down; the teeth set on the stepped drum are moved into position to engage with the final result gears. Performing the additions one place value at a time is potentially slower than a Leibnizian machine that performs them in parallel. Serial addition has one major advantage: by banking the carries to be calculated at a later point, the machine obviates the need for mechanical solutions to deal with carries as they arise. Any time a stepped drum moves a result wheel from 9 to 0 in a digit, a "warning" mechanism is triggered to be added at a later stage. The user powers the carry process by rotating the drive shaft, so the force does not diminish in a series of successive carries, and the continuous rotation of the drive shaft resets the springs of the warning mechanism and performs the carries themselves.

Hahn's machines emerged from a well-organized and well-equipped site of coordinated inventive practice. Hahn had a clockmaking workshop that operated outside the guild regulations of Stuttgart; in addition to traveling journeymen and local apprentices, he employed his brothers and sons in the manufacture of clocks, watches, and—most famously—astronomical clocks.[80] Making parts for calculating machines belonged within the workshop routine of making parts for clocks. When Hahn ran out of work, he had his workmen take apart trial calculating machines and use the parts to build clocks and watches.

In his 1779 publication, Hahn stressed the limits of his best design at that point. Although his current design functioned well enough, Hahn was still dissatisfied. He had reduced his machine to practice but not to any kind of regular manufacture, since regular journeymen could not build it. In particular, he found the use of springs and "forks" unsatisfactory. So he set out to create a revised version that an "average worker" could complete, a design "so thought-out" that the machine could be produced more quickly and would thus be "simpler and more durable."[81] For Hahn, a fully perfected machine required a design optimized for middling journeymen—not simply a machine that only a master could build. The bringing of a machine to perfection cannot be understood absent a work context or a recognition of the skilled labor available.

Müller's Emulation

In 1783, the Darmstadt "Engineer-Captain" Johann Müller, in the process of inventing his own calculating machine, heard from eyewitnesses that Hahn's machine often produced errors.[82] Hahn apparently blamed these on bad workmen. "In thinking about my machine," Müller explained in a letter written just after he had completed some trial sections of a machine, "I particularly reflected upon these circumstances, in which [Hahn] had the workmen in his house under his constant supervision." Given Hahn's level of supervision over his artisans, it was fair to guess that any machine designed in similar ways would produce similar errors.[83] In his process of emulation, Müller sought to produce a carry mechanism less subject to failures of skill and precision, one capable of being produced by average journeymen. Hahn and Müller shared these goals, but Müller insisted that Hahn could never achieve them.

Müller quickly learned of the need for close supervision and collaboration in creating a calculating machine. After coming into some money upon his marriage, he gave his drawings to a clockmaker, who worked unsupervised for three months, to no avail. While Müller was traveling on official business, another artisan, a journeyman clockmaker, worked six-and-a-half months "without completing anything more than the crude housing and the mechanism for the first four digits," a device very prone to error.[84]

After the two-digit mechanisms were mutilated under the supervision of a master clockmaker, Müller explained in a letter written in the middle of the process, "I could no longer entrust him with the supervision and I needed to bring the journeymen into my house."[85] In this new domestic work setting, the mechanism was so "improved in six weeks, that one could compute with 4 digits." Given his success in creating a partial trial piece, Müller felt confident enough to start announcing his invention to the learned world. With the trial piece finally working well enough, he removed the "too often patched-up [*geflickte*] inner mechanism" and replaced it with a finalized, many-digit version. Implementing the final mechanism of his machine took two journeymen five-and-a-half months and then a journeyman working alone another eight months, all of them working within Müller's household.[86]

Müller began his effort to have a machine built by attempting to have a design implemented; he succeeded only when he created a place of coordinated practice in his own house: a place of supervision, design, trial and error, and microadjustment over many months. To describe the messiness of his trial mechanism, Müller used the sewing term *geflickte*—patched up, darned.[87] As an idea and as a realized thing, the machine emerged from the collective production of "too often patched up" parts.

For all their outward similarities, Müller's machine—in the eyes of many commentators, contemporaneous and modern—has a better-realized mechanism than Hahn's.[88] Instead of moving rods up and down to set the digits to be added, the user of Müller's machine turns dials on the side of the machine. Instead of Hahn's circular orientation of upright stepped drums parallel to the sides, Müller used a radial orientation of drums parallel to the bottom. The drums themselves differ, for they play far different roles in the process of carrying. The stepped teeth of Hahn's drums are oriented radially around the *entire* drum. The teeth of Müller's drums are oriented only around a part of each drum. Hahn's drums play no role in giving the machine time to perform carries, whereas the toothless arc on Müller's drums allows the brief but necessary time for the machine to complete a carry without jamming immediately following an addition.

Just as their drum designs reflected their different solutions to how to effect carries, their springs expressed differences in how they thought about the reliability of these elastic components. In his 1779 description, Hahn stressed his worries about the use of springs in his machine—both for the durability of the machine and for the ability of ordinary clockmakers to build it and fix it. Müller put different concerns about springs at the center of his efforts. In a letter written during the development of his machine, Müller explained that the practical inability to ensure that mechanisms would be sufficiently precise motivated his improvements: "Through my improvements, the digits, when they do not exactly come beneath the openings, snap by themselves right to the middle, as soon as the crank begins to move, so that the teeth all necessarily must mesh correctly."[89] His existing machines have a carry mechanism that converts the rotation of the primary crank into movements that constantly ensure that the elements of the carry are pushed and locked into proper position. Whereas Hahn's machine requires carefully calibrated springs to move carrying arms into position, Müller's draws on large band springs to move the lever into places, thus rendering the precise of calibration of springs far less important. In both cases, relatively successful carry mechanisms emerged only through extensive trial and error in sites of close collaboration among inventors, master artisans, and journeymen.

Inventing Collectively and Distributing Credit: Hahn vs. Müller

Hahn suspected Müller of plagiarism: both machines are circular and, though Hahn could only suspect it, both use stepped drums. In reacting to Müller's machine, Hahn chose to damn Müller with faint praise. By creating his machine, Müller revealed that he possessed the knowledge to capitalize on Hahn's limited disclosure of his machine: "Even if [Müller's machine] does only what mine does: that's proof enough of Herr Müller's mechanical knowledge, if he had a solid introduction before him, in the form of my description of my invention, just as I had in Leibniz. How easy Herr Müller had it with my candid description of the history of a discovery; someone with such a mechanical understanding can be led on the right trail of thinking."[90] Hahn expanded on Müller's advantages. First, details omitted in incomplete disclosure were readily discernable to a mechanical mind—a charge Leibniz had levied against Robert Hooke a hundred years before. Second, Müller knew that it was possible to succeed. During the many years Hahn labored before producing something reasonably functional, he had no guarantee that the project would yield a working machine. Even "if the alleged inventor did not himself look into my machine," or had a friend look into it, "clues came to him" that in time "all would be in order"—that one *could* perfect a machine after all.[91] Third, in his early description, Hahn had narrated the major technical issues he had slowly come to recognize and subsequently overcame. Hahn argued that many problems do not appear until well into the inventive process—and Hahn had laid out the entire series of difficulties for Müller. At a minimum, Müller had a series of important hints for the creation of the machine, even if he never saw one, and he had sufficient technical acuity to build upon them.[92]

Hahn further mocked what he understood to be Müller's fundamental innovation—a bell that would ring when the machine was going to fail. Far from a decisive improvement, Hahn argued, the bell was but a "little play thing" that served as a "fog" that obscures the vision of the unknowledgeable.[93] This fog had deceived the editors of the *Teutsche Merkur* into celebrating the greater reliability of Müller's machine. Unacquainted with actual technological development, the bookish editors were duped into believing that a small technical add-on could substitute for a lengthy process of invention and iterative implementation necessary to produce a reasonably error-free machine. According to Hahn, Müller's innovation only precluded a small class of failures; the flashy device was a feeble substitute for the sustained work in the workshop required to perfect a carrying mechanism. Those without knowledge of real technical development, including the editors of the *Teutsche Merkur*, accepted that Müller's preposterous little bell could eliminate errors. In contrast to the conclusions of the commonsensical reasoning of a reading public and editors, long-term mechanical experience justified Hahn's skepticism: "I know from mechanical experience, that one has higher hopes about it working when caught up in the initial delight around a happy idea, than one subsequently discerns in thinking about it dispassionately or undertakes a genuine inquiry."[94] According to Hahn, Müller had the emotive rush of a simple idea, but not the subsequent, essential, dispassionate iterative development of materialized designs and designed materials capable of leading to reduction to practice.

Müller acknowledged that Hahn's description of the external appearance of his machine and its use helped abridge his inventive efforts: "Here is everything, that I have Hahn's invention to thank for: I would have, perhaps, vainly worked upon some other ideas without his description, before I would have come upon the most expedient." Müller rejected that being led in helpful ways entailed any detailed revelation of the internal structure of Hahn's machine. The box remained opaque: "The inner mechanism can be produced in various ways of bringing it into reality, albeit not equally good and easy or convenient ones; each causes a different outside disposition."[95] If Hahn's argument held, Hahn would be denying himself credit, since he would be little more than an imitator of Leupold's circular machine: "Hahn errs greatly, in arguing in the *Teutsche Merkur*, that the description of his machine gave rise to a similar invention, or that I got more from it, than he could have drawn from Leupold's description about the inner mechanism of the sketched out machines of Leupold and others; or from the outer disposition of Leibniz's machine."[96] To justify his own inventiveness, Hahn needed an account of invention that permitted multiple inventors—one that would almost certainly exonerate Müller.

This debate centered on how to weigh claims of mechanical ingenuity against the various forms of disclosure that arguably made it possible. Müller cast himself as emulative and thus virtuous. Hahn cast Müller as merely imitative; even worse, Müller's machine had a dishonest appearance of an improvement—the warning bell—designed to deceive the unknowledgeable. Müller's machine was neither independent of Hahn's, nor an improvement upon it, nor further "realized." Müller was no second Prometheus.

In focusing on Müller's small technical innovation of a warning bell, his defender, Merck, misunderstood the processes necessary to bring machines to practice. And why should Merck have grasped this? After all, he was a literary critic and philosopher of sorts, not a craftsman. Hahn was acute in his doubts about what Merck *professed* to be the superiority of Müller's machine: the bell and preemptive stopping mechanism offered marginal improvements at best. They were incidental to the drawn-out process of reiterative designing, fitting, and fine-tuning necessary for carrying mechanisms. Hahn did not know, in contrast, about the likely superiority of Müller's machine. It had a more robust carry mechanism and was less dependent on the vagaries of springs; the machine emerged from a drawn-out iterative process of design, refitting, implementation, and redesign. Hahn suspected Müller had failed to pursue the lengthy process necessary to move from an idea to a realized machine when in fact Müller and his craftsman had come to a design more capable of reduction to regular manufacture. Müller's machine is best understood as emulative: something perhaps superior to and inspired by Hahn's machine, but not a slavish copy of the original. Müller had structured his own inventive process around what he took to be Hahn's failure to produce a design capable of being built by "simple" workmen: those who could only mechanically "ape," not freely "imitate."

Müller presented his machine to the Academy of Sciences of Göttingen in 1784. The academicians praised his manner of demonstration: "He personally explained the use of [his machine] with an order, clarity and understanding that showed that he has not only the ingenuity of a good inventor, but also, can impart instructively the perfectedness of his discovery—something hardly typical of mechanical artisans." Far from hiding the inner mechanism, Müller explained it: "Since he showed the inner construction of the machine to me and to other members of the mathematical Class [of the academy], we can well explain as well as testify in detail about the ingenuity of his invention as well as his talent."[97] Neither Müller's machine nor his talent remained a black box after his display. The academicians equally admired the planned durability of the machine. Built of steel and brass, the machine had been crafted "with great care" into "a simple, certain arrangement." Should it break, moreover, "an ordinary clockmaker can fabricate it again."[98] For the academicians, as for Hahn and Müller, producing a machine ready for regular use meant reducing it to a form that ordinary artisans, not simply the original makers and their very best journeymen, could repair: a device that could be made through the craftsmanship of certainty—of regular production—not simply the craftsmanship of risk.[99] Müller hoped his machine would soon be readily produced by regular artisans at considerably lower cost, as "workers no longer need to work on flaws," previously a heavy burden on them.[100] Once artisans no longer needed reference to sketches and designs, once making a machine became habitual manufacture, they could produce the machine in larger numbers.

Müller acted more like an academician or member of the republic of letters in disclosing his machine in private to the academy. He did so in pursuit of markets for the machines—and likely better employment for himself. While the machine would probably cost no less than 1,000 Guineas, that "would not be too much for a great Lord or a rich Englishman."[101] The publicity following his displays brought him inquiries from numerous German princes and even from St. Petersburg. Müller had a yet bigger market in mind: "I wish to be able to bring the machine to London." The Prince of Mecklenburg, who had seen the machine, promised to write a letter of introduction to King George III.[102]

Before long, the Göttingen academician Georg Christoph Lichtenberg had bad news: Müller was unlikely to be able to sell machines in London, for "Lord Stanhope, called Mahon, had completed two calculating machines."[103] Stanhope and Müller were both understood to have finally completed what Leibniz had begun: "How peculiar it is, that, just as Newton and Leibniz discovered the differential calculus at the same time, a German, Müller, and [the Englishman] Stanhope have brought the Leibnizian Calculating Machine to perfection at the same time. One of the most remarkable phenomena in the history of the discoveries of the human mind!"[104]

Beyond Hylomorphic Creation

In December 1784, Johann Paul Bischoff, an engineer in the employ of the Margrave of Ansbach, visited Hahn. On a classic tour of technical inspection on behalf of the Margrave, Bischoff came to see Hahn's various machines, including the calculating machines and, in exchange, to regale Hahn with details of machines he had encountered. Bischoff later celebrated Hahn's efforts in an impressive manuscript history of calculating machines.[105] Although the visit was flattering to Hahn the famous mechanician, it profoundly upset Hahn the pastor. Bischoff's views, documented in Hahn's diary, embodied the atheism, skepticism, and freethinking that dismayed Hahn about the naturalists of his time: "Human beings have no soul, for we cannot know, where it is, and because people, who have died and woken up again, know nothing of it." The world was older than 6,000 years. God revealed nothing to humankind: "We do not know, if God is a specific being or if everything taken together is God." Convinced of Bischoff's honesty, Hahn tried to answer him, as he saw how darkness had spread in the world.[106] Bischoff soon declared himself a follower of Lessing, Voltaire, and Rousseau, not of Jesus. Hahn duly recorded Bischoff's failure to attend Sunday services.

Bischoff had deceived Hahn. Within a few days, Hahn judged Bischoff true to the ethical implications of his depraved views: he was engaged in industrial espionage. "I was greatly annoyed, for I noted that my guest had snatched up everything from me," from calculating machines to a design for pendulums. "He attacked with a ruse, for he offered to copy a machine for me." He "hindered my work eight days"—a grievous affront to the industrious mechanician-pastor.[107]

The incident with Bischoff confirmed Hahn's fears about the growth of atheism attendant upon new technologies and new philosophies: "In the evening, I disputed with Bischoff about religion. I see ever more, how naturalism spreads all around."[108] Hahn's famed calculating and astronomical machines could be construed as emblems of the worst sort of atheism—the world as mechanism, not the world as revealed through mechanism! Like many of his contemporaries, Hahn sought to reconcile his faith—however unorthodox— with reason and its products, to overcome the apparent trajectory by which mistaken philosophical reasoning would lead to materialism and atheism.[109]

A breakthrough came several years later in a short book by Johann Herder, whose works Hahn had been reading since the mid-1770s. At night, after working long days in February and March 1788 to improve his calculating machines, Hahn read Herder's radical reworking of Spinoza—a recasting based on contemporary physics and natural history, later important to Schlegel and Hegel.[110] Herder transformed Spinoza from a dogmatist to a more epistemologically modest natural philosopher: "I wish that Spinoza had been born a century later, so that he might have philosophized far from the hypotheses of Descartes, in the freer and purer light of mathematics, natural science and of a truer natural history."[111] Herder told the history of eighteenth-century sciences that would have greatly benefited Spinoza: "The more corporeal matter was physically investigated, the more active or interactive forces were discovered in it, and the more empty the concept of extension was."[112] Leibniz's monads presaged these empirical discoveries. These active forces inherent in matter eliminated Spinoza's separation of the modes of God. However, God's two modes of thought and extension could be replaced with the following: "that the Deity reveals Himself in an infinite number of forces in an infinite number of ways."[113] Life replaces extension. This revised Spinoza allowed one to be simultaneously a theist and naturalist—to be an exponent of Spinoza's toleration and liberality while abandoning his dated, sterile, and dangerous forms of reasoning. With the shedding of dogmatism comes the shedding of a dogmatic materialism: no longer a materialist monist, Spinoza becomes a spiritual pantheist.

As he read and reflected on Herder's Spinoza in the nights of March 1788, Hahn became more and more enthusiastic; these natural philosophical conclusions proved a great resource for reconciling Hahn's controversial theological views within contemporary natural philosophy. For Hahn, as for numerous other contemporaries, Herder's work suggested a path toward a reconciliation of reason and faith by overcoming the problems of separating the mechanical and nonmechanical: "Our old philosophy of the nature of the soul and of the body, which Wolff and Leibniz established, greatly hindered me and many thousand others for it made many things in Scripture incomprehensible and unbelievable. It is just the idea of pure spirituality and simplicity of the human soul, which has fostered great disbelief about the Scripture."[114] Philosophical dualism precluded a proper appreciation of Scripture—and particularly of Revelation, which had long been Hahn's controversial focus.[115] Herder's work revealed that the best natural philosophy meshed easily with Hahn's vision of the progressive revelation of God's plan—a revelation understood as the corporealization of the divine to be found in all things.

In his diary, Hahn recast Herder's Spinoza within his own idiom. For Hahn, God was not removed from the world but immanent within it: "His powers, his spirit is in all things. The world is no merely mechanical work that moves itself. He is the mover." The laws of nature in all their uniformity and regularity emerged from God's wisdom. Miracles and wonders "are no exception to the divine laws, that is, the most perfect order; rather, we do not see the means of operation: just as in the highest art works everything follows the common laws of motion."[116] Power and mechanism were not separate: "Life, movement, and being: that is not only a mechanism \a clockwork/, that stands outside of God; rather the forces of a mechanism are God's unmediated powers, through which he maintains all things."[117]

Creation was no more idealistic than dualistic—understood as revelation, it was no hylomorphic process of a spiritual idea imposed onto recalcitrant, brute matter. God created from within: "He is no otherworldly God. He is through all, and in all, and above all."[118] Hahn argued that understanding the Creator in this manner could improve one's understanding about human making. To create was to share in the process of revelation: "Since there are philosophers, physicists, theologians, and artists, there is certainly a God, who is the highest philosopher, physicist, lawmaker, and greatest artist. For where should philosophy, physics, knowledge of the law, and the art of humankind come but from God?" Hahn portrayed the creative act as an embodying revelation of living force: "Should the natural soul—beautiful, full of order, full of art—through which we think and know something, arise by itself? Yet we learn all these things, though human beings are raw by nature and lacking in knowledge. But we can learn this; a capability and capacity must lay hidden: a higher spiritual spark, which can develop and be revealed in the flesh and in the corporeal world."[119] Both the divine creation and the creation of ourselves as human beings came through the revelation of that spark—through its *corporeal* revelation.[120] A development of the seed within, not a negation of the human by the imposition an external idea or form, was necessary: "The final state of human beings is 'as spirit-infused Materiality.'"[121] Self-revelation was to create oneself as an ever more spiritualized organism through mind and matter alike.

Hahn worked on his account of the divine creation as revelation in matter over time using a less literary material; in reworking Herder, he drew upon his long experience with workmanship as a process of making that tacks continually between design and materialization, as he had described in his autobiographical portrayal of the long path of the development of his calculation machines. Only after beginning to materialize his first visions of a design, as we saw above, did Hahn realize "how dark and chaotic the picture of the machine remained in me."[122] Emergence from darkness and chaos came in the corporeal revelation of the design over time, a process of creation far from a hylomorphic model. Any real maker, he insisted in his polemic against Müller, knew better than to conceive making as design followed by implementation.

In a foundational passage for the late eighteenth-century conception of genius, Edward Young argued, "An *Original* may be said to be of a *vegetable* nature; it rises spontaneously from the vital root of genius; it *grows*, it is not *made*."[123] The vegetative model of originality undercut the anxiety that the act of creation itself might be mechanical. The model explained how a long train of sense impressions could be ingested but then utterly transformed into something original; it revealed how genius was a creator of forms unified all the way down, as Leibniz had taught.[124] A maker of his time, Hahn portrayed God as the ultimate creative plant: "He is the root of the tree of the universe, from which all the branches and the entire stem of the creating lift-fluid and powers rise up."[125] For him, divine creation was not a hylomorphic one where God imposed a finished design onto matter; the incipient Romantic account of genius in which one makes things ex nihilo mistook the nature of creative activity. Divine creation was the revelation of seed within and through the flesh. In Hahn's account, God's revelation is captured by the image of a well-coordinated workshop, such as Hahn's, where iterative invention took place and where the trivial bifurcation of idea and matter had no place. Despite his evident heresies, Hahn's prince protected him not because he was inspired, not because he was some original genius, but because he could bring extraordinary devices to completion. Unlike a thousand projectors, Hahn melded together design and matter, people, ideas, and things, to produce the goods. In a world of ignorance about people and things, that was a certainty.

1111   Fourth Carry

Babbage Confronts Prior Art

Charles Babbage's claims to the novelty of his Difference Engine were suspect from the very start, even to himself. His earliest manuscript "demonstrating" the possibility of the Difference Engine advanced a fairly humble attitude about the possibility of predecessors: "Commencing then with the first action which it will be necessary to execute, either it is one which is known and to be found in books or in machines, or it is novel in its kind; in the first case it is needless in the present instance to choose among the received methods of accomplishing it . . . ; If on the other hand it is new in its kind we must . . . \contrive/ some method of executing it or reserve it after consideration."[1] Like many inventors of calculating machines, Babbage, during his initial investigation, appears to have been unaware of previous efforts. Soon Babbage became concerned about establishing the priority and distinctiveness of his Difference Engine—a claim of novelty he deemed necessary to secure fame as well as funds.

The originality of the machine came under scrutiny soon after news of it began to spread. An unknown Dr. Church with a patent for a printing letterpress began making inquiries about whether Babbage's machine violated his property.[2] Babbage replied that Dr. Church "need not be in the least alarmed at my infringing on his patent; both the object and the mechanism of his engine are as far as I am acquainted with his totally different."[3] However frivolous Church's complaint—patent trolling has a long history—Babbage had real concerns about the novelty of his invention.

Concerns about priority clearly intensified whenever Babbage angled for governmental support as a *philosophical* inventor, using the forum of the Royal Society repeatedly to justify government financing of his invention. Babbage's friend Francis Baily wrote delicately that a committee member of the Royal Society, due to produce a report on the engine, "wishes to suggest to you whether it would not be proper to get all the information you can, on the subject of former machines which have been constructed with a view to calculation." Baily suggested, "I would advise you to collect all the information you can on this point; and endeavour to show how far they went, and why they failed."[4] Babbage's manuscripts testify that he undertook a project of seeking out "prior art," including a manuscript he acquired later in 1823 of an otherwise unknown "Machine for multiplying any number of digits by any number" by Basil Lord Daer.[5] Over many years, his book collection came to include the rare works of Morland and Poleni; Babbage also purchased numerous machines, which later came to the Science Museum in Kensington.[6]

In 1828, his friend and collaborator J. W. F. Herschel conceded in a newspaper, "Such an invention is by no means new, two such machines have been produced by . . . Samuel Morland; and two more, to answer the same ends, by the late Earl Stanhope, about forty years since."[7] In response to Herschel's article, A. F. C. Kollman, the organist at the Royal German Chapel at St. James, responded with news of Leibniz's long failure and Müller's great success: "Neither Mr Babbage nor Mr Herschel mentions that a complete and perfect machine of that kind has existed these forty four years past which has been examined by the mathematical class of the Royal Society of Sciences at Goettingen and the use and effect of it been shewn by the inventor before an extra meeting of the whole society and numerous other men of science an account of the success."[8] Kollman provided precise bibliographic citations to the report of the Göttingen Academy given above and to Müller's pamphlet: "In that work Captain Muller shews the successful process by means of which he attained, in a short time, the great object which Leibnitz had sought for in vain so long and with such great expense."[9]

Leibniz had failed, but Müller hit close to home.[10] Müller had focused sharply on the need for mechanisms to ensure the security of the machine, as did Babbage. Even worse, as his machine neared a reduction to practice sufficient for manufacture, Müller envisioned connecting a fully functioning calculating machine to a mechanism for printing tables—that is, Müller envisioned turning his calculating machine into nothing less than a sort of difference engine:

> If the calculating machine sells well, I would in the future make a machine, which would simultaneously print in printer's ink on paper any arbitrary arithmetical progression in natural numbers . . . and which would halt of its own accord, when the side of the paper was full up.

After setting the first figures, all one has to do is to turn a handle and after stopping to turn the paper or to put another sheet in its place. In this way, a sequence of sixty terms can be delivered in a minute.[11]

In his short book on his calculating machine, Müller described the idea of automatically calculating the arithmetical sequences and connecting them to a printing mechanism. At some unknown point, Babbage had Herschel translate and paraphrase Müller's printed account of his calculating machine, with no shortage of humor about German bureaucratic titles: "Offers for money to disclose interior construction of his Engine & to make them for sale. & to make others to do more—such as (among other things) 'to deliver whole series of numbers which proceed according to Complicated Arithmetical relations'—each number to be produced by from [*sic*] 2 to 3 turns of the handle."[12] Whether Babbage "plagiarized" Müller remains unclear. He had many years to do so, and his extant machines and parts show his deep concern with the issues of security Müller saw as essential for reduction to practice.

As Herschel publically admitted in 1828, a great threat to Babbage's originality came nearby, in the late eighteenth-century machines of Charles Mahon, the third Earl Stanhope, and fellow of the Royal Society. Babbage and Herschel had heard rumors of Stanhope's machine soon after they began work on the engine. In the spring of 1822, the fourth Earl Stanhope, the son of the late inventor and radical member of the House of Lords, received a circumspect query from the assistant secretary of the Royal Society: "I have been informed, but correctly or incorrectly I know not, that your father the late Earl Stanhope, a short time before his death, invented a machine for performing arithmetical operations. A very ingenious Member of our society . . . is apprehensive, from what I have told to him, that your father may have anticipated him in this discovery." Apologizing for bothering the earl, the letter writer explained, "My only motive is to correct a statement if untrue, which may have the effect of preventing the publication of an ingenious & useful invention, and robbing the Inventor of his due share of fame & profit."[13] The earl replied to this inquiry several months later: "I never considered these machines as objects of utility but of \mere/ curiosity, & you will perceive that they have no connection with the very ingenious Machine which you mention."[14] A year later, amid his first sustained effort to gain state patronage for his Difference Engine, Babbage met personally with Stanhope's son. A diary entry from April 21, 1823, explains that the earl "confirmed [for] me that his father's mathematical machine performed its operations with more difficulty than they could be done by" Babbage's prototype machine. Stanhope's "multiplying was merely addition repeated."[15]

These reports were far too sanguine, for Stanhope's machines had important mechanical lessons for the discerning eye and hand. In the late nineteenth century, an authority on calculating engines noted, "The process employed by Stanhope is precisely that performed mechanically in Babbage's celebrated Difference Engine."[16] A visitor to the Science Museum in London will readily recognize the justice of this claim, as both the reconstructed Difference Engine 2 and Stanhope's multiplying machine dated 1777 have elegant helical or spiral arms as the centerpiece of their carry mechanism. Both designs involve "warnings" that record the need to perform carry. Far more distinctively, both use a spiral wheel—to power and to time successive carries to avoid jams—the very contrivance that Babbage's publicizer Dionysius Lardner called "a mechanical arrangement of singular felicity."[17] The two Stanhope machines in the Science Museum belonged to Babbage—though we do not know when he purchased them; his son Herschel Babbage later bequeathed them to the Science Museum, along with the machines of Morland, as "relics of his celebrated father."[18]

Our evidence about the development of the Difference Engine in the later 1820s does not reveal when Babbage and Clement made spiral arms such a crucial portion of the apparatus. An undated letter, likely from 1830–31, suggests that they were something new in the design. Clement's assistant draftsman Charles Jarvis wrote, "I have thought of the spiral barrel you were speaking of the other day and no difficulty of any consequence has presented itself. I am anxious to name this because you might mention the barrel to others who might see some trifling objection and condemn it at once."[19]

Given Babbage's and Herschel's efforts in the early 1820s to find out about other machines, along with the fact that they did get Stanhope's machines and Müller's book at some time, we might with some probability think that Babbage had substantial access to earlier arithmetical machines. Ultimately, we simply do not know whether Babbage drew upon Stanhope's and Müller's machines either for inspiration about building safety mechanisms, as in the case of Müller, or for the specific mechanism for carry and for safety, as in the case of Stanhope. Less important than proving or disproving a borrowing is noting the framework for our concern. Our anxieties around authorship rest upon an understanding of inventiveness that leaves too little space for appreciating the cumulative nature of most technical activity and that too often treats such cumulative production as tantamount to denying the contributions of individual inventors.

When the first dangers of his priority arose in 1822, Babbage offered Herschel a lucid plan for dividing credit if necessary:

If any one has found out other principles of applying machinery to arithmetic and has produced such results as my engines will procure he deserves to be amply rewarded and I wish him success and what is much more to the purpose if he has contrived such and if they put the type together and if he is a Frenchman I am quite sure he will be well rewarded. Supposing the worst for me this is a case of two independent inventors and each must be content with that portion of fame due to the engines he produces. It will be quite otherwise with any future contrivers however ingenious.[20]

Here as elsewhere, Babbage straddled the world of the genius inventor and the older one of the emulative craftsman. Had Babbage operated in an artisanal moral economy of emulation seventy years earlier, worries about authorship and plagiarism would not have been so pressing. As a philosophical inventor working early in the nineteenth century, an heir to the economy of philosophical glory, Babbage was inclined to prove and trumpet his absolute originality, insofar as possible.

Babbage's ally David Brewster defended Babbage's originality in light of predecessors: "Some of that class of individuals who envy all great men, and deny all great inventions, have ignorantly stated that Mr. Babbage's invention is not new. The same persons, had it suited their purpose, would have maintained that the invention of spectacles was an anticipation of the telescope; but even this is more true than the allegation that the arithmetical machines of Pascal were the types of Mr. Babbage's engine." They only performed manual arithmetic, so no genuine connection is possible. The Difference Engine, in contrast to all calculating machines, "is to embody in machinery the method of differences . . . and the effects which it is capable of producing and the work which in the course of a few years we expect to see it execute, will place it at an infinite distance from all other efforts of mechanical genius."[21] True to his moment, Brewster denied any meaningful anticipation and stressed the infinite gap between Babbage's genius and other mechanical contriving.

The refusal to accept any connection, falling upon a dichotomy of absolute invention or mere anticipation, is the fundamental problematic issue. Insofar as Babbage and his allies chose to portray his invention as a philosophical idea that was absolutely new, something springing from his own genius alone, an accusation of plagiarism is meaningful and deadly. From the perspective of emulative invention, Babbage can only be seen—even if he had detailed knowledge of Müller's and Stanhope's innovations—as someone inspired by and quite rightly developing and improving on the works of predecessors. Only a narrow picture of inventive activity and its associated vision of individual intellectual property could—and constantly—neglect the inspirational power, moral as well as epistemic, of the innovative copy, the improving emulation.

Teething Problems: Charles Stanhope and the Coordination of Technical Knowledge from Geneva to Kent

If ever you should bring your Reasoning Machine to the Degree of Perfection you sometimes speak of—you may not unaptly call it the Telescope of the Mind.
NORTH TO STANHOPE, December 1, 1801

There really is a difference between a mere scientific man, and one who is employing
several <u>Workmen</u> and <u>Artists</u>, who would be obliged to stand still, if he did not attend
to <u>more</u> things than one, in order to keep them all a-going in their several Departments.
STANHOPE TO NORTH, September 3, 1812

Stanhope was having problems with his teeth.[1] A series of short notes to his
workmen details things to be fixed: "All Teeth of the Horny Piece to be cut
<u>flat</u>, like the 3 teeth." Many of the gears of his mock-up calculating machine
had many teeth with the wrong shape: "To examine Length of teeth of Horny
Piece & centers of ditto."[2] The extant machines show that the horny piece got
its flat teeth, just as Stanhope requested (figures 5.1 and 5.2).

The working model in question appears to have been at a fairly advanced
stage of construction.[3] Stanhope's numbered comments to his workmen reveal his calculating machines in the middle of design and making. The comments range from the engraving on the box to the polishing of parts and from
the size and disposition of parts to the shape of teeth. A series of diagrams
shows each part "as it is" followed by one "as it ought to be."[4] Some note outright errors in implementation; many reflect the process of designing amid
the process of making. The existing model in the Science Museum in London
bears numerous markings from this process of implementation.

Stanhope's comments on teeth shape are found among a disordered series
of sketches, draft treatises, detailed plans, and patterns for iron casting, as well
as documents soiled in the workshop. A valuable index of the texture of his
collaboration with the clockmakers and other artisans building the machines,
the rich store of documents Stanhope produced in the course of devising his
machines allows us to reconstruct the interplay of sketches, mockups, and
working models of machines—an interplay as social as it was technical. The
process of invention was profoundly material: form in the designer's mind (or

FIGURE 5.1. "Horny piece": detail of carriage mechanism, Chevening Machine. © Trustees of the Chevening Estate.

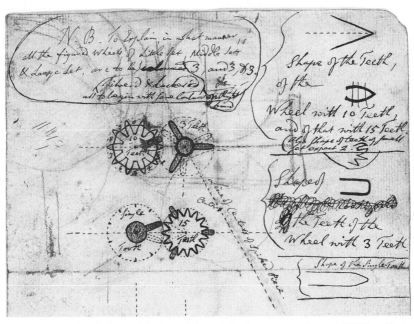

FIGURE 5.2. Stanhope's design of the shape of the teeth for the carriage mechanism. CKS U1590/C83/1. © Trustees of the Chevening Estate.

eye) was not simply impressed upon matter or drawn perfectly on paper. Read alongside his three existing machines, these documents reveal the intimate intertwining of design and making in Stanhope's effort to create calculating machines. A complete mental form did not precede material realization; neither did material realization happen independently of the envisioning and depiction of form.

This chapter focuses on form making, not final forms.[5] Embodied form came only through a nonlinear process of sketching, mocking up, tearing apart, reusing, redrawing, and modifying—a mixing of work on paper and work with metals and wood. His process was gradual and halting, with the constant creation of different solutions, in metal, wood, and on paper, given the current embodiment of the machines. Stanhope's efforts do not reflect eternal, presocialized truths about the making of artifacts in the preindustrial age; his materialized design practice emerged from late eighteenth-century ways of forming materials, of coordinating different practitioners, of representing forms and matters—all linked to political economies of innovation devised to encourage the dense intertwining of design and making. A close examination of the collective process of devising and creating these machines illuminates a distinctive late eighteenth-century moment in the coordination among practitioners and ways of designing and implementing. Stanhope's realized calculating machines emerged from the distinctive social and technical order in this transformation in the relationship of mind and hand in modernity (see figure 5.3).[6]

The documentation generated in the course of realizing Stanhope's machine undermines a series of false dichotomies: between histories of technology focused on inventive geniuses and those focused on workshop production, between histories of technology and material culture focused on social relations and those focused on the qualities of materials, between histories of drawing focused on sketching as creative and constitutive of ideas and those focused on sketches as communicative of already extant ideas. Each dichotomy presupposes anachronistic divisions of mental form and matter that were the product—not the cause—of social and technical transformation. This chapter looks at the historical emergence of each dichotomy within Stanhope's creative efforts and his theorizing about them. This chapter chronicles the abandonment of a vision of how philosophers and elite artisans might better coordinate and collaborate to make new things and improve older ones. In its stead arises a more familiar social dichotomy between mind and matter, between envisioning mental designs and implementing them in matter—a move from collaborative making to segregated manufacture. Moving between the process of making and Stanhope's philosophical reflections

FIGURE 5.3. Stanhope's calculating machine, made by Bullock, 1777. Inventory no. 1872-0137. (Science Museum, London.)

upon invention and thinking, the chapter reveals this abandonment both in the evolution of the paper designs and in Stanhope's account of the inventive process.

Stanhope's working manuscripts show his intense concern with the selection of the right form of teeth and the processes of their production (see figure 5.2): "Now all those that are conversant in the motion of Wheels and the proper shape \of their teeth/ will know that the \teeth of those/ wheels that lead, ought (if well made) to be of a different shape from the wheel that is lead."[7] The flat teeth were essential for the proper functioning of the key innovation of Stanhope's machine: its delayed carriage mechanism with its essential "horny piece." The distinctive helical form of the horny piece is his great solution to the problem of secondary carries. Each digit must have two arms: one to perform the primary carries, another to perform the secondary carries. Those arms must be spaced relative to the arms of other digits to allow the carries of previous digits to complete as the horny piece makes its rotation. The plans for Stanhope's machine, including the horny piece, were extremely detailed in many ways but indeterminate in others. Even the fine

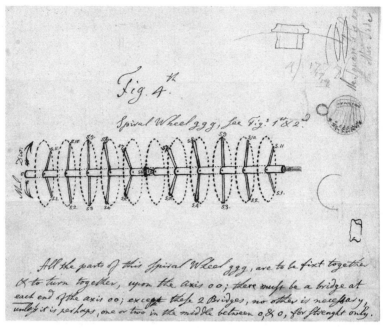

FIGURE 5.4. Design of the carriage mechanism, with "spiral wheel" or "horny piece" ggg. CKS U1590/C83/1. © Trustees of the Chevening Estate.

drawing, likely dirty from use in the workshop, indicated nothing about the flatness of the teeth (figure 5.4).

The drawings and diagrams for Stanhope's machines suggest the dramatic shifts in technical design since Leibniz struggled to have his machines made. Stanhope inherited a mathematical theory of shape and number of gears; he was an initiate into a new level of precision technical drawing and specification; his workmen could draw upon a new array of tools and metallurgical processes permitting the creation of more finely made and more precise instruments. More than all that, he was a member of a culture of technical collaboration that offered novel ways of working together to encourage innovation. At his estate in Chevening, Kent, he created a workshop to embody this culture—a rich environment of technical emulation and factive collaboration.[8]

A Technical and Political Education

In June 1774, Stanhope wrote to his former teacher and collaborator George Louis Le Sage in Geneva about a new idea for a calculating machine. Born in 1753 in England, Stanhope spent his formative years in Geneva—where he imbibed in equal measure republican politics, skill in technical drawing, mechanistic philosophy, the quantifying spirit, and mechanical ambitions— from 1764 until he left with his family in 1774.[9] He studied under Le Sage, a mathematically impoverished natural philosopher, *Encyclopédie* contributor, and logician known for his mechanistic account of gravity that gained important support in the nineteenth century among skeptics of action-at-a-distance.[10] For much of his life, Le Sage sought to devise "Baconian" procedures of systematic logical discovery in mechanics as well as in philosophy. He likewise sought to bring the new sciences into technological practice, and he lectured on mechanics to clockmakers. Stanhope collaborated with Le Sage to reduce friction in marine chronometers.[11] By 1769, Stanhope had invented a "mathematical instrument," which Lady Mary Coke was "assured [was] better for the purpose it was intended than any other of the kind."[12]

Stanhope's experiences in Geneva beyond the academy were likewise crucial. As an employer and unlikely apprentice, he involved himself with various groups of artisans and other skilled workmen—so much so that he was said to speak "French using the words and the accent of a butcher of Geneva or a boatman of Morges."[13] He worked closely with Genevan artisans in creating a number of inventions, from new escapements to wind mills. Fluent in the technical language of clockmakers, not just their French, he designed clocks and other machines using the common repertory of mechanical components of a cosmopolitan Genevan clockmaker of the 1770s.[14]

Stanhope, one contemporary wrote, "has a surprising genius; . . . [E]very day he invents something and he instructed and exercised himself in popular movements."[15] Not simply a collaborator or employer of artisans, Stanhope got himself elected to the Geneva governing council of "two hundred," where he sought to resolve the split between the politically empowered aristocracy and rest of the population through the creation anew of the original, founding republican virtues of Geneva. In so doing, he made himself into a major force for the lowest level of mechanicians and journeymen, who had neither political rights nor the right to form their own firms. The ruling aristocracy responded by closing down their private societies and forbidding their public exercises.[16] Stanhope, by all accounts, became a hero to the disenfranchised artisans.

While Stanhope's political agitation failed to resolve the divisions of Genevan society, his efforts reflected one major axis of a new social and cultural organization of knowledge and skilled manufacture in Geneva—a reorganization that served to secure its horological hegemony in the nineteenth century. In a letter of 1774, he called for universal education of all classes of Genevan society, with content appropriate to each level of society.[17] He was echoing the practical and pedagogical views of his close Genevan friends and colleagues among the savants. At that moment, they were attempting to implement just such a pedagogical vision to achieve a new political and economic foundation for Geneva in late mercantilism through the reform of the College at Geneva and the creation of a new "Society of Arts."[18] In defending his efforts to reform the Collège de Genève, H. B. Saussure explained that the "general purpose of national education is therefore, not making the people into savants, but making the work of Savants useful for the people."[19] Just such a pedagogical, social, and political transformation, to make philosophical work useful to practical technological and economic development, was needed for getting the shape of teeth right in Stanhope's calculating machines and clocks alike.

Practice and Politics in Embodying Theoretical Knowledge

Choosing the shape and number of teeth for gears was a central locus in the eighteenth century for the assertion of scientific authority over artisanal activity. While this assertion involved the domestication of new theoretical approaches into concrete artisanal activity, in the Genevan case, it came with calls for a more applied focus for natural philosophical inquiry. Before explaining the proper shape of teeth, the entry on "tooth" in the *Encyclopédie* explained, "Although machines using teethed wheels have been made for several centuries, Mechanicians have entirely neglected these considerations and left to workers the care of this aspect of the execution of machine, in which they observe no other rule than the teeth of the gears and of the little gears, so that the gearings engage with freedom and without stopping. . . . Although skilled clockmakers have solid enough notions about this question, the true figure of teeth of wheels still remains a sort of problem for them."[20] The *Encyclopédie* presented the gulf between the mathematical theory of gear shape and the techniques of producing them as an iconic example of the breakdown between natural philosophers and skilled artisans.

In 1778, Stanhope published, in English translation, an account of a machine "for determining the perfect Proportion between different Moveables acting by Levers and Wheel and Pinion" by a Mr. Le Cerf, a Genevan watchmaker. Le Cerf offered a "compass of proportion" that would allow watchmakers to determine the proportions and dimensions of their machines without the complex calculations the best theory required. The device would enable skilled workmen to align themselves with the best theory without necessarily understanding it. "It is to be regretted," Stanhope wrote, "that the author does not enter into a greater and more minute detail upon the shape of the working-teeth, &c. as this, in my opinion, will, in several cases, materially affect the very simple, general rules which he has elegantly laid down."[21] While Stanhope pointed here to the limits of the theoretical discussion in Le Cerf's account, the overall point of his translation was to supplant pervasive "arbitrary" practices for calculating the number, size, and shape of teeth with a mechanical device embodying the best sort of theory.[22]

The instrument and its accompanying treatise stemmed from the effort to reconfigure the relationship between propositional knowledge and practical know-how in Genevan horology. The theoretical superiority of gears with epicycloidal teeth was well known by the late eighteenth century.[23] Turning this theoretical fact into knowledge capable of being put into practice epitomized the more general problem of the gap between theory and practicable manufacture. As the author explained, "However beautiful this discovery," its worth was far less because "it is far easier to recognize the advantages" it makes one hope for "than it is to procure them through execution."[24] Making the theory practical required new instruments that would "make the formation of this curve absolutely mechanical."[25] The treatise and instrument embodied a new social formation bringing together skilled makers and natural philosophers.

Soon after the Stanhope family left Geneva, savants and watchmakers in the circles he had frequented formed a Society of Arts, bringing together natural philosophers and skilled artisans.[26] The Society of Arts offered an institutionalization of the sort of experience Stanhope had as a student in Geneva. It was an effort to embody in practical processes the latest philosophical work, to push experimental and scientific inquiry in directions useful to scientific work, to institutionalize collaboration between natural philosophers and artisans, and to create emotive frames to stir Genevan industry and improvement. These efforts sought to orient the production of propositional knowledge around issues of industry and production as well as to make explicit ample swaths of artisanal knowledge. The society likewise sought means for turning formal propositional knowledge into know-how, for making the latest techniques open and public, to encourage innovation and improvement through "emulation": "At issue is exciting, by means of emulation, the minds of Artists to observations and discoveries; to call forth all the new forms of industry to naturalize themselves among us; to favor useful establishments; and, above all, to support and to accredit our already established manufactures, and to join, for the glory and good of the Arts, insights from outside to those produced from our own inquiries."[27] Affective transformation, industrial espionage, and the movement of enlightened ideas and practices were to come together.[28]

The new forms of collaboration, forums for printed dissemination, and machines effecting these transformations were products of conscious effort on the part of citizens, not the state, to transform high philosophy, artisanal practice, and the economy as a whole. The Society of Arts took as its mission the reflection upon the failures of current practice and the seeking of practical schemes for remedying them: "Among our manufactures, clock making holds the first rank. . . . We know how much our skilled Masters distinguish themselves today in this art and they know better than anyone all that is lacking still in its perfection." From basic scientific claims to detailed procedures, the society sought to perfect the workplace: "Principles, instruments, materials, execution: these all offer vices to correct, difficulties to overcome." The goal was nothing less than to reach the limits of the "imperfection itself of our organs."[29]

Stanhope's publication of the description of the horological tool pointed to a key Genevan pedagogical and organizational solution that helped lead to its horological predominance in the nineteenth century. This article, and the Society of Arts that supported the work, rested on a conviction about the insufficiency of formal propositional knowledge of the number and shape of teeth: knowing *that* teeth might optimally take one shape according to theory did not translate easily to knowing *how* to make them. High propositional knowledge needed to be translated into an explanation accessible to skilled workmen. Even more important, propositional knowledge had to be embodied in working processes, including tools and new skills in using them, that were capable of putting that knowledge into use. Such embodiments were far from easy or accidental: they were the product of active collaboration and mutual correction between savants and machinists. New scientific work was not simply "applied" to practical problems: it was fundamentally transformed in becoming something capable of being embodied in a process of manufacture. An "industrial enlightenment" came from a simultaneous transformation in natural philosophical practice and allied artisanal practices. Mere access to formal propositional knowledge was deemed insufficient without the complex of social, intellectual, and material practices capable of making formal propositional knowledge something possibly used in making things.[30] Stanhope sought not the implementation of a formal proposition but the network of devices and practices permitting such an implementation to become materialized. This demanded appropriate forms of social organization and communication capable of bringing novel scientific knowledge and new mechanical analogies into practice. The efforts in Geneva to transform both philosophers and artisans alike highlights the danger for historians in projecting an anachronistic division between creative natural philosophers capable of using propositional knowledge and thinking analogically and artisans capable of, at best, minor improvements. Societies of emulation were devised to reorient natural philosophy and equally to encourage artisans to become more oriented by propositional knowledge.

Formed in this Genevan crucible, Stanhope embodied its social, epistemic, and mechanical solutions to the problem of intertwining propositional and prescriptive knowledge of nature and machines. His thirty years of experimental effort at technical improvement testify to his upbringing within this culture. Stanhope developed his knowledge and skills within an economic system striving to discern how best to introduce new natural philosophical and mathematical principles into artisanal practice alongside, and as part of, a massive industry-wide effort to imitate and counterfeit English clockwork.[31] For all his formal mathematical and natural philosophical knowledge, Stanhope turned more to learned manuals for clockmakers than to the latest mathematical treatments by the lights of Euler or Lagrange. Even when he contemplated contributing to that high mathematical tradition, he did so with his knowledge of the practice of horology and with the goals of practical implementation; the point was not simply to apply but to create a theory properly bridged to the world of practice: "I am working to reestablish . . . the theory of Huygens and Newton on small vibrations, and to show the way of applying the theory to practice, which neither mathematicians or clockmakers have done yet."[32] Stanhope sent the Royal Society of London a half-burnt timber as a token of his effort to fireproof buildings, illustrating the concreteness of his efforts at improvement and his insistence on such concreteness as necessary for inventive development.[33] In clockwork, he dedicated himself to an improved escapement and particularly to a multiyear effort to improve pendulums that united a concern with the shape of points of suspension to a theoretically motivated, quantitative experimental program to distinguish the friction of the means of suspension from air resistance.[34]

"Cylinder A": Stanhope's Initial Model

In his letter of July of 1774 to Le Sage, Stanhope described his best current vision for a calculating machine, and he announced plans to publish in the *Philosophical Transactions*. Not specifying the "details of execution," Stanhope set out the fundamental processes he then deemed necessary: carrying and shifting by one digit repeatedly to effect the additions.[35] While Stanhope abandoned this form of the machine before long, tracing its fortunes allows us to track the improvisational nature of his inventive practice that eventually led him to create a machine with stepped cylinders and a delayed carriage

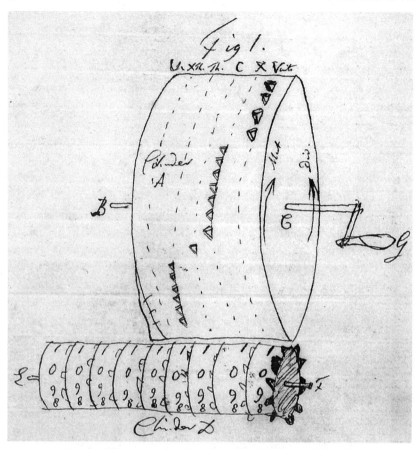

FIGURE 5.5. Cylinder A, "Figure 1": performing additions serially, not in parallel, to allow carries. CKS U1590/C83/1. © Trustees of the Chevening Estate.

mechanism. Stanhope's papers include the draft of a treatise on a calculating machine following this early model (figure 5.5). As described in his draft treatise, Stanhope's Cylinder A serves as a general model for fulfilling several major requirements for all multiplying machines: the multiple addition problem, the sufficient-force problem, and the simultaneous carry problem—that is, performing multiple additions of the same digit for a set of digits and allowing for the propagation of carries, including "secondary carries," without jamming the machine.

With the model Cylinder A, Stanhope envisioned machines capable of performing the addition of a number with multiple digits automatically but not simultaneously. In Leibniz's machine, the multiple digits of a number such as 6174 would be added in parallel and simultaneously to the result wheels. In

Stanhope's model as depicted in the diagram, the machine has just added 4 in the ones place, then 7 in the tens, then 1 in the hundreds, and finally 6 in the thousands (though the result cylinder D does not record these). While such a machine would still need a robust carriage mechanism capable of quickly propagating a carry throughout higher digits, the machine would not have to be able to add a carry and a primary addition at each place value as did Leibniz's. By performing carries from one place value before performing the addition of the digit in the next higher place value, a machine with Cylinder A performed all the carries needed that would occur before the next primary addition from the multiplying Cylinder A. Since the primary additions would be completed in sequence, such a machine was likely to be slower and larger but would involve a considerably simplified carry mechanism.

In these papers, Cylinder A offers a desideratum, nothing like a specification or even design of a buildable solution. In a slightly later set of considerations on the nature of invention, Stanhope distinguished the "Nature of the Invention [that is] Scientific Principles of it," what we might take as a basic model for a mechanism, from "The \Particular/ Mode of practical execution including \the/ Parts of the Mech," a fuller, concrete mechanical means for implementing the model. The model of Cylinder A provides the "Principles, I mean those qualities of it, which are essential, either to its acting at all as required, or else to its acting well."[36] To fulfill these principles, Stanhope envisioned and began to build at least two ways of practically realizing this Cylinder A.

In the first mode of practical execution, Cylinder A comprised a series of stacked decagons (figure 5.6). He explained in a set of instructions accompanying the diagrams that the "present great circular Wheel without teeth, to be cut into a decagon, as in Fig. [5.6] the angles of which decagon, are to fall into the intervals between the teeth of wheel bb."[37] The sides of each decagon had their teeth in a simple arithmetical progression from zero to nine. Each decagon was to be offset a set number of degrees. The user of the machine would have to be able to rotate each decagon to set the digit desired. Stanhope began working out some details of the locking mechanism for this scheme—in one diagram, he set out the rough numerical dimensions and sketched a piece "E" that was to lock the decagons into place after they had been set (figure 5.7).

Stanhope soon abandoned the realization of Cylinder A using stacked decagons in favor of stepped drums (figure 5.8). He hit upon the idea of mounting a series of stepped drums around the edge of his envisioned cylinder in a stepwise fashion (see figures 5.9–5.11).[38] No evidence suggests that Stanhope knew of Leibniz's drums, though it is certainly possible that he gained knowledge about them in Geneva, where several of his friends had

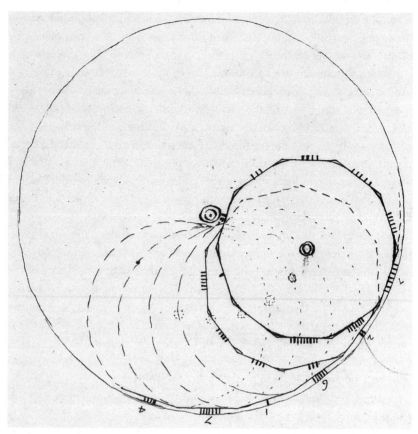

FIGURE 5.6. Stacked decagon realization of cylinder A. CKS U1590/C83/1. © Trustees of the Chevening Estate.

connections to Göttingen, where the corpse of Leibniz's machine rested.[39] Unlike all the German machines, Stanhope's drums used a parallel series of increasing numbers of teeth rather than a parallel set of increasingly long metal bars; his drums adapted this feature from his large pinned barrel and offered him greater flexibility in thinking about their use. He sketched the design in much greater detail and began calculating the size and proportions of the various parts. Figure 5.10 reveals that he was coming to believe that the right-angled version of the drums might better be realized as curved along circular arcs. (Compare the single stepped drum in the top left with the curved components of the main diagram.) Stanhope appears to have retained the notion of drums, and circular ones at that, but thought of putting them to some new use.

At the bottom of figure 5.12, we find sketches of a far different way of using stepped drums. No longer placed radially as a way of implementing Cylinder A, they are placed parallel to one another and admit the possibility of additions happening simultaneously in every digit. Stanhope appears thus to have independently arrived at something like Leibniz's rectangular configuration after considering something like the Leupold/Hahn/Müller organization. With this design came a new fundamental model for the machine, one with two distinct temporal stages: a primary addition stage, where no carries would be performed, and a carry stage, where primary and secondary carriages are performed. The entire carriage of stepped drums would slide to admit carries without jamming.[40] In the course of attempting to realize Cylinder A, Stanhope had moved from seeing the stepped drums as a means for implementing Cylinder A to a model of the machine organized with a parallel series of stepped drums.

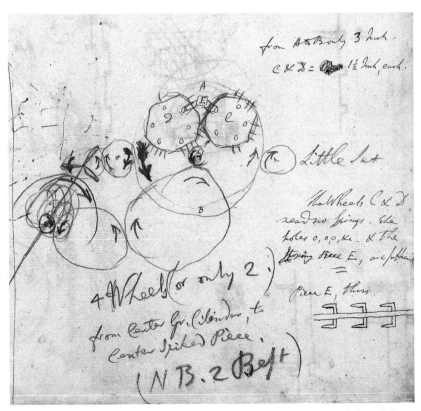

FIGURE 5.7. Stacked decagons; working out of the mechanism for holding in place; holes with "Stopping piece E." CKS U1590/C83/1. © Trustees of the Chevening Estate.

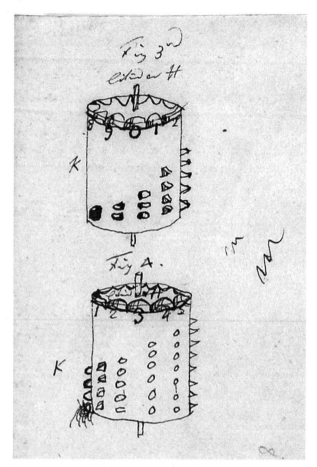

FIGURE 5.8. Stepped drums (or "cylinders," as Stanhope calls them). Note the steps are teethed, unlike in the drums of Hahn, Müller, or Leibniz. CKS U1590/C83/1. © Trustees of the Chevening Estate.

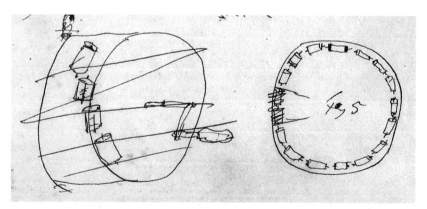

FIGURE 5.9. First crude sketch of the circular mounting of the stepped drums on cylinder A. CKS U1590/C83/1. © Trustees of the Chevening Estate.

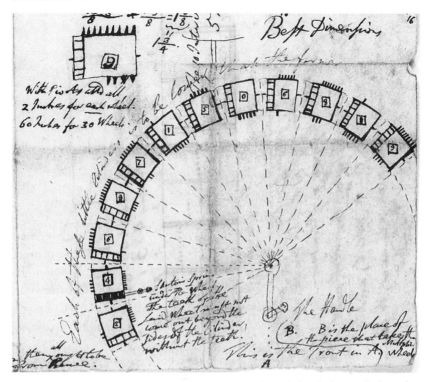

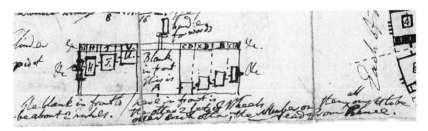

FIGURE 5.10. Circular disposition of stepped drums: "Each of these little cylinders is to be lower than the former." Set to multiplicand 245971506819. CKS U1590/C83/1. © Trustees of the Chevening Estate.

FIGURE 5.11. Circular disposition of stepped drums. CKS U1590/C83/1. © Trustees of the Chevening Estate.

For whatever reason, perhaps from difficulty in producing the parts, perhaps from a desire to produce a more compact machine or a vision of something better, Stanhope abandoned the model schematized by Cylinder A, which was to solve at the same time the multiple addition problem and the simultaneous carry problem. Instead of abandoning all the component parts that had been physically produced along the way, he noted how each piece was to be modified, replaced, or, more rarely, scrapped altogether. He soon fit

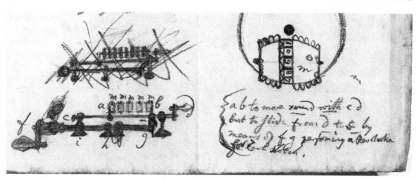

FIGURE 5.12. Stepped wheel with early postcylinder A design. CKS U1590/C83/1. © Trustees of the Chevening Estate.

the stepped drums, now a standard mechanical representation for him, into a new configuration. The design ramified into an addition part and a distinct mechanism to accommodate the need to permit multiple, parallel additions while performing the carries. Stanhope's breakthrough was to recognize that the carries could be stored and performed *after* all the additions had taken place; his mechanical breakthrough was his "horny piece" and the accompanying mechanism to effect it.

Stanhope's Theory of Invention

In the process of contriving a machine centered on the schema of Cylinder A, Stanhope came to distinguish far more sharply the problems various parts of the mechanism had to solve. Stanhope's development process fits poorly into two extreme poles within the history of technology. Stanhope was neither simply tinkering and developing prior mechanisms in an evolutionary way, nor was he implementing a design first conceived idealistically—a design whose novelty was only made possible through propositional scientific knowledge. Grounded in the mechanical vocabulary of Genevan clockmaking, Stanhope drew heavily on mechanical analogies to earlier devices; like the best of craftspeople, he improvised and departed from those analogies.

A few years later, Stanhope articulated the factors enabling such an inventive practice, at once grounded in extant mechanisms and free to improvise on them. His theory of invention distinguished three parts:

1. Object of a mechanical Invention. End proposed by the Inventor.
2. Nature of the Invention. <u>Scientific</u> <u>Principles</u> of it. By <u>Principles</u>, I mean those <u>qualities</u> of it, which are <u>essential</u>, either to its acting <u>at all</u> as required, or else to its acting <u>well</u>.

3. Manner in which the same may be performed. The \Particular/ Mode of practical execution including \the/ Parts of the Mech.

He gave an

Example of the Operation of Sawing Wood.

1st. Object. To divide the Particles of the Woods.
2. Nature of the Invention. Relative motion between the wood and the saw.
3. Manner in which same may be performed.
{ 1. The Saw Moved
2. The Wood moved
3. Both moved.[41]

For Stanhope, the heart of invention consisted in the enumeration of different manners in which to perform the "nature of the invention." A philosophical approach to invention was important in realizing the middle level of the process: the articulation of the "qualities of [an invention], which are essential, either to its acting at all as required, or else to its acting well." In the case of cutting wood, it was important to recognize that the crucial requirement is a "relative motion between the wood and the saw" that allows the inventor to enumerate all the different "manners" of producing that relative motion; this process of enumeration allowed the inventor to escape the trap of envisioning moving either the wood or the saw, but not both. The more philosophical inventor must abstract from both the absolute motion of a saw moving through wood and the absolute motion of wood moving over a stationary saw to the relative motion of saw and wood. Rather than envisioning the action of cutting as one or another fully realized apparatus, the inventor should attempt to abstract to the motion necessary, which every physicalized setup must ultimately produce. With this abstraction, unnecessary constraints are distinguished from real constraints, and new solutions become possible: one could imagine a more effective sawing apparatus in which wood and saw both move.

Using his numerous mechanical inventions as examples, Stanhope often claimed that grasping the middle level of abstraction between a goal and particular mechanical solutions enabled him to envision alternative, and often superior, implementations capable of reaching desired goals.[42] In Stanhope's account of invention, philosophical knowledge of intermediate-level principles of design allowed an inventor to break out of limited sets of lower-level mechanical realizations by breaking apart assumptions and mechanical habits. This escape from constraints allowed him in practice to tinker in novel ways, to apply mechanical analogies without being limited by the ordinary uses for them. With this account of invention, Stanhope was attempting to discern the "freedom" master craftspeople demonstrated in the creation of derivative but transformative developments of existing inventions. He was codifying the cognitive side of emulative practice: the break between mere aping and productive emulation that he experienced in Geneva and attempted to recreate in England.

However inadequate as a general theory, Stanhope's normative picture of invention instructively refracts Stanhope's inventive practice as seen in his working papers. Stanhope's extensive manuscripts reveal his repeated efforts to enumerate cases in order to break out of the dilemmas presented by the limited contemporaneous means of achieving various technical goals. He credited his most famous and important invention, the Stanhope iron press, to his ability to pull back from too narrow a focus on extant solutions to mechanical problems to grasp the problems themselves more generally.[43] According to Stanhope, his work on rudders and his trials with steamships likewise rested on the same move away from current mechanical solutions.

Understanding the conditions under which artisans and inventors draw upon mechanical analogies creatively and not just slavishly or mechanically has long intrigued theorists of innovation.[44] Historians of invention have sought to grasp concretely, for example, "to understand how Edison borrowed from" the analogy of the clock escapement in creating motion picture technologies, "while at the same time . . . not [being] constrained by it."[45] Recent historians of the early Industrial Revolution disagree about the importance of the abilities of artisans in drawing on analogies in creative and nonconstraining ways to explain the explosion of invention in the eighteenth century. Following historians of technology generally hostile to the "linear" model of technology as applied science, John Harris, Liliane Hilaire-Pérez, and Maxine Berg point to the incremental microtechnological inventiveness of artisans in the eighteenth century, supported by their mobility and culture of open technique. Hilaire-Pérez, for example, claims that the "part of artisans" was "the crucial factor in the industrial Enlightenment." In her account, "artisans were not just skilled. . . . [T]hey combined different kinds of cognitive resources to contrive new products and processes."[46] While praising the increasing skills of artisans in the period, Joel Mokyr denies that the capacities of artisans could account for the rapid increase in macrolevel inventiveness of the late eighteenth century: most artisans "were good at making incremental improvements to existing processes, not in expanding the epistemic base of the techniques they used or applying state of the art knowledge to their craft." Not only poor at drawing on formal knowledge, most artisans were "not well-positioned to rely on the two processes of analogy and recombination, in which technology improves by adopting or imitating tricks and gimmicks from other, unrelated, activities."[47] Opposing an older view that saw the revolution as a product of "illiterate tinkering" with little connection to the new sciences, Mokyr focuses on the pedagogical, institutional, and economic conditions under which formal propositional knowledge served to promote major technological transformations and to overcome the stasis characteristic of even the most skilled artisanal communities.[48] In Mokyr's account, sustained inventiveness requires competencies above and beyond artisanal practice to move toward creative analogy making and novel recombination; it also demands creative competencies in drawing upon the state of the art (or natural science) in novel ways.

The division between artisans and philosophers in this recent debate flattens out the much richer eighteenth-century appreciation of the range of creativities in mechanical and craft production. The hierarchy among artisans we've seen in previous chapters echoed this range of creativities. What distinguished artisans from philosophers in the eighteenth century? No sharp division existed in practice or, remarkably, in most Enlightenment theory. To be sure, the lowest sort of artisan was thought capable only of reproducing something, without recognizing its imperfections or finding means for improving it. Other artisans created more autonomously, as theorists well understood. As discussed in chapter 4, Johann Sulzer set out a hierarchy of more and less free imitators. Content to follow others, base imitators seek only to replicate a work with no independent reflection upon the fit between its overall design and its components. Rather than aping the choices of earlier craftspeople, "free" imitators constantly assess how each part contributes to whole. By replacing ill-serving elements with ones that better execute the desired function, this independent judgment enables the free imitator to produce superior works.[49]

With his theory of invention, Stanhope attempted to isolate the freedom inventors ought to have; his theory captures his efforts to proliferate solutions to the problems of carry remarkably well. He attributed his ability to do so to his mastery of a syllogistic logic of discovery, as we will see. We can attribute his enumerative abilities rather to his rich set of new technical and scientific competencies: a background in an experimental approach to invention, not exclusively Baconian; a grounding in the quantifying spirit of the late eighteenth century; a particular understanding of the need to produce numerous distinct solutions to technical problems, inherited from Le Sage; skill and fluidity in precision drawing; and dexterity in collaborating with natural philosophers and skilled artisans alike.

If studies of invention must empirically describe new technical practices and the new social organizations connected to those technical practices, they would be remiss to dismiss the actors' theories of technical and social innovation as merely mistaken ideology or theoretical posturing.[50] While normative accounts of invention do not correctly or completely describe inventive practice, they are not incidental to it, as actors often struggle to put their normative theories of innovation into practice. Likewise, while normative accounts of the relation between savants and mechanics do not correctly describe their interaction, neither are they entirely incidental to it. Normative accounts of invention and its social forms condition the possible forms of technical and social organization. Stanhope's theoretically inspired practice of enumeration and exclusion was central to his inventive practice; just as his practically inspired technologies of coordination and mediation were central to that practice.

Spiral Wheel: The Delayed Carriage Mechanism

Any good calculating machine must be able to perform an addition where a carry required in any digit can be propagated through all the higher-order digits if necessary, as in 9999 + 1 (see box 5.1). A simple adding machine could perform the addition on one digit, then perform the carry on the next digit, and then do the addition for the second digit with any carry. Such an arrangement is more difficult for multiplying machines, which need to be able to accommodate an addition of up to 9 and a reception and/or communication of a carry all at the same time for each digit of the machine. If both operations were performed on gears mounted upon or driven by the same axis, the apparatus would get stuck and cause the carry or the addition mechanism to jam, if not both. Separating the process of carrying from that of adding, however, creates its own difficulty—namely, the need to perform any new carries produced in performing a first set of carries, as in 9989 + 111. If a machine adds 111 to 9989 place value by place value without performing the carries as it goes along, two carries will remain to be completed. Stanhope resolved this problem in two opposing ways: in the first, by performing all additions and carries for each place value individually as in Cylinder A, and in the second, by deferring all carries to a two-stage process.

All the Cylinder A designs were devised to solve two problems with one mechanism: in a single rotation, Cylinder A would add a digit in each column and it would do so sequentially; carries would follow immediately after the addition at each digit to prevent the machine from jamming when carries occurred. For unclear reasons—perhaps the size necessary or the sheer diffi-

---

Box 5.1. The Challenges of Carry

Example of Propagated Carry

A ' indicates a carry needing to be performed in the next digit to the left.

$$
\begin{array}{r}
9999 \\
+ \quad\quad 1 \\
\hline
999'0 \\
99'00 \\
9'000 \\
'0000 \\
\hline
10000
\end{array}
$$

Example of Simultaneous Addition and Carry

$$
\begin{array}{r}
49 \\
+ \ 41 \\
\hline
8'0 \\
\hline
90
\end{array}
$$

If the additions occur simultaneously in every column, then both the carry and the four need to be added at the same time in the second column.

Example of Secondary Carry

$$
\begin{array}{r}
9989 \\
+ \ 111 \\
\hline
9'09'0 \\
'00'00 \\
\hline
10100
\end{array}
$$

9'09'0   After addition, two carries to perform.
'00'00   After primary carry, two *new* carries to perform.
10100   After secondary carry, done.

---

culty of constructing Cylinder A in any of the ways he envisioned—Stanhope decided to separate the process of the primary addition of the digits of the multiplicand from the process of producing the delay necessary to allow carriage. With this decision, he produced the most innovative aspect of his calculating machines: their "delayed" carriage mechanism and the "horny piece" essential to it.

Stanhope was clearly very worried about getting the carry right and devised mechanisms to test his machine. On the bottom of a detailed page setting out changes in the current mockups, Stanhope describes a "Temporary wooden Cilinder . . . 3 times size of aaa. 7 Figures 8996527."[51] With this temporary cylinder, Stanhope was working to perfect the carry mechanism. The listed default values for this wooden cylinder are well chosen for testing out

---

Box 5.2. A Value for Testing Carry

8996527	
4500	
8990'0'027	Addition before carries added
89'01027	
8'001027	Secondary carries with propagation
9001027	Final result

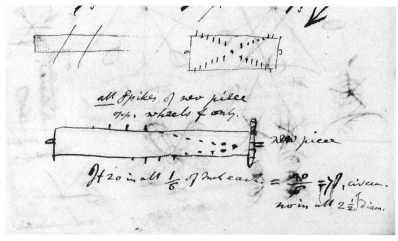

FIGURE 5.13. Early sketch of spiked wheel f, with calculations. CKS U1590/C83/1. © Trustees of the Chevening Estate.

and modifying carrying mechanisms: adding a figure such as 4500 to 8996527 would involve primary and secondary carries as well as propagation through the machine (see box 5.2).

A hastily scribbled working page, replete with deletions and revisions, offers the first working sketch of what he eventually called his "horny piece" or "spiked wheel" (figure 5.13). The continuity of mechanical forms from his Cylinder A is clear: he retained the notion of using teeth arranged radially around a cylinder to perform carries sequentially. Drawing on one function of Cylinder A, but not both, his insight was to place the teeth in a helical pattern around the cylinder. There was considerable continuity of materials as well; Stanhope's notes to his artisans detail the reworking of pieces of parts already produced and fitted in the current mock-up of the machine.

On the working design where the "spiked wheel" first appears, Stanhope set out the general disposition of his new piece with "spikes" and sketched several different sets of gears, with different number of teeth, to receive carries when necessary. He produced a set of stop-motion diagrams to show the different possible states of the new parts of the carry mechanism accompanying the spiked piece (figure 5.14).

As is typical of his design practice, he reworked his carry mechanism using existing pieces and carefully noted what an artisan would have to do with each extant piece—retain it, modify it, or toss it (figure 5.15).

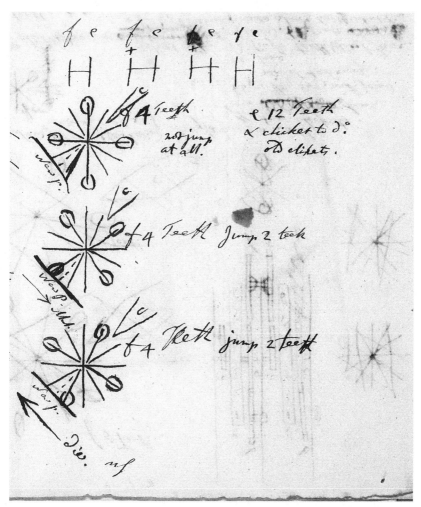

FIGURE 5.14. Stop-motion figures of carry wheel f, c, and spiked wheel. CKS U1590/C83/1. © Trustees of the Chevening Estate.

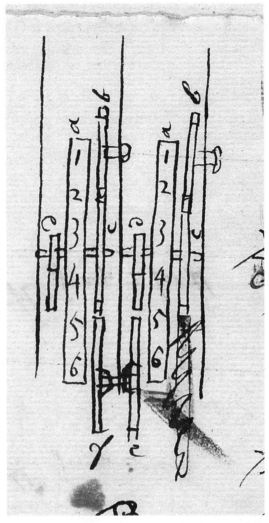

FIGURE 5.15. Reworking the carry mechanism to accommodate the new spiked piece. CKS U1590/ C83/1. © Trustees of the Chevening Estate.

aa   same
bb   as they are with joined wheel taken off
cc   same
dd   single tooth Wheel same
e   wheel of 12 teeth
ef   joined
f   4 teeth
ef   1/20 of circumference above horizontal line
e   teeth out towards single tooth, which ½ way between 9 & 0
e   to turn independent of d not of it

The "spiked piece" was soon rechristened the "horny piece" in a series of elegant drawings illustrating its construction. He set out what was to be retained and what was to be changed in the current mockup machine (figure 5.16). He instructed his workmen to rework most of an earlier, already constructed carriage mechanism into a new carriage arrangement capable of performing the carriage from the "spiked piece." Piece kkk, for example, the "present catch & spring," was to be retained, though the "end of catch [was] to be cut off." He devised several ways to indicate the motion and timing of

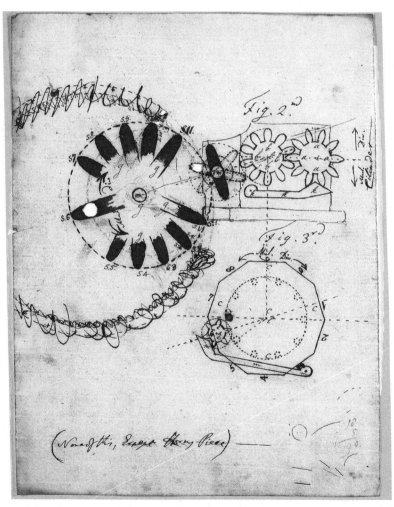

FIGURE 5.16. "None of this, Except Horny Piece." Rejected design of carriage mechanism with horny piece ggg cross section. Note that the teeth are not flat. CKS U1590/C83/1. © Trustees of the Chevening Estate.

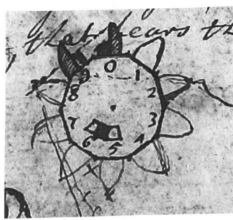

FIGURE 5.17. Revision to the placement of the pin in instructions to an artisan. CKS U1590/C83/1.
© Trustees of the Chevening Estate.

the "horny piece" ggg.[52] In two of his figures, he explained, "ff has only 2 teeth, which are the two <u>black</u> teeth at f, in the Fig 2nd the <u>dotted</u> teeth, are only to shew the different <u>positions</u>, of the two <u>black</u> teeth of f."[53]

While elegant, this set of diagrams did not get heavy use in the shop. If initially written to instruct workers to modify parts and machine new ones, they appear in that process to have led Stanhope to yet more alterations of his designs instead. As he revised them, he wrote "None of this at all" and "None of this, except horny piece."[54] The envisioned single tooth is soon replaced with a pin in the 10-wheel.

And yet the teeth still needed fixing. And the crucial pin necessary to communicate the need for a carry was simply wrong (figure 5.17): "In small set, the <u>Pin</u> to be in the center of the Flat, that clears the Figure 5—Then the <u>Stop Pin</u> is to be in the Line between the Center of the Wheel, and the Corner that is between the Figures 5 & 6, when the 0 is upper most."[55] Before long, Stanhope had the working exemplar now in the Science Museum in London, which bears the material evidence of the many modifications in its production.

Tacking between Paper and Metal

Skill in technical draftsmanship as a technical and social practice for collaborative invention was central to Stanhope's Genevan apprenticeship. In his well-known account of technological design, Eugene Ferguson argues, "As the designer draws lines on paper, he translates a picture held in his mind into a drawing that will produce a similar picture in another mind and will eventually become a three-dimensional engine in metal." Such drawings "enabled others to build what was in their minds."[56] Historians of technology have long criticized a linear model where ideational designs on paper are then produced in metal; such a vision, and the metaphors used to describe it, "leaves out any social relations, any mismatch of sender and receiver, any translation, any feedback."[57] In his study of eighteenth-century French engineers, Ken Alder has shown the social, pedagogical, and instructional work necessary for the creation and enforcement of a regime where drawn artifacts could be then simply implemented in metal: French engineers devised "novel techniques for drawing artifacts on paper and for defining them with physical tools. And implicit in these new representations was a new conception of work, and a new vision of the social order."[58] Before the creation of such a division, drawings involved far more collaboration. David McGee argues, "The properties of early machine drawings actually point more immediately to what might be referred to as a social process of design in which the second person is the real expert, the person who actually determines the final dimensions of the machine that is made." The traditional craftsperson, designing with and through the elements of the machine, constantly adjusts the configuration, size, and shape of parts and paper designs.[59] The tone of Stanhope's notes to his workers reveals a sharp hierarchy, but not one simply of the mind and the hand—at least not until the machine had been thoroughly developed. His earlier sketches leave much up to his artisans, who, as trusted collaborators, could determine final dimensions and shapes, as in his early design of his spiral wheel in figure 5.4.

Sketching rough ideas was central to Stanhope's very process of envisioning; the coming into the mind of a specified *idea* of the machine was something that happened through sketching, not a prelude to sketching.[60] Like many engineers, he often appears to have had a notion that required a mechanical process of elaboration through sketching (figure 5.18).[61]

Stanhope did not design something on paper at an initial discrete stage and then proceed to order it implemented in metal: complete ideational design did not precede implementation.[62] His drawings—as well as the existing machines themselves—testify to a constant tacking back and forth between pieces already produced in wood and metal and new paper designs. There was tinkering, to be sure: tinkering on paper as well as in metal and wooden prototypes; tinkering on paper about things already in metal; tinkering in metal based on instructions in metal. Even moments of radical redesign, such as the rejection of Cylinder A, involved a steady reuse and transformation of pieces already built; they involved thinking about the machine through

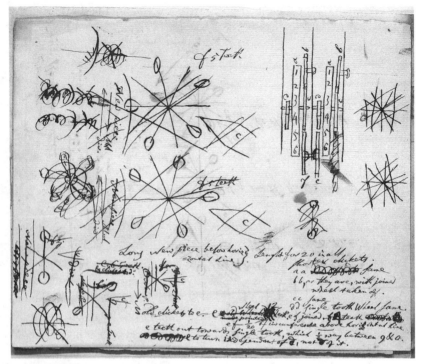

FIGURE 5.18. *Left*, canceled sketches of the carry mechanism's interaction with the spiral wheel; *right*, reworking of an older carry mechanism. CKS U1590/C83/1. © Trustees of the Chevening Estate.

the physical parts already produced. The existing Stanhope machines in the Science Museum in London are precious material monuments to Stanhope's process of making. Michael Wright's technical description notes, "There are some odd, perhaps irrational, points in the design . . . which suggest that its construction may have been rather tentative. These are: the use of the un-finished iron base; the way that the fork is fitted, with the frame pillar fitted over it; and the way in which the top plates are set. There is also evidence in the mistake made in engraving the numerals on the rings of the main figure drums."[63] Stanhope's order to remedy this last mistake is number three on his list of orders for his artisans. The spiral wheel carriage mechanism emerged from just such a back-and-forth movement in Stanhope's design practice.

Inspirations from Genevan Mechanics

Cylinder A and the stepped drums and spiked pieces that then replaced it im-mediately call to mind the familiar drum of a music box that encodes its tune, formally termed a "pinned barrel" (see figure 5.5).[64] Such barrels belonged to the mechanical vocabulary of Genevan clockmakers. One such clockmaker, Antoine Favre, revealed the details of the first "music box" to the *Société des Arts* some years later in 1796. When Stanhope turned to inventing calculating machines, he had absorbed a rich mechanical vocabulary from the Genevan clockmakers. A fluent speaker of the French argot of the clockmakers, he was likewise a fluent user of their mechanical vocabulary.

Stanhope's basic technological vocabulary developed through his immersion in the innovative world of learned clockmaking, which was concerned with ever greater accuracy, along with his close interaction with clockmakers in Geneva, who were largely concerned with innovation through the imitation—and often counterfeiting—of successful English and French watches and clocks. Time and again, he sought to improve the elements of that vocabulary—whether escapements or suspensions—and to combine them into novel forms.[65] At the heart of his Genevan mechanical education was learning to seek improvements to extant mechanisms—whether through innovative shapes of pendulums and their mountings or the choice of metals—by trying out various mechanical solutions and testing them. He tested those improvements in the quantifying spirit of the time.

While Le Sage's mechanistic physics and logic of invention were the focus of Stanhope's natural philosophical education, the improvement of marine chronometers was the focus of his technical education. An early letter to his father explains, "I am indeed ashamed of having stayed so long without writing, but as I was inventing a way to hinder the variations of the Pendulum by heat & cold, I was very loath to write till I could send you an idea of it."[66] Stanhope soon won a prize competition for the improvement of pendulums from the Academy of Sciences in Copenhagen; his work appears to have focused mostly on the shape of pendulums and the design of their points of suspension.[67] As he was preparing to leave Geneva, he and Le Sage were working on diminishing friction in marine chronometers.[68]

Upon arriving in Paris in 1774, Stanhope made the rounds of *philosophes*, mathematicians, and elite clockmakers while continuing to reflect on how to improve his new clock escapements. He visited Jean-Jacques Rousseau, the Académie des sciences, and the famous clockmakers—and bitter rivals— Pierre Le Roy and Ferdinand Berthoud.[69] Le Roy entertained some of Stanhope and Le Sage's schemes to improve escapements, but Berthoud dismissed them outright. Stanhope found the Bibliothéque Publique du Roi astonishing but shockingly deficient in the standard texts of contemporary clockmaking, lacking even the works of Harrison.

Stanhope's youthful apprenticeship in Genevan horological culture led him to seek out a fruitful, nonunidirectional collaboration among the competencies of philosophers, skill masters, and implementing journeymen. When Stanhope returned to England from Geneva with his family, he was, his father-in-law noted, "yet very new to our vile world, indeed quite a traveller in England."[70] He no longer was among the Genevan artisans whom he had patronized, befriended, collaborated, and politically supported. In England, he set upon finding skilled workmen to recreate this context and to make his calculating machines and his improved clock escapement. After dining at the Royal Society Club, someone suggested that Stanhope seek out Jesse Ramsden, whom he soon engaged to build the machine.[71] At that moment, Ramsden was becoming famous as the best producer of bespoke mathematical instruments in England.[72] Delays quickly set in: most of Ramsden's workers got sick; month after month, the work still had not begun.[73] Queried by his mentor Le Sage about the progress of the machine, Stanhope wrote that Ramsden "had amused me for an infinite time" with all his promises, but all for naught. But Stanhope also had brighter news. He had turned to the most famous clockmaking firm in London, that of Benjamin Vulliamy.[74] The letters of the time describe a machine of twenty to thirty digits requiring only a turn of a crank.[75] In May 1778, Stanhope announced, "My arithmetical machine will be done in eight days, that is, my first; I've already tried multiplication, division, and the rule of three on it. It succeeds \very/ well."[76] Whether Vulliamy was the maker after the long delay is unclear.[77]

When he began to attempt to construct his machines in England, Stanhope sought the help of the most reputed artisanal firms in an effort to recreate something of the working relationships he had developed in Geneva. Busy and increasingly successful, Ramsden and Vulliamy failed to prioritize Stanhope's work; even worse, neither offered the collaborative environment he had experienced in Geneva, so he turned elsewhere.

Stanhope's existing machines all bear the name of James Bullock, a clockmaker active in the 1780s and 1790s. Ramsden and Vulliamy were famous; Bullock was little known—much like Leibniz's artisan Ollivier. In the absence of other details, we can discern a pattern apparent from the construction of earlier machines: calculating machines seem largely to have been built by master artisans who could focus exclusively on them; the large, prosperous firms of Vulliamy and Ramsden had too much business to take the risk of monopolizing their efforts with such a difficult project that had a limited probability of success.

After Stanhope's death, his son remarked, "There are some particulars in the mode of working it . . . which were not fully understood even by a Mechanician who lived many years with my father."[78] Seeking initially something more like collaborators as found in the community of artisans and philosophers he knew from Geneva, Stanhope seems rather to have found employees who could implement his designs—at least until around 1800.

Stanhope's desire for artisans capable of closer collaboration continued until his death. While we know little about his workmen during the apparent period of greatest effort on the calculating machine, Stanhope's most important collaborator in the final years of his life exemplified the highly educated and highly skilled workmen he had known in Geneva and sought with difficulty in England. Like Philippe Vayringe from chapter 4, Samuel Varley (1744–1822) was an autodidact watchmaker and jeweler who made himself into a natural philosopher offering lecture series. His London lectures became the basis in 1794 for the London Chemical and Philosophical Society, a typical society of emulation. "Earl Stanhope ultimately made it worth [Varley's] while to leave his prosperous career in town" and move to Stanhope's estate at "Chevening where he assisted the Earl in all his works, particularly his two important ones the perfecting the Stereotype and the Printing Press."[79] For sixteen years, Varley explained, he "was his lordship's constant attendant in all his experiments, provided the greatest part of his apparatus, and superintended the construction thereof."[80] In an 1805 will, Stanhope left Varley (or his heirs) the extraordinary sum of "£1,000, and all my tools, machines, machinery, and instruments, mathematical and astronomical, chymical and mechanical."[81] For much of his life, Varley focused on the creation of new precision instruments for manufacturing, such as new lathes for screw making—the material embodiments of the culture of emulation so central to Stanhope. Samuel's apprentice—his nephew, the artist Cornelius—published details of many of these instruments and other secrets from Chevening many years into the nineteenth century, after his uncle's death; Cornelius became one of the foremost makers of precision drawings of improvements to tools of the period, mostly published in the *Transactions* of the Society of Arts, the model for all other societies of emulation.[82] Another publication detailed the novel cement made by the "truly-scientific mechanic" Varley: "exceeding useful in the various mechanical arts, and particular in turning."[83] By bringing such artisans to Chevening, Stanhope turned his estate into a space not just for the production of new schemes of improvement but also for the perfection of the machine tools and draftsmanship necessary to construct other machines.

Varley came after the machines were built. Stanhope's drawings bear witness to a shift toward a hylomorphic design—one devised to be implemented, not to serve as the basis for creative interchanges between philosophers and artisans. The greater technical specificity of his drawings capture, certainly, the maturing of his design, but they also point to a less bidirectional relationship in design and making. A series of soiled, far finer diagrams with measured, multiple view plans deploys a new set of conventions to capture the delayed carriage and parts of the "horny piece." A series of carefully drafted cross sections set out the angles for subsequent digits of the "horny piece" (figures 5.19 and 5.20).

If a Ramsden or a Varley could easily visualize an entire device based on a sketch, the later builders of Stanhope's machines seem to have needed ever-greater specification. From his time in Geneva, Stanhope was more used to working closely with top-level artisans as collaborators; the evidence suggests he ended up largely with artisans capable of producing what he had set out in ever more specified terms, what Joel Mokyr calls "competence."[84]

Stanhope appears to have had reasonably well-functioning models of his calculating machines by 1780. The existing machines, though prone to break because of design problems and some poor choices in materials (notably ivory dials), still work reasonably well; their innovative carry mechanisms function today. Stanhope clearly planned on manufacturing the machines in greater numbers and not just making them as bespoke luxury items: "I will try to reduce the price of the machine so that businessmen and bankers can purchase them."[85] One existing copy of his machine reveals a robustness in design and method of construction unnecessary for a one-off machine: "The Chevening machine was built with the benefit of greater experience, and the use of a cast base plate, involving the trouble of pattern-making and foundry work (and in practical terms quite unnecessary) suggest that some sort of production was envisioned even if in modest numbers."[86] Among his papers are the used patterns for the foundry work and finalization of the design. A

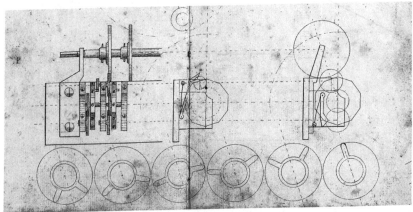

FIGURE 5.19. Fine production drawing for carriage mechanism. Along the bottom are carefully drawn angles for the "horny piece." Note the changed location of the other gears. CKS U1590/C83/1. © Trustees of the Chevening Estate.

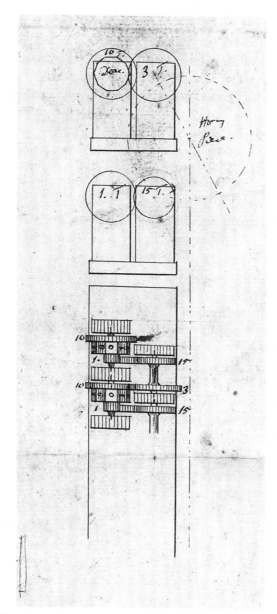

FIGURE 5.20. Plan and cross-section of location for ten-, two-, and fifteen-teethed wheels with "horny piece." CKS U1590/C83/1. © Trustees of the Chevening Estate.

few years later, in the early 1800s, Stanhope would cast his innovative presses in iron. Manufacture of his machines, not just their making by master craftsmen, was possible. Only with manufacture could the machine be understood as ideational in essence, the mix of form and matter integral to its genealogy dispensed with. Design and matter, form and implementation, could be ontologically separated.

Historians of clockmaking have long recognized how entrepreneurial clockmakers at the end of the eighteenth century, such as Eli Terry, created systems for producing interchangeable parts that eliminated the need for great skill: Terry's "greatest achievement was to destroy the tradition and skill which, as a master clock artisan, he valued so highly."[87] Such transformations allowed manufacture to replace skilled making. The creation of instruments for manufacture was necessary for higher levels of precision, which then created conditions where a bidirectional interaction of artisans and philosophers gave way, increasingly, to engineers and implementers.

And yet Stanhope did not manufacture his calculating machine, despite much encouragement; he neither published their designs nor patented them. In 1812, his friend North upbraided Stanhope for failing to bring any of his numerous schemes to fruition: "Still I cannot but judge that Society would profit much sooner by any one of your Lordship's mechanical or logical Devices if you could find yourself disposed to turn an undivided Attention into one Channel—whether the preferred Subject should be Logic or Harpsichords—Engraving or Ship-Building—we then arrive at some pleasant Land of Promise—& might enter with enjoyment of useful Discoveries—at present we must content ourselves if some new Delays should occur."[88] In his reply, Stanhope thanked his friend for his "kind and friendly scold" before explaining, "There really is a difference between a mere scientific man, and one who is employing several Workmen and Artists, who would be obliged to stand still, if he did not attend to more things than one, in order to keep them all a-going in their several Departments."[89] The social arrangements of skilled labor necessary for the production of technological projects precluded the quick and easy bringing to practice of any one machine. Projecting devices was easy; fostering the social and technical environment capable of allowing them to be perfected and produced was altogether more challenging.

North was not to let his friend off so easily, however. If the machine couldn't be perfected at the moment, why did Stanhope not disclose its secrets, as he had long promised? "I know your Lordship has ever entertained Principles favourable to the Argumentation of human Happiness & consequently I cannot easily discovery any Motive that can so long have prevented you from disclosing a Discovery which you have conceived to be generally useful. . . . I cannot see what you could do more useful in the arithmetical and logical Lines than to give them freely to the world."[90] North knew Stanhope well and was happy to highlight little hypocrisies. The earl often praised the public disclosure of inventions tending to human happiness, and yet he held many of his secrets very close.[91]

From Mechanical Invention to Mechanized Reasoning

Stanhope's friend North was incredulous. Stanhope claimed his numerous inventions stemmed from his use of the syllogistic method: "I mean to shew, in my work, how [logic] can be applied to making Discoveries. And if I have succeeded in making a great Number of mechanical & other inventions, it is by means of Logic applied to certain Facts, mechanical, chemical, physical, &c. & by means of Logic only."[92] He gave numerous examples, including his way of increasing the power of printing presses: "I then logically analized practical means, of mechanically the said means, & by Logic, I discovered it. The same as to the burning of Lime &c &c. Vive la Logique."[93]

In a typical eighteenth-century dismissal of logic, North replied, "Has not Invention & Communication made its most rapid Progress since the Syllogistic Forms have been laid aside, as they have been for a Century past?" A sociologist of scientific knowledge *avant la lettre*, North challenged Stanhope's insistence that a rational reconstruction of past discoveries would reveal hidden syllogisms at work: "If the syllogistic Form of reasoning is not in the Mind of an Inventor, at the Time of Invention, in what sense is it said that the Syllogistic Form aids Invention; if the Syllogistic Form is not used by the Teacher at the Time of Teaching, or if the Teacher can Teach as easily without it; in what sense is it said that the Syllogistic Form assists in the communication of Knowledge?"[94] The earl held firm that he discovered through logic, whether he was improving clocks and calculating machines or making discoveries in electricity.

Since such discovery was logical, it could be mechanized, Stanhope elsewhere explained: "Those sublime discoveries of Newton . . . are found, as of necessity, and without mental Effort, by a mechanical operation, most easily performed in my Analytical Machine."[95] The heart of invention was analysis: "The Method of Analysis is properly termed the Method of Invention. Therefore, Invention, upon whatever subject, is a mere Mechanical Operation."[96]

In 1802, the Swiss philosopher Marc-Auguste Pictet published his travelogue of a visit to the United Kingdom, including a stop to see the many inventions of Stanhope, with whom he had been a friend since their adolescence in Geneva. Among the many "productions from this fertile brain," he

FIGURE 5.21. Charles Stanhope, Logical Demonstrator, square model, c. 1805. Object number 1953–353. (Science Museum, London.)

wrote, Stanhope "most values a system of logic accompanied by a machine for reasoning."[97] Another report of 1801, which quickly found its way into French and German, illustrated the breathtaking quality of Stanhope's claims: "We have often heard him assert, that with his reasoning-machine, he shall be able on all subjects, to draw true conclusions from any given premises; . . . he shall be able to ascend, by regular steps, from the first definitions to the higher and most sublime speculations of Newton."[98] Having more or less perfected his calculating machine, Stanhope produced an array of simple machines to help display logical relationships and assist in making reasoning mechanical (figure 5.21).[99]

The potential implications of mechanizing reason worried a few readers. A letter in the *Christian Observer*, a mouthpiece for the abolitionist Anglican evangelical Clapham Sect, explained: "We live, Mr. Editor, in an age of astonishing improvement; and I should not wonder if by establishing, in the first place, those doctrines of materialism which prove the soul to have all the properties of body; and then, by giving the material soul the proper impetus through the means of mechanical instruments, we should learn to dispense with the services of the clergy, and to spare the necessity of all moral exertion."[100] The letter concluded that exertion of thinking too was no longer needed: "Nor should I be much surprised if we were at length to arrive at the happy point of being able, through the largeness of that stock of 'machines for reasoning by,' . . . to exempt our posterity from the burthen of thinking upon every topic; . . . it is by the progressive and indefinite substitution of the operations of mechanics in the place of the exertions of mind, that we shall exemplify the modern doctrine of man's perfectibility."[101] In drafts of his never-completed logic text, the earl neither shied away from the implications for human reasoning of his claim nor retreated into a domain of genius separate from simple scientific discovery: "The \Mechanical Instrument/ Machine above described for drawing Consequences, will shew to Man in what consists his boasted faculties! . . . For what is the Man of Common Sense, who will not feel Humility, when he reflects, that Pieces of Metal and Glass can be so arranged, as to produce Conclusions in Logic with (at least) as much accuracy, as the ablest of the Human Race."[102] This self-reflection will occasion an affective transformation: "May he feel also, that Men are brethren; and that his utmost ability ought to be exerted to extend their Happiness. . . . With these deep Impressions on his Mind, he will become anxious to perform, that sacred Duty, toward his FELLOW MEN. With a virtuous Zeal, enlightened by Reason, he will courageously, and steadily, pursue that Object. He will pursue it, with unremitting Perseverance."[103]

11111   Fifth Carry

Babbage's Collaborators Emulate

Leonardo imagined the first one.
The next was a pole lathe with a drive cord,
illustrated in Plumier's *L'art de tourner en perfection*.
Then Ramsden, Vauconson, the great Maudslay,
his student Roberts, Fox, Clement, Whitworth.
[. . .]
I listen to the clunk-and-slide of the milling machine,
Maudsley's art of clarity and precision: sculpture of poppet,
saddle, jack screw, pawl, cone-pulley,
the fit and mesh of gears, tooth in groove like interlaced fingers.
I think of Mozart folding and unfolding his napkin
as the notes sound in his head.

B. H. FAIRCHILD, "The Art of the Lathe"

In 1831, the central organ of emulation in England, the *Transactions of the Society for the Encouragement of Arts, Manufactures, and Commerce*, published a lengthy description of a new metal planing lathe. The publication involved two advances. The first came in the content: Joseph Clement's new means for making metal planes. The second came in the form of the artist Cornelius Varley's pictorial depiction of these machine tools, one "fully adequate to the complete illustration of this beautiful machine."[1] Clement was the key engineer in the ultimately failed effort to produce Babbage's Difference Engine; Varley was a draftsman, lens-maker, and later a famous watercolorist who served as an apprentice to his uncle, Stanhope's collaborator Samuel Varley, and published several of Stanhope's and Varley's improvements in machinery and other processes in the same journal.[2] Publication in the journal celebrated Clement while disclosing his innovation for others to improve upon. In the spirit of emulation, the lathe was neither kept as a trade secret nor patented; he relied "for protection mainly on his own and his workmen's skill in using it."[3] The knock-on effects of emulation intersected in this publication of the lathe—a lathe that contributed greatly to Clement's celebrity and success as a maker of precision tools. Planing lathes and the new technologies of depiction helped make the severing of design from execution into more than a polemical dream of philosophers. Such tools of materialization helped make it a reality that would reduce the need for practitioners like Clement and Varley—nonmechanical emulators who produced the tools of more mechanical imitation.

The long processes of tool and depiction making at once deferred the production of Babbage's engine and were an essential material condition of the possibility for its physical becoming. There is no better testimony to the quality of the work of Clement's shop. The still functioning Difference Engine 1 demonstration piece in the Science Museum in London stands as testimony to the quality of the work of Clement's shop and serves today as a reference point for understanding achievements in machining by the 1830s (see figure C1.1).

Clement's Workshop in Action

On February 12, 1828, Herschel wrote to Babbage, then in Italy recovering from the deaths of his wife and father, about the state of the work on the Difference Engine: Clement "tells me he keeps 10 men constantly at work on the Engine. I <u>saw</u> 7 there in the workshop and <u>one</u> man working on the drawings when I last called. He seems to have been chiefly at work on the drawings."[4] Herschel's notes and letters let us glimpse Clement's workshop at the moment it was mostly dedicated to the production of the paper and metal tools necessary before the full-scale effort to produce the parts of the engines commenced. The letters make vivid the labor entailed in the interface between design and implementation.

Clement's early success came as a superior draftsman and then maker of improved tools. In 1822, before Babbage engaged him, Clement was known as "that ingenious Mechanical Draughtsman."[5] His biographer notes, of some of his drawings for the Society of Arts, "He reached a degree of truth in mechanical perspective which has never been surpassed."[6] When Babbage turned to him to superintend the project, Clement had a reputation as someone to whom inventors might turn to see if their ideas could be realized. Speaking at the foundation of the Institution of Civil Engineers in 1818, Henry Palmer argued, "For the Engineer, being a mediator betwixt the Philosopher and the working Mechanic, must, like an interpreter, betwixt two foreigners, understand the language of both."[7]

In the letters and notes, Herschel detailed, all the while admitting his incompetence, the changes to the drawings that Clement brought to his attention. An important "Great Drawing" had "a vertical of the whole on the large drawing board, all the bottoms of the axes are drawn in."[8] In his next letter, in addition to numerous smaller drawings, Herschel describes a depiction of the engine machine, an often reworked, "large new Drawing" that offered a "general plan about 4 feet by 1½."[9] Herschel's descriptions demonstrate that Clement and his assistants worked on drawings of smaller contrivances, paper patterns, and sample parts in conjunction to and with an ever-evolving plan and elevation of the whole (see figure C5.1).

Working drawings, once they had reached a certain level of development, were elaborated on the general plan and general vertical drawing of the engine. As sites for reasoned improvisation, the drawings were a means for thinking through and working up the architectonics of the machine. For example, Clement and his assistants had evidentially been working on the apparatus that controlled how many units were being added, including the carry mechanism. They did so in miniature and in the overall plan and elevation: "Another New Drawing 2 feet by 3 is filled with details of a working drawing. It contains a plan, elevation and end view of the manner of making the 'spiral axis' apparatus pump by means of an excentric, so as to move the 'drop-pin' apparatus." These working drawings developing the spiral axis were put into plan, elevation, and view. "In the large new Elevation drawing on the Great Board," Herschel explained, "is inserted all the elevation of the work of which the plan is the last mentioned drawing."[10] In this process, preexisting ideas in the mind's eye were not simply projected in drawing; sketching as well as drawing in highly formal ways came together to produce more precise ideas of the parts of the machine in relationship to the whole. Through the process of elaboration on paper, the design of a machine capable of materialization emerged.

The work of drawing took much time—too much for Herschel's taste. Conversion from drawings into metal proceeded slowly. Elsewhere in the shop, patterns were being made "for a good number of wheels but the preparation of tools [was] still the main business."[11] Herschel distinguished all this preparation from the "working department" of actually making parts.

At the end of 1828, the drawings covered "about 400 square feet."[12] As detailed above, the 1828 report of the Royal Society that encouraged continued government support for the engine focused less on Babbage's big ideas of the machine than on the detail of care in the process of moving the machine toward practical implementation. The Treasury had asked the Royal Society about the "adequacy" of Babbage's progress.[13] "Adequacy" was a standard

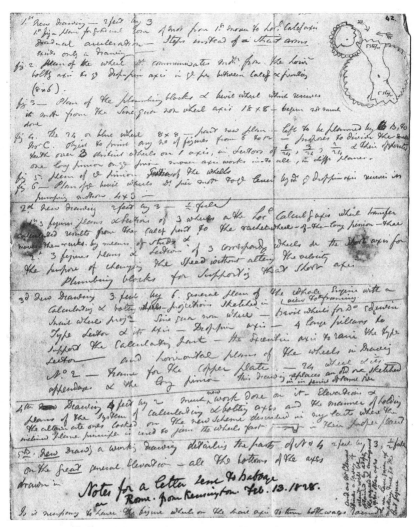

FIGURE C5.1. J. W. F. Herschel, Notes for a letter sent to Babbage, Feb. 13, 1828. Royal Society HS 27.42. © Royal Society.

legal term for judging the sufficiency of disclosure in patent applications. Mere ideas of machines, the province of the projector, were inadequate—they needed more detailed specifications, the province of the mechanic and the engineer, makers of adequacy. The proof of Babbage's progress in 1828 was precisely the working stuff—the drawings—of Clement's shop, which demonstrated "all the skill and system which the most experienced workmanship could suggest."[14] If in his private correspondence Herschel refrained from considering the drawing as "work," the report of the Society noted, "The

FIGURE C5.2. "Joseph Clement's Drawing Board of Large Area," *Transactions of the Society for the Encouragement of Arts, Manufactures, and Commerce* 43 (1825). (Courtesy of Columbia University Library.)

drawings form a large and most essential part of the work: they are executed with extraordinary care and precision, and may be regarded as among the best that have ever been constructed."[15] Not too much time had been wasted on materialized trial and error. The project, then, deserved funding precisely because Babbage had engaged someone who moved it toward the possibility of implementation. The project deserved continuing support because Babbage had engaged the social process of working together with someone who could do far more than merely implement his crude macrolevel ideas mechanically.

Making the drawings for the engine led Clement to devise new technologies to facilitate the large-scale draftsmanship. The workshop had a novel, very large drawing table, worth some fifty pounds, according to Clement.[16] He described his table in a publication: "Having been engaged in making some large mechanical drawings . . . their dimensions being seven by six, I found it impossible to execute them in the usual way; necessity obliged me to contrive some plan of manoeuvring the board so as to get to the middle of it without getting upon it"[17] (figure C5.2).

Babbage and Herschel complained about the apparent desultory quality of Clement's workshop. Babbage, however impatient, knew better than to desire that the workshop start making parts before the drawings and tools were ready. In his account of the Exposition of 1851, Babbage wrote, "The successful construction of all machinery depends on the perfections of the tools employed, and whoever is a master in the art of tool making possesses the key to the construction of all machines. . . . The contrivance and the construction of tools, must therefore ever stand at the head of industrial arts."[18]

Emulation and Precision Toolmaking

In his preparatory notes for the second Royal Society committee to examine the Difference Engine, Herschel commented, "A large number of new tools and admirable ~~contrivances~~ \methods/ for executing work have with perfect precision have been contrived and made."[19] Clement's workshop was among the most important midcentury sites for the creation of machine tools and training of machine tool makers.[20] Clement's employees went on to emulate his practice, further developing his tools and concerns in new directions central to the development of precision engineering in Britain. In 1834, a former employee of Clement wrote to Babbage: "Much talk about the Machine here so much so that A Man who has worked at it has a greater chance of the best work and I am proud to say I am getting more wages than any other workman in the Factory."[21] He was employed by Joseph Whitworth, subsequently a major improver of metal tools of the century: "Mr Whitworth was possessed of a special aptitude for that minute accuracy of detail in mechanical work which necessarily must have been a marked characteristic of the skilled workmen engaged on Babbage's machine, and the experience which he gained in Clement's workshop, Mr Whitworth in after life certainly made the most of."[22] Whitworth introduced a uniform system of screw threads that came into general use by the 1860s.[23] A much-celebrated Victorian hero and maker of armaments, Whitworth in time was made a Baronet and a member of the Royal Society and was given an honorary doctorate by Oxford.

Babbage collected testimonies about the great impact of his effort on machine tools and their makers, as Bruce Collier has shown. The engineer James Nasmyth explained in 1855 that he had "no hesitation" in affirming "that the admirable contrivances and tools which the late Mr. Joseph Clement in conjunction with yourself designed and constructed . . . furnished such ideas to the mechanists of the world as gave an impulse towards the perfect [*sic*] in mechanism, such as has had no small share in bringing about the wonders which modern machinery has enabled us to realize."[24] Clement and Babbage had spurred modern tool making and thus modern industry. The Earl of Rosse, president of the Royal Society, wrote to Lord Derby, then head of the government, to comment on the value of the British state's investment in Babbage's Difference Engine: "This Country has received an equivalent many times over for the expenditure on the Calculating engine, in the improvements in tools and machinery directly traceable to the attempt to make it." This transformation would make the production of machines, so expensive in the 1820s, far more precise, expedient, and inexpensive.[25]

From Clement to Jarvis

On November 24, 1835, Babbage's housekeeper witnessed a contract of employment between Babbage and Charles Jarvis, a former assistant of Clement's. Jarvis assumed the position of chief draftsman and superintendent of the difference and analytical engine projects. The two parties agreed to a clause for noncompliance: "And whereas the said Charles Babbage has explained to the said Charles G. Jarvis certain inventions which could not without much trouble and loss of time be communicated to another person." If either Babbage or Jarvis "shall neglect or refuse to fulfil and perform the agreement on his part herein contained, the party so neglecting or refusing shall pay to the other party, the sum of Five hundred Pounds Sterling by way of liquidated damages."[26] Contractually dedicated only to the engine, Jarvis was to be richly compensated £1 1s. a day for his part in the workmanship of risk and to keep him from taking up employment on the continent.

In earlier letters and conversations, Jarvis had set out the individual competencies needed for designing and building the engine: "Although a man who does not know enough of the Science of Mechanics to ascertain the magnitude and direction of a force resulting from two other forces whose directions are different may be a very good tool-maker," Jarvis noted, "we call upon him for too much when we expect him qualified to arrange and construct a new and complicated machine."[27] Unquestionably, he cast Clement as a mere toolmaker. Jarvis had the qualities needed, and such people deserved better pay: "Such an assistant will not be content with the salary of a mere mechanic."[28]

Jarvis was adamant about the social, spatial, and temporal organization necessary for the effort. The physical distance between Clement and Babbage had slowed decision making and collaboration—which led to the failed effort to move Clement to Babbage's premises. Jarvis outlined an "alteration" in the organization of the work on the machine that would more fully integrate design and implementation: "The designs and drawings be all made on your premises and under your immediate inspection: working drawings made from them, and the work ordered and directions given in writing to various persons to construct various parts of the machine which parts might be going on at the same time and the entire machine be speedily completed."[29] Bringing the machine quickly to completion required a better-controlled space with better organization of time and clear lines of authority, as Simon Schaffer has argued.[30] Jarvis described an improved process for contending with problems involved in implementing designed pieces: "Whenever any difficulty occurred you might be at once appealed to whenever it was found very difficult to produce a given effect, but comparatively easy to produce <u>nearly</u> that effect."[31] The design and implementation of the project would be brought together in a physical space with a clear and expeditious procedure for making judgment calls about redesign and adjustment entailed in the course of production. Transformations in design would be seamlessly attendant upon materialization. Babbage did not need someone who merely implemented his designs; he needed someone who could communicate the pushback of material in implementation. Jarvis claimed to be that person, and Babbage paid him a guinea a day to be that person. Together they produced hundreds of careful drawings of analytical and Difference Engines, but never did they produce a working version of either engine.

Difference Engine 2 Vindicated

The first fully error-free calculation of a full-size Difference Engine was completed in 1991, a month shy of Babbage's two hundredth birthday, at the Science Museum in London. The effort to build Difference Engine 2 rested on the remarkable set of high-quality drawings produced by Jarvis and Babbage in the final attempt to rethink and simplify the engine. Although built using modern machinery, the parts were produced to tolerances within those achievable in Clement's workshop and using metals closely approximating those such as the gunmetal Clement had used. The high quality and clarity of the drawings did not mean that the late twentieth-century builders could proceed without constant judgment calls: "Building the engine was not the slavish physical implementation of an abstract ideal with doom attending even the smallest deviation from the master's instructions."[32] In picking up the project, the reconstructors were confronted with a fundamental flaw in the design for the carry mechanism that would have made the machine jam and would have left it little more than a piece of "Victorian sculpture."[33] The curators followed what they took to be Babbage's mechanical approach in tweaking his design: "Had Babbage proceeded from design to manufacture, the deficiencies in the drawings would have become evident. There would have been no way of avoiding them, and he would have sought solutions. We were simply picking up where he left off, and the task for us was to find solutions with conscientious regard for contemporary practice while using the best knowledge available of Babbage's style of working."[34] The reconstructive effort "vindicates" Babbage—but what precisely is vindicated?[35] Not simply the design as given, for it is faulty; not just some underlying essence or spirit of the machine, for many of the detailed mechanical contrivances are rightly celebrated; not even his design plus some general sense of his approach to machine design, for that misses the play of implementation and redesign. Rather, what's vindicated must be something like Babbage's overall design with its architectonic principles as well as its strategy in contrivance, combined with the entire package of processes of transformation of design in the process of implementation. In other words, what's vindicated is his design only when coupled with the process of working with engineers and mechanics capable of working in nonmechanical ways, just like the team at the Science Museum. What's vindicated is the hypothetical Babbage who got to work with Jarvis in creating the machines they had drawn up collectively. The curators, in some sense, recreated something like Jarvis's reorganization of the workshop under Babbage's quotidian supervision. The curators recreated the possibility of implementation as a free activity subject to the constraints of an underspecified design. The curators emulated.

Swedes Emulating Babbage

The London team was not, however, the first to emulate a difference engine. The completion of the first prototype difference engine capable of automatically computing tables and setting type took place in 1843, and 1855 saw the first sale of a difference engine. The machines came from Sweden, not England, from a classic nineteenth-century emulator and his son, not a wealthy gentleman. Lardner's written description of Babbage's machine inspired the printer, translator of Shakespeare, and journalist Georg Scheutz to attempt to build an engine, despite his ignorance of the technical detail.[36] Scheutz was known for his liberal views and his proclivity for improving a variety of inventions. His son, the engineer Edvard Scheutz, largely responsible for the prototype, studied at the Royal Technical Institute. Their sequence of difference engines, first a wooden-framed prototype and later fine finished versions in metal, emerged from two different ways that engineering practice and natural philosophy converged in the early nineteenth century. The wooden prototype came from Edvard's training at a technical institute that sought to teach students "to acquire and communicate skills and knowledge which are necessary for proficiency in the exercise of handicrafts" or "a trade or industrial occupation."[37] The finished metal version came from workshop owner Johan Wilhelm Berström, another nineteenth-century Vayringe. Starting as a glassmaker's apprentice, he studied physics and chemistry with "the help of eminent professors," was sent by the king of Sweden to study workshops in Europe, and became a noted purveyor of daguerreotypes.[38] A grant from the Swedish king, obtained with pressure from the Swedish parliament, financed the reduction to practice of the wooden prototype into the high-quality, precision-manufactured machine. Knowledge of Babbage's desultory process led the monarch to impose a strict deadline, and upon satisfactory completion of the machine in 1853, the king deemed it to be "of special benefit to science and an honor to our native land."[39] Benjamin Gould at the Dudley Observatory in Albany, New York, purchased the machine; it now belongs to the Smithsonian Institution.[40] The British government purchased another, though a larger market never emerged. When the Scheutzes came to England to sell their machines, Babbage energetically supported them, regardless of his fiery early talk of property, and helped them publicize their machine. They dedicated the first printed examples in English of "tables calculated, stereomoulded and printed by machinery" to Babbage.[41] For all their efforts to implement the engine, the Scheutzes treated Babbage as the inventor of the machine, and Babbage was pleased to be vindicated as an intellectual inventor, able to act free from pecuniary interest.

Freedom and Mechanics

Unlike mechanics, philosophical inventors were supposed to be free of pecuniary interest in their projects. Babbage needed state support precisely to allow him to become disinterested. In setting out his vision of a reorganization of the workplace, Jarvis articulated the interests that divided an inventor and his mechanic:

> It must be very agreeable to construct a newly invented machine the cost of the parts of which cannot be taxed; and still more agreeable to be able to charge for time expended upon arranging the parts of that machine without the possibility of the useful employment of that time being disputed, and to doze over the construction year after year for the purpose of making one thing after another . . . ; but it is not to the interest of the inventor that this should be suffered. . . . [H]is interest is any thing but promoted when his invention is placed at the disposal of persons who do not even know the names of purposes of the several parts without appealing to better heads than their own.[42]

Clement, in this view, was systematically self-interested never to complete the machine in a timely fashion—but also incapable of grasping the machine as a whole. In proposing his reconfiguration of the work, Jarvis recognized that his views might be dismissed as self-interested.[43] Precisely because he was not interested, Jarvis could fairly gauge the workshop and offer Babbage advice. Freedom from interest in reflection could help create the conditions of the engineer's freedom necessary for creative and disciplined implementation. Clement and Jarvis were expensive because they were highly skilled in the free imitation of the master engineer, heirs to the imitative freedom of the master of the workshop. They were craftsmen of risk, not of certainty. This freedom, central to emulation, required no illusory liberation from preexistent models—the vegetable model of authorship—but rather the constant creative use and improvement of models and contrivances. Philosophical aesthetics had badly eroded the freedom accorded to the mechanical artist badly in the late eighteenth and nineteenth centuries. Countless treatises from Kant onward told a similar tale: "The principle of trade, and the principle of the arts, are not only dissimilar but incompatible. Profit is the impelling power of the one—praise, of the other."[44] Profit precludes genuine freedom. In the late Enlightenment and Romantic account of genius, the true artist was said to be doubly free: free from external causes, including those of self-interest, and free to create with no slavish imitation.[45] The tradesman was doubly enslaved, ruled by self-interest and caught in imitation. This division, of lasting cultural significance, obliterated the hierarchy of artists with different levels of freedom in interpretation that had long figured within artisanal practice and reward systems. Mechanical models of mental activity, including calculation, worked to narrow the range of potentially free activity. New mechanical contrivances—including calculating machines—helped erode the space of human creative freedom. Freedom came increasingly to be understood as comprising precisely those activities beyond the merely mechanical. This legacy undergirds much of the history of technology.[46] Caught between deterministic models of unfree craftspeople and exaltations of heroic geniuses, the history of technology has long struggled to account for the freedom of the inventor in building upon antecedents.

Calculating Machines, Creativity, and Humility from Leibniz to Turing

As soon as someone gets a computer to do it, people say: "That's not what we meant by intelligence." People subconsciously are trying to preserve for themselves some special role in the universe.

MICHAEL KEARNS, 2004[1]

In 1844, the satirical periodical *Punch* carried a series of testimonials about the power of one J. Babbage's "New Patent Mechanical Novel Writer" (figure 6.1). An E. L. Bulwer of Lytton, Bart., proclaimed himself "much pleased with Mr. Babbage's Patent Novel-Writer, which produces capital situations, ornate descriptions, a good tone, sufficiently unexceptionable ties, and a fund of excellent, yet accommodating morality." While ridiculing Charles Babbage and his Difference Engine, the parody lambasted the mechanical quality of the production of popular fiction as a product of the unthinking agglutination of existing bits. For his part, "LORD WILLIAM LENNOX, *Author of Waverley*" expressed his "pleasure in stating that he finds the operation of the Patent Novel-Writer considerably more expeditious than the laborious system of cutting by hand. Lord W. has now nothing more to do than to throw in some dozen of the most popular works of the day, and in a comparatively short pace of time draw forth a spick-and-span new and original Novel." His lordship felt machines could do yet more: "Lord W. would suggest the preparation, on a similar plan, of a Patent Thinker, to suggest ideas; in which he finds himself singularly deficient."[2]

Debates about the originality of machines were in the news.[3] In 1843, Babbage's collaborator, Ada Lovelace, publically denied the possibility that machinery might be original: "The Analytical Engine has no pretensions whatever to *originate* anything. It can do whatever we *know how to order it* to perform."[4] A century later, in his epochal "Computing Machinery and Intelligence," Alan Turing labeled this claim "Lady Lovelace's Objection" to the possibility of a machine fully capable of creative thinking. Turing avowed that Lovelace and Babbage misunderstood the implications of what they had

FIGURE 6.1. "The New Patent Novel Writer," *Punch*, 1844, p. 268. (Courtesy of Rare Books and Manuscript Library, Columbia University.)

wrought: "The Analytical Engine was a universal digital computer, so that, if its storage capacity and speed were adequate, it could by suitable programming be made to mimic the machine in question. Probably this argument did not occur to the Countess or to Babbage." Turing noted that a variant of this claim "states that a machine can 'never do anything new.'" His response was a brief, but telling, knife twisted in the hubris of originality: "Who can be certain that 'original work' that he has done was not simply the growth of the seed planted in him by teaching or the effect of following well-known general principles."[5] More than overrated, originality might be a figment of human pride. Like much, if not all, reasoning, invention might in fact be mechanical, something entirely understandable within a materialist psychology. In expanding on her claim, Lovelace had explained that the machine "can *follow* analysis; but it has no power of *anticipating* any analytical relations or truths.

Its province is to assist us in making *available* what we are already acquainted with."[6] At most, machines aid in the putting together of parts; their province is entirely dispositional, not creative. Like the "Patent Novel Writer," they can produce congeries, not intellectual unities. Turing's point was that this very act of combining parts could generate those things we perceive as original.

The early nineteenth century saw a hardening of the divide between original creation and creation as combining or agglutination; reinforcing and protecting this divide required minimizing the salience of earlier influences for original cultural production. The Patent Novel Writer epitomized the absence of originality. The earlier theory and practice of emulation, discussed in the two previous chapters, offered a vision of creativity made possible through the imitation and improvement of completed things. Higher forms of imitation, practiced by master craftsman, whether goldsmiths or artists, were anything but machinelike: they were precisely practices of making that could not be automated. In accounts of technological change as well as aesthetics in the nineteenth century, the protection of novelty from the dangers of external machinelike causation became paramount. Denying absolute originality has long been central to the professional history of technology, and yet denying the salience of imitation to the creative process has long been crucial for attributing credit to many of the central actors figuring in the history of technology. Accounts of invention contain—usually implicitly—accounts of human mind.[7]

From the seventeenth to the nineteenth century, calculating machines provided grounds for reflecting upon the originality thought to be distinctive to human beings. The distinction between origination and execution had long been central to debates around the sufficiency of mechanism for explaining the phenomena of the world. Nowhere is this clearer than in the tensions within aesthetic theory provoked by materialist psychologies of invention and creation. Policing the distinction between origination and execution could protect the creative potential of the creator and isolate originators from mere executors. Policing this distinction could protect the genuine creative abilities of a divine creator and perhaps a few geniuses in imposing form onto nature and art.

Makers of calculating machines who opined on their philosophical significance saw themselves as contributing less toward an atheist, monist materialism than toward a refined conception of human beings, the nature of the material world, and their distance from the creator; they were all theists, though typically unusual theists by the standards of their time. Defending the creative capacities of a divine creator was central in debates in early-modern natural philosophy and mattered to creators of calculating machines.

Material Minds, Perhaps; Calculating Ones, Not So Much

Around 1800, Charles Mahon, the third Earl Stanhope, noted in the manuscript of his logic textbook, "Those sublime discoveries of <u>Newton which</u> illustrate the Name of one of the greatest Men who ever lived, are found, as of necessity, and without mental Effort, by a <u>mechanical operation</u>, most easily performed in my <u>Analytical Machine</u>."[8] Not simply claiming that the machine could reproduce the discoveries of Newton, Stanhope claimed his machine did, in fact, reason: "I have, therefore, proved, \beyond all dispute,/ that a <u>Machine</u> can be made <u>to reason</u>; inasmuch as the specifick Machine above-described does actually perform the Argumentative Operations." The machine could synthetize and analyze. To discover or invent, according to Stanhope, was to analyze: "The Method of <u>Analysis</u> is properly termed the Method of <u>Invention</u>. Therefore, Invention, upon whatever subject, is a mere Mechanical Operation."[9] A mechanized procedure does more than replicate aspects of reasoning, Stanhope argued; that mechanism actually reasons. The ways the machine discovers shows how human beings do so. Invention requires no genius—it is a mechanical procedure capable of automation, a procedure whose mechanical nature has long been obscured. A century and a half before the general-purpose digital computer, Stanhope connected logic to calculating machinery and used the capacities of his machine to make strong claims about human intellect.[10]

Whatever our views on the merits of reducing mental activity to forms of computation, Stanhope's arguments hardly surprise us. Conceiving of the mind as material leads naturally to consideration about whether machines are capable of the operations of reason and whether our reason is in fact algorithmic in a way susceptible to mechanization; conversely, the success of calculating machines in imitating reasonable activity lends considerable weight to the possibility that the mind could be material. Nothing was natural or even compelling about such inferences in the eighteenth century, even among those willing to conceive of the mind as potentially comprising only organized matter.

Debates about whether matter could think punctuated the eighteenth century, so we might expect to find calculating machines regularly invoked either to illustrate the limits of mechanical reasoning or to demonstrate its wide scope and to help motivate the possibility of reducing thinking to mechanism. That expectation is wrong.[11] Barring a handful of references, calculating machines do not appear in eighteenth-century discussions, in radical materialist treatises, in attacks on materialism, or much in between. Calculating machines do not figure regularly in famous philosophical texts or in obscure periodical publications. Even the most radical materialist treatises do not invoke the calculating machine to strengthen their case. The materialist tract *Le Philosophe*, for example, defines the *philosophe* as just a thinking machine capable of self-reflection: "The *philosophe* is a human machine just like other human beings; but it is a machine that, by its mechanical constitution, reflects upon its own movements."[12] The radical quality here is the claim that self-consciousness could be material. Many nonorthodox Christians shared a commitment to the material nature of the human body with freethinkers and atheists.[13] Opponents and Whiggish historians alike have tended to lump them all together as godless Spinozists. Neither atheists nor radical theists, however, focused on calculating machines.

For all their venom, the well-known vibrant debates about the possibility of thinking matter rarely drew upon claims that machines could perform aspects of reasoning to bolster an argument for a materialized mind.[14] The case of Stanhope suggests that such inferences were *possible* within the major strictures of eighteenth-century thought about logic, material minds, and mechanical devices. Such inferences by and large simply were not made. While more examples certainly can and likely will be found, it is nonetheless the case that the obvious materialists and their opponents did not generally draw inferences about human minds based on the qualities of calculating machines.

Calculating machines were *not* particularly good to think with in the eighteenth century to make sense of human reasoning. Several central facets of eighteenth-century philosophy mitigated against such use.[15] First, formal and symbolic reasoning notoriously found few strong advocates outside Germanic universities. Historians of logic, with some reason, bemoan the neglect of their subject in the eighteenth century. Rather than being seen as the fundamental structure underlying thought or an essential normative framework for proper thinking, formal logic was typically portrayed as an artificial propaedeutic to thought, something quite dangerous. For philosophers from Locke to Hume to Condillac, all following in the footsteps of Descartes, formal logic was antithetical to good reasoning. Theirs was an age committed to the grasping of ideas and their connections, not to assenting to formalism. Even if machines *could* perform arithmetical operations roughly analogous to symbolic reasoning, such reasoning was not generally thought to capture much of human reasoning and its creative potential. Scholastic Aristotelians and Leibnizian-Wolffians, who remained much invested in the importance of logic, had no truck with visions of thinking matter by virtue of their commitment to substantial forms.

Second, most major eighteenth-century versions of materialism involved some addition of forces and powers, if not of outright sensibilities, to brute matter.[16] By midcentury, the fundamental model for the materialistically inclined was more vibrating strings or sensitive oscillating fibers than material particles in motion, analogical to machines in metal. Only if the forces, powers, and affects of matter could be reduced to matter-in-motion would mechanical calculating machines have the salience they briefly did in the seventeenth century—or our own. Such mechanical reductions of gravity and forces belonged, however, to precisely the epistemologically suspect Cartesian hypothesis building that Newton and his followers had inveighed most strongly against.

Materialist and quasi-materialist accounts of mind proliferated in the eighteenth century—but they were based neither in formal logic or arithmetic nor merely in brute matter analogical to metal calculating machines. Physiological theories of mental activity, such as those of David Hartley and Charles Bonnet, grounded a nonformal, explicitly associationist account of mental activity in a neurophysiology of sensitive vibrating fibers.[17] These radical accounts suggested matter *could* account for the mind. Any machine capable of emulating such thought would necessarily be less a metal, calculating automaton than a bundle of sensitive, interlinked, vibrating strings.

Calculating machines were simply the wrong sort of matter to think with. Calculation was seen as but a small, albeit important, part of human reasoning. Thomas Hobbes claimed, "By Ratiocination, I mean computation."[18] This infamous reducing of thinking to reckoning found precious few followers. When someone tried, at the beginning of the nineteenth century, the reaction was telling.

Mathematics and Mechanization

The first known use of "calculating machine" as a term of opprobrium appears in 1803, as part of a castigation of an effort to reduce thinking to an arithmetical logic. The post-Kantian philosopher Jacob Friedrich Fries denounced a reduction of thinking to calculation in a major new logical text: "The first paragraph . . . states, *whoever calculates, thinks,* and the entire work following contains only the erroneous inversion of this claim, whereby, *whoever thinks, calculates.* Whereby it becomes understandable, that, then, when one wants to begin at the axiom, all thinking is calculating, all understanding and reason in earth as in heaven becomes an endless repetition of a calculating machine in a calculating machine."[19] Fries's target in this paragraph was the philosopher and sometime teacher of Hegel, Christoph Gottfried Bardili. In the remarkable passage that scandalized Fries, Bardili cast thought as a form of refined calculation: "Whoever calculates, thinks. But he thinks, without describing anything other than his thinking itself in thinking. At first he describes his thinking through written calculation in an object outside himself."[20] For Bardili, what we can know about the world includes, most fundamentally and significantly, those truths known through logic, where logic is understood along computational lines. Neither a collapse of thinking to logic nor a collapse of logic to calculation was congenial to most contemporaries. The luminaries of early nineteenth-century German philosophy—Hegel, Fichte, Schelling, and Fries—all attacked Bardili vehemently, as did the editor of some of Kant's logical lectures.[21] Fries was insightful in pointing to a calculating machine: Bardili had articulated a conception of reasoning in which the most significant part of mental activity could be understood as calculation. A calculating machine could, in this view, know nearly all that was to be known. And that was clearly quite wrong. Not even good mathematics, after all, was mechanical.

Mathematics Is Not Calculation

For many eighteenth-century writers, the more mathematics seemed primarily a form of symbolic manipulation, the more suspect it was as an authoritative source of knowledge about the world and as a form of mental discipline. The great successes of clever calculation using the new tools of algebraic analysis in rational mechanics induced profound ambivalence. The natural philosopher and aphorist Georg Christoph Lichtenberg claimed that Leonhard Euler was "one of the greatest mathematicians that ever lived; and certainly the greatest calculator, that has ever lived; but he was no physicist." "Calculator" was damning praise. Without turning to the laboratory, Euler worked "to develop through reckoning a bunch of statements." Newton, in sharp contrast, "was the greatest mathematician and greatest physicist at once. The two were one."[22] Lichtenberg's distrust of the overreliance upon mathematical formalisms and calculation was common. The potentially automatic, mechanical quality of the calculation and algorithmic processes profoundly disturbed all those committed to a conception of knowledge as the possession of ideas. The automatic quality of algorithms seemed likewise to negate the self-possession, the liberality, of the mathematician: to be a calculator, after all, was to be a rule-following mechanic. In his *éloge* upon Euler's death, Condorcet noted that the Swiss mathematician "seemed sometimes to occupy himself only in the pleasure of calculating and regarded the point of the mechanics or physics that he was examining as an occasion only to exercise his genius and to deliver himself to his dominant passion."[23] Far from a self-conscious engagement of reason, calculation could lend itself to an impassioned move away from a free inquiry after truth. The many fruits of algebraic analysis—the heir of the calculus—showed calculation to be a powerful tool, but one in need of careful surveillance.

Throughout the eighteenth century, an insistence that real mathematical knowledge comprises ideas clashed with the algorithmic and symbolic fecundity of contemporaneous mathematical practice.[24] Numerous writers raised the possibility that mathematics might be merely the manipulation of signs, that it might merely be something mechanical. Such a position suggested that mathematics might not contribute in a genuine way to knowledge worthy of the name. A century earlier, Hobbes had deemed the arbitrary nature of the axioms and principles of mathematics as precisely the source of its certainty. In the eighteenth century, critics of mathematics used such an account to diminish it. The natural historian, and mathematician manqué, Buffon articulated the position that mathematics was but tautologies based on arbitrary suppositions: "Since definitions are the only principles on which the whole is built, and given that they are arbitrary and relative, all the consequences are equally arbitrary and relative."[25] Denis Diderot went further, to denounce such mathematics as metaphysics in the pejorative midcentury sense: "Games and mathematics are much alike. . . . [T]he object of a mathematician has no more existence in nature than that of a player. Both are simply conventions."[26] If reason was often considered as epitomized by mathematics, the automatic quality of calculation suggested higher reason might be unnecessary: "Is investigation an art so mechanical, that it may be conducted by certain manual operations? or is truth so easily discovered, that intelligence is not necessary to give success to our researches?"[27] The argument was most popular among opponents of mathematics as a model for reasoning and as a higher discipline. The rejection of a formalist view of mathematics was understood as essential to rebutting claims that mathematics was a mechanical process, fruitless, tautological, and devoid of significance.

While attacking the new, infinitesimal calculus, George Berkeley took aim at its lack of "evidence."[28] Traditional mathematics, he explained, epitomized good reasoning by providing "a perpetual well-connected chain of consequences, the objects being still kept in view, and the attention ever fixed upon them."[29] In this view, good mathematics is both formally certain and "evidence" preserving, meaning roughly that underlying ideas and their connections remain ever present to the mind. At the very least, mathematics begins with evident principles and ends with evident conclusions, in which the connections among ideas are grasped as ideas and not merely assented to thanks to their form. In the broadest sense, evidence seems to have captured a widely shared sense that the formal nature of mathematical reasoning tracked a grasp of underlying ideational referents. Common to many followers of Locke, Leibniz, and Descartes alike, a preference for such an idea-focused account of mathematical knowledge undergirded the century-long animus against formalized reasoning in much of Western Europe. Mere assent to the form of argument was widely seen as inferior to seizing the connection of ideas, even among those sympathetic to the value of syllogistic and other sorts of formal reasoning.[30] This framework of evidence provided norms, often violated to be sure, for mathematical practice and reflection upon mathematics itself. In his first *Critique*, Kant argued that, through the concrete process of construction, the mathematician "arrives, by a chain of inferences but always guided by intuition, at a completely evident and at the same time universal solution of the question." In contrast, no such intuition can guide philosophical inferences and render them evident.[31] Evidence rested on an implicitly physicalist vision of mathematics in an uneasy alliance with the manipulation of symbolic systems of ever greater power. Admiration of the fecundity of formal and symbolic reasoning fit only with some difficulty with the view insisting that mathematics ultimately provided ideas and evidence. Symbolic reasoning clearly was fruitful, yet "the name of reasoning cannot be given to a process in which no idea is introduced."[32] Few philosophers maintained that all mathematical knowledge was evident, especially when symbolic reasoning was involved.

Berkeley, for example, argued that the new infinitesimal calculus worked as a sort of mechanical procedure, but not as a means for connecting ideas or as a general technique for teaching to reason and judge.[33] For Berkeley, the advances of the new mathematics undermined its evidence, if not always its certainty.[34] While granting some utility to symbolic calculations in the study of physical appearances, Berkeley and other critics of the new analysis took the absence of evidence to undermine the suitability of symbolic mathematics, on the one hand, for reasoning about fundamental physical and metaphysical truths and, on the other, as a training for the mind to reason more generally: "You may operate and compute and solve problems thereby, not only without an actual attention to, or an actual knowledge of, the grounds of that method, and the principles whereon it depends." In computing this way, "you may pass for an artist, computist, or analyst, yet you may not be justly esteemed a man of science and demonstration."[35] The calculus was no medicine for the liberal mind—it was a crutch for mechanics.

In his *Critique of Pure Reason*, Kant worked hard to insulate mathematics from the threats of automaticity. His challenging account of Euclidean geometry as comprising synthetic a priori truths served to show that geometry rested on a nonalgorithmic appeal to intuition. Practicing mathematics was not just rule-following, logic alone could not explain it, not all mathematical knowledge comprised tautologies, and mathematics was no more automatic than art. Crucially, mimicking the form of mathematics without the ability to appeal to intuition would get philosophy nowhere: "By means of his method the mathematician can build nothing in philosophy except houses of cards."[36] Critics of mathematics were not wrong about the dangers of formal reasoning: they were wrong to reduce mathematics to such reasoning, and they were likewise wrong to think mathematical reasoning could bring certainty to all domains.

For all his political radicalism, Charles Stanhope seemed a throwback to the bad old days of syllogistic logic. His remarkable claim that a machine could invent and could reason rested on the celebration of reasoning as the manipulation of signs: "Reasoning is Nothing more than an Arrangement of Signs. . . . the Arrangement of Signs is a mere Mechanical Operation." He concluded, happily, "Therefore, it follows, that Reasoning is Nothing more than a mere Mechanical Operation."[37] Stanhope celebrated this mechanical quality as the fundamental characteristic of mathematics to be pursued in other domains. Stanhope's views on the mechanical quality of reasoning, so rare for an Enlightenment savant, made far more possible his conviction that machines could reason.

Materializing the Mind in the Seventeenth Century

Commentators on Blaise Pascal's machines from the mid-seventeenth century marveled that they could perform operations previously considered an exclusive province of reason. His sister remarked, "This work was considered a new thing in nature, as it reduced into a machine a science that resides entirely in the mind, and having found the means to perform all the operations with perfect certainty, with no need for reasoning."[38] Pascal's machine illustrated that some parts of reasoning could be done mechanically—but no one claimed that Pascal's machine made it obvious that reason might be entirely susceptible to mechanization. He noted, "The arithmetical machine produces effects that approach thinking more than anything animals do. But it does nothing about which it could be said that it has a will, as is the case with animals."[39] Reflection upon calculating machines made material minds more thinkable, to be sure; just as often, such reflection delimited the scope of explanations possible within a purely mechanical philosophy.

In an undated manuscript, Gottfried Wilhelm Leibniz reflected upon some lessons his calculating machine could impart:

> The machine ... shows [*faire voir*] that the human mind can find the means of transplanting itself in such a way into inanimate matter that it gives to [matter] the power of doing more than it could have done by itself: to convince via the senses those who have difficulty conceiving ~~that~~ \how/ the Creator could house [*loger*] the appearance of a mind a little more generally in a body, however furnished with many organs; since even brass can receive the imitation of an operation of reason which concerns a particular or determinate truth, but [also] more difficult ones, especially as the Pythagoreans believed one can distinguish a human being from an animal ~~and to give the definition [of man] as an animal capable of using numbers~~ and to place as part of the definition [of man] the faculty of using numbers.[40]

Leibniz cast the calculating machine as a palpable intervention in the major early-modern debate about what sorts of causes were philosophically licit in explaining the creation, emergence, and ordinary phenomena of nature. The envisioned calculating machine provided tangible evidence of what matter in locomotion, acting entirely though efficient causality, could—and could not—achieve. A goal-achieving artifice, the machine illustrated that matter in motion, suitably well organized, could exhibit some mind-like behaviors. Complex organized phenomena required neither some immaterial soul or "plastic nature" directing the motion of matter nor the direct and particular concurrence of God himself. The machine likewise suggested that matter in motion could never be so well organized except through an initial, creative organizing activity of a rational mind, human or divine. The *appearance* of intelligence did not require the continuous activity of an intellective substance. The *creation* of the appearance of intelligence in matter, on the other hand, certainly did require the activity of an intellective substance.

Using the example of fetal development, Pierre Bayle, late in the seventeenth century, argued that explaining the phenomena of a complex being requires more than an intelligent cause creating and organizing its material form at the start. The production of activity over time requires continuous regulation by an intelligent cause: "In order for a little organized atom to become a chicken, a dog, a calf, etc., it is necessary that an intelligent cause should direct the movement of the matter that makes it grow." This cause behind the growth of fetus must have "an idea of the work and the means of making it larger, as an Architect has the idea of a building and of the means for making it larger."[41] Organized growth required, in this view, something capable of having an intentional cognitive grasp of a plan for the growth;

such a master plan for development could not be created in a material system organized to grow according to that plan by itself. Bayle, however, rejected that God himself was this regulator: "To present God as the entire reason in this investigation is not to philosophize," for it offers little of an explanation at all.[42] A proper philosophical investigation, then, required more than describing the efficient and ultimate causes of the phenomena.

Robert Boyle had argued a few years earlier in 1666–67 that the efficient causation of matter in motion could not *explain* the initial organization of matter, and yet we must have recourse to matter in locomotion alone in understanding the movements and development of organized things: "In this great Automaton the World, (as in a Watch or Clock,) the Materials it consists of, being left to themselves, could never at the first convene into so curious an Engine: and yet, when the skilful Artist has once made and set it a going, the Phaenomena it exhibits are to be accounted for by number, bignesse, proportion, shape, motion, (or endeavour,) rest, coaptation, and other Mechanical Affections of the Spring, Wheels, Pillars, and other parts it is made up of." Our difficulty in understanding how organized matter in motion could possibly produce certain effects should not lead to a precipitous turn to nonmechanical forms of explanation: "Those effects of such a Watch, that cannot this way be explicated, must, for ought I yet know, be confess'd, not to be sufficiently understood."[43] Appreciating the wisdom and skill of the divine artist requires, in this view, humbly accepting the limits of human understanding of divine mechanisms without committing the error of positing alternative, nonmechanical, causal explanations. The "World being once fram'd, and the course of Nature establish'd, the Naturalist, . . . has recourse to the first Cause but for its general and ordinary Support and Influence." Like Bayle, Boyle insisted that turning to the creator to directly explain the ordinary course of quotidian nature was to philosophize improperly.

In offering a tangible analogy for God's creative capacities in ordering matter, the calculating machine made a mechanical philosophy such as Boyle's far more plausible. Leibniz worried that appeals to the wrong sort of entities to explain everyday phenomena would undermine a proper philosophy of nature and appreciation of the divine. In a well-known letter to Damaris Masham, Leibniz argued that, since human beings can produce a machine that could calculate completely mechanically, who could reject that God could similarly produce organic machines capable of acting as if they possessed reason? "One can conceive that God has from the beginning given [matter] a structure appropriate to produce over time actions conforming to reason. And since our workmen, whose talents are so limited, can give examples of [such things] in certain situations [*rencontres*], by means of machines that imitate reason, it is easy to deem that whatever is most beautiful in the artifice of men ought to be found by the strongest reason in the works of God."[44] Once the structure and the principles of nature have been established, matter can, without external intelligent cause, work to reach its programmed goals. No additional entity with a cognitive grasp of the plan is required.[45] Leibniz demonstrated further the everyday appearance of reasonable behavior without recourse to a reasoning agent by relating the calculating machine to skilled human behavior: "Habitual actions (such as those one draws on in playing the harpsichord, while not thinking at all about what one is doing) confirm what I have just said, namely, that the Machine is capable of acting reasonably without knowing it."[46] So skilled is God in design and construction that He can—and does—easily set up matter to perform apparently intentional and goal-driven actions. The appearance of some kinds of reasoning does not require reasoning, much less an intelligent supervising cause.

Leibniz insisted on the falsity of a metaphysics or natural philosophy that maintained that matter in motion sufficed to explain the ultimate causes and emergence of organized natural things and their activity toward their various ends. Matter operating according to the laws of motion alone could never organize itself into a structure capable of achieving ends: only the organizing activity of an intelligent cause, be it God or human, could create such a structure. Not accepting the difference between the need for a nonmaterial creator and a nonmaterial supervisor leads to errors in natural philosophy, for it entails a failure to recognize the scope of efficient causation. Anyone making this error will likely fail to sufficiently investigate the domain of that causation on its own terms. To fail to investigate efficient causes is to indict the creator.

If skilled human beings could create machines appearing to be able to reason through efficient causation, why could God not produce machines that in fact reason through matter in motion? And if mind-like appearances could more or less be implanted into well-structured matter, why could human minds not themselves be material? John Locke's infamous speculations that thinking matter was not inconceivable soon found stronger advocates among those who sought to defend the infinite artistry and power of the creator. Omnipotence made a spiritual mind unnecessary. "Man is a mere piece of Mechanism, a Curious Frame of Clock-Work, and only a Reasoning Engine," William Coward, a major early English materialist, argued in 1702. Coward was no atheist: the complexity of the mechanism necessary for thinking indicates that only God could have fashioned it. Man "is such a curious piece of mechanism, as shews only an almighty power could be the first & sole artificer, *viz.* to make a reasoning engine out of dead matter, a lump of insensible earth to live, to be able to discourse; to pry and search into the very nature of heaven & earth."[47] Coward's opponent Richard Bentley accepted that to protect the spiritual quality of human beings required nothing less than to circumscribe divine skill: "If brutes be supposed to be bare engines and machines, I admire and adore the divine artifice and skill in such a wonderful contrivance. But I shall deny then that they have any reason or sense, if they be nothing but matter. Omnipotence itself cannot create cogitative body. And 'tis not any imperfection in the power of God, but an incapacity in the subject."[48]

Materialists such as Coward claimed they were in better position to defend God's existence and distinctive creativity than those positing nonmaterialist accounts of the universe: "It is rather a magnifying [of] his Power, that is able to make Accidents (to speak Philosophically) that is Life or Motion, with Sensation to distinguish it from meer Motion, to execute such Operations."[49] The more skillful the God, the less reason need be something immaterial. Well-structured matter could think—and could create.

And yet the eighteenth century did not follow the mechanical slippery slope made possible in the seventeenth century. However many countenanced that minds might be material, few thought them mechanical or anything like calculating machines.

Materialist Reduction: A Noble Dream Overcome

Few people knew more about the variety of eighteenth-century calculating machines than the Göttingen physicist Georg Christoph Lichtenberg. The remains of Leibniz's machine were in Göttingen: Lichtenberg followed and helped mediate the dispute between Philipp Hahn and Johann Müller, and he informed Müller about Stanhope's machine. In an oneiric composition of 1770, Lichtenberg divided an envisioned history to come into two moments: "We had not yet at that time made the most noble part of a living being into a calculating machine or into a syllogism box, it was yet long before the time, that we can begin to call thick Blood, the animalistic fear of humans, and nervous diseases Genius,—in short, before the time of *Night-Thoughts* and the *Lunar Tables*."[50] The moment in which calculating or logical machines might be thought capable of standing in for the human mind (or soul) would precede an age characterized by a nonmechanical genius. To encapsulate this change, Lichtenberg linked a crucial aesthetic text to a major astronomical one. He connected Edward Young's *Night-thoughts*, that central text of German Romantic aesthetics and the cult of genius, with Tobias Mayer's lunar tables, which served to bolster a Newtonianism that eschewed unempirical and hypothetical causal mechanisms.[51] If a mechanical philosophy of brute matter without other forces were indeed true, and the mind could be understood through a mechanistic psychology, then the calculating machine might indeed aid reflection about the nature of thinking. Lichtenberg was not averse to the possibility of the mechanization of the world. He called it a noble dream: "If it is a dream, then it is the greatest and most sublime dream ever dreamt, which can fill a great gap in our knowledge—a gap which can only be filled by a dream."[52] Lichtenberg spent years reflecting upon the works of the Genevan physicist Georges Le Sage, the Earl Stanhope's teacher. Le Sage's decades-long quest to mechanize universal gravitation epitomized the dream of a mechanical universe, with no unmediated forces such as gravitation, fermentation, and so on.[53] Lichtenberg ultimately denied Le Sage's "effort to explain mechanically weight, attraction, and affinity" while recognizing the historical importance of having passed through a moment of the search for materialist explanation.[54] Those heady days of the mechanical hypothesis had been necessary and, subsequently, were surpassed. Like most cutting-edge natural philosophers in the later eighteenth century, Lichtenberg doubted the reduction of forces and affinities to mechanism through untested—and perhaps untestable—hypotheses: "It is not as irritating to explain a phenomenon by means of mechanics and a big dose of the Incomprehensible, as to do so entirely by means of Mechanics." The honest espousal of a lack of knowledge trumps reductionist materialist dogma.[55] He had imbibed the central lesson of an epistemological Newtonianism: "Modesty and caution is what is wanted above all things, in philosophy but especially in psychology. What is matter as the psychologist thinks of it? Perhaps there is no such thing in Nature. He kills matter and then says it is dead."[56] A mechanistic psychology would not do: "I am quite convinced that we know precisely nothing about that which we can grasp conceptually—so how much more may there not be left behind, which the fibers of our brain cannot image?" Around 1790, Lichtenberg commented on the utility of building machines for determining whether various operations could be produced mechanically: "And what is Calculation, other than something like this machinery? It would be a calculating machine. Note this well."[57] Lichtenberg stated the salience of the calculating machine as a model for thinking and for the constitution of nature—just before firmly rejecting both a fully materialist universe and a conception of reasoning that would allow mechanization.

For Lichtenberg, the age of the plausibility of a mechanistic materialism and mechanistic psychology had rightly been supplanted by the age of *Night-thoughts*—an age that resisted the dangers that mechanistic determinism posed to the possibility of activity in nature and especially to the possibility of human creativity. After a moment of stressing the power of the mechanical and the automatic, the time had come to stress the creativity of the free artist in ethics, art, and mathematics. As discussed in chapter 4, Young, the author of *Night-thoughts*, reoriented thinking about the nature of creativity in his *Conjectures on Original Composition*: "An *Original* may be said to be of a *vegetable* nature[. . .]; it *grows*, it is not *made*."[58] Manufacture and creativity were to be divided. Imitation, long a dimension central to all creative production, was to become mechanical.

Creativity in Materialist Psychology

What did genuine creation require? The wrong views could easily lead to atheism. God seen only as an artisan was not sufficient, or so Immanuel Kant argued in his *Only Possible Argument in Support of the Existence of God* of 1763. Proofs of the existence of God based on the harmony and order evident in created contingent things, he argued, could at best reveal his existence as a "skilled workman," but not as the far more fundamental "creator" of the world ex nihilo: "According to this brand of Atheism, God is strictly regarded as the Master Artisan [*Werkmeister*], not its Creator: He orders and forms matter, but He does not produce or create it."[59] Kant's theological criticism registered a growing divide between the account of making deemed appropriate to liberal artists and that deemed appropriate to mechanical artists. The classical rhetorical tradition, perhaps the central site for early-modern aesthetics, had predicated creativity upon the reuse and reconfiguration of existing natural and artificial things.[60] "Invention" was the stage of accumulating relevant materials before combining them anew. An artist's or writer's impressions of the world and other artifacts were central to the process of imitation; those impressions were a necessary but not deterministic cause of her own production. The freedom of the will—an incorporeal faculty—made it possible to accept the intrinsically imitative quality of all creative activity while leaving open the freedom to create, drawing upon those materials. Skill came though imitations, but not all imitations reduced entirely to rote copying. The theory and practice of emulation allowed for a hierarchy of imitators, with ascending degrees of freedom allowed to different sorts of makers; the reward systems in many workshops accorded well with—and certainly preceded—this account.

A trickle—and later a flood—of critics rejected this rhetorical conception of creative activity. The Earl Shaftesbury insisted that the highest sort of human creator ought to replicate the divine ability. No poet worth the name simply combined preexisting parts. "Like that Sovereign Artist or universal Plastick Nature," a true poet "forms *a Whole*, coherent and proportion'd in it-self, with due Subjection and Subordinacy of constituent Parts."[61] This poet "is indeed a second *Maker*, a just Prometheus, under Jove," who produces genuine wholes. In his account of the organic machine discussed in chapter 2, Leibniz offered a mechanical explanation for the created unity at the heart of aesthetic reflection since Horace, if not earlier. Whereas human makers always had to draw upon preexisting parts, however small, in making, God manufactured his parts all the way down. An organic thing comprised parts all perfected and fitted to the whole. Leibniz's physical account offered a rich model for rethinking the creation of artistic unities, one resounding in criticism through the twentieth century.[62]

"Organic" views of making were central in reconceiving the nature—and hierarchy—of the arts. Such romances of organic creation remained minority views for most of the Enlightenment. Parallel to these exaltations of autonomous organic creators was the effervescence of materialist accounts of mind, which threatened to deny creative freedom to the maker with their deterministic accounts of creation. Material brains surely might sense, reason, associate, and compute—but could they create in any meaningful way if all these were caused by external processes? Without a disembodied mind outside the realm of material causation, was all creative activity externally caused? Stanhope claimed his analytical machine could invent, could discover. The danger that originality might be a prideful illusion, reiterated a century and a half later by Turing, had a powerful precedent in Enlightenment discussions about the possibility of creativity in natural philosophy, literature, and the arts. These determinist visions of the fundamental causes of making still figure in opposed histories of technology and the arts and opposed visions of the legal protection of creative activity.[63]

The run-up to the Romanic conception of genius saw the rhetorical model of creative activity strengthened, even as new materialist psychologies made accepting the imitative quality of making while maintaining the freedom of the creator or inventor difficult. A central intellectual problem animating eighteenth-century aesthetic reflection was how to explain the creation of something new while upholding in some way the importance of imitation; how to demarcate the truly inventive from merely mechanical imitation; and how to contend with materialist, potentially determinist models of the mind while preserving some freedom to create de novo. The threat that invention might be a mechanical and determinate process of making associations among previous sense experiences resonated through all materialist accounts of mind.[64]

The struggle to integrate imitative creation within new materialist physiologies marked British associative psychology prominently. This psychology envisioned mental processes as building from sense impressions, which the mind connected in large part automatically through a process of natural association, often treated as a psychic analog to the inherent, unexplained quality of gravity in physics. In giving a physiological account of the associative picture offered by Locke, for example, David Hartley argued, "The powers of generating ideas, and raising them by association, must also arise from corporeal causes, and consequently admit of an explication from the subtle influences of the small parts of matter on each other, as soon as these are sufficiently understood."[65] Imagination worked to put together ideas in novel combinations. Under this account, creation was precisely the putting together of previously existing ideas in novel, but not entirely original, forms. In science and art alike, judgment is required to control the imaginative combination of new forms. Both, however, rest on recombination, properly disciplined.[66] Genius rests upon imitation—creative, to be sure, but imitative nonetheless. Creation is thus a bringing together of the already extant, not a coming to be of an entirely organic new form. The combinatorial view extended beyond advocates of associationist psychology in England and Scotland.[67]

If these accounts allowed for the creation of new combinations of extant things and ideas, explaining the production of genuinely novel things that were nevertheless unified proved a challenge that provoked much development of aesthetic reflection. Could organic wholes emerge through the bringing together of parts? In his 1774 study that sought to encompass scientific discoveries within a unified account of making, Alexander Gérard explained that the "fancy" allowed the association of ideas necessary for new unified forms: "Every work of genius is a whole made up by the regular combination of different parts so organized as to become altogether subservient to a common end."[68] Genius required mechanism but never reduced to it. Both continuing and breaking with the associative model, Gérard sought to insulate the creations of genius from most artisanal invention. In the higher arts, an artist conceives the unity to be made from the component parts all at once and then implements them in time.[69] Gérard drew upon Edward Young's metaphor of creation as a vegetative process: "When a vegetable draws in moisture from the earth, nature, by the same action by which it draws it in, and at the same time, converts it to the nourishment of the plant: it at once circulates through its vessels and is assimilated."[70] This vegetative model found success initially and above all in Germany.[71] Combined with a theory of unconscious influence largely from Leibniz, the vegetative model helped account for how genius made use of unconscious knowledge of sensations without being determined by them. The freedom of the creator was secured—by assimilating that freedom to flora.[72] The vegetative model of originality served to undercut the anxiety that creation itself might be mechanical. The model suggested how a long series of sense impressions could be ingested and thereby transformed into something original; it distinguished the mere artisan, one who merely assembled ideas or parts, from the true creator of artificial organic unities.

Before imitation became debased, machines were seen as unable to appreciate unities, whether artificial, natural, or mathematical. In an essay of 1759, Abraham Kästner, the Göttingen mathematician, teacher of Lichtenstein, and sometime restorer of Leibniz's calculating machine, explored the value of mathematics "as a pastime." The best poet among mathematicians, and best mathematician among poets, Kästner praised the delight in unity experienced by anyone disposed to mathematical truths:[73] "The mind, which finds its pleasure in knowing truths and concluding them from one another, forms itself through a taste, in which nothing outside of mathematics pleases, for outside of it the mind never finds truth, coherence, and reason." Kästner built on the long tradition of mathematical evidence: becoming mathematical involves far more than assenting to formally true statements. The mind develops a taste almost exclusively for the pure, unified truths of mathematics. This taste can, however, reach outward, should it encounter genuine unity elsewhere. The mind "experiences, in a place with no geometrical truths, an agreement of parts that compose a whole."[74] The experience of agreement, of unity, common to mathematics and the best of arts, is spiritual or mind-like, not machinelike.

A mathematical mind—quite unlike a calculating machine—thus can indeed take pleasure in the fabrications of a poet who, as Horace said: "In such a manner forms his fictions, so intermingles the false with the true, that the middle is not inconsistent with the beginning, nor the end with the middle."[75] Imitation in mathematics and poetry alike fashioned one's taste for unity—a capacity of taste unavailable to any calculating, logically reasoning, or music-playing machine. Sharing the Romantics' celebration of artistic unity, Kästner remained a partisan of the traditional rhetorical conception of imitation to shape the virtuous self and the capacity to taste. A generation later, imitation had itself become mechanical. The mid-nineteenth-century satire of the Patent Novel Machine dramatized the debased imitative artist with a machine destined always to assemble, never to make or to admire a true unity.

Autonomy and the Eclipse of the Imitative Model

Not long after praising imitation in the arts and science in a standard eighteenth-century way, Kästner's younger contemporary, Immanuel Kant, changed his mind. He rejected imitation as antithetical to genuine autonomy in morals, knowledge, and creativity. Kant provides an important index of the transformation, within philosophy and criticism, whereby imitation came to be seen as machinelike and opposed to freedom. Kant came to distinguish art from mere handicraft sharply; he rejected the hierarchies among different sorts of crafts and among different practitioners of those crafts, consigning them all to "remunerative" arts.[76]

In attempting to make headway in thinking about humans as creators, Kant wrote copious, detailed notes while working through his concerns with skill, genius, and freedom. For Kant, superior arts go beyond habitual, mechanistic activity: "In all arts and sciences one can distinguish mechanism from genius; for the former, only skill is required, for the second, spirit."[77] The classic Christian divide of the letter and spirit of the law was projected onto the activity of making: "Everything that introduces mechanism . . . brings genius down."[78] Despite his castigation of the mechanical, Kant could not so easily ignore imitation: "Imitation is more than mechanism: for in the latter, we do not have only an example, but also guidance through someone else, e.g. a model. Free imitation."[79] Kant struggled to reconcile human freedom in creating with law-governed and, in many cases, necessary constraints: the apparent contradiction was powerfully generative in his reflections on mathematics, morality, and aesthetics. Yet Kant still insisted on training and discipline, and on rules: "Genius without discipline is crude."[80] Spirit is not enough: design without implementation is not art. Hylomorphism requires rules for embodiment—to move from *hyle* to morphism: "In all liberal arts there is . . . something compulsory or, as it is called, a mechanism, without which the spirit, which must be free in the art which alone animates the work, would have no body at all and would entirely evaporate."[81] The coming-into-being of any piece of art rested fundamentally upon habituated skill in following rules, though it is in principle independent of those rules.

Creativity needed to be at once free and bound by rules. In the *Critique of Judgment* that emerged from these notes, Kant captured the apparent paradox clearly: "That the *imagination* should be both *free* and yet *of itself conformable* to law, that it should carry autonomy with it, is a contradiction."[82] To ensure human autonomy required an account that captured freedom to create with a form of self-imposed, necessary law. With this self-imposition came a potent divide between design and implementation. Imposing that law requires becoming little less than a second creator, to animate the creation of a whole through new laws.[83] Kant had long before feared the atheism that followed from a physico-theology that viewed God as a workman who assembled rather than the giver of form and of laws. Now he reaffirmed human freedom with a picture of human creation as mirroring the divine. An idea was a design for the whole instantiated into all the parts: "All the parts are

there for the sake of the others, and all for the sake of each, as in an animal."
The parts "are not associated and sought together, but generated thereby. The
spirit is entire in the whole and entire in every part."[84] Kant imposed the
Leibnizian vision of the organic—a machine entirely and perfectly machined
to coordinate—as his vision of the highest form of human creation. Human
beings can create organic forms and not just agglutinate preexisting forms. Thus
they can be understood as free. No patent-novel writing machine could ever
compare. Kant's artist had an interlinked, twofold freedom from necessity: a
freedom from outside aesthetic rules coupled to a freedom from economic
necessity. A liberal art "is agreeable in itself"; a handicraft or remunerative art
"is disagreeable . . . in itself and is attractive only because of the effect (e.g.
remuneration)."[85] The analogy with his ethics is clear: a liberal art is categori-
cally agreeable to the maker; the other is only hypothetically agreeable.

When Kant shifted his views on the value of imitation, he retained his
older conception of Newton as an imitator; this was a clear demotion. To
show that Newton could be imitated was largely to show that his work required no genius. Human beings could do yet more. Kant wrote his comments on creation in the wake of defending mathematics from any claims
that it might be a mechanical procedure blindly following logic. Yet in his aesthetic works, he came to make mathematics decidedly second-class: "**Mathematics is not the land of ideas**, but of concepts made intuitable: it does not
go from the whole to the parts, but from universal to particulars."[86] While not
mechanical, mathematics did not belong to realm of genius, of real creativity.

Materialism and Property in Ideas

Arcane aesthetic debates about the relative contribution of different mental
and corporal faculties had important legal ramifications. In 1765, the economist Louis-Paul Abeille bitterly denounced those who sought exclusive privileges—patents—on the basis on their inventiveness. Privileges depended on
a false account that prioritized the individual inventor in the production of
knowledge and of technologies: "The instruments necessary for culture are
not the work of the cultivator; in no art is the effort and exercise of a single
man sufficient. The existence of society depends, then, on the communications of forces, of insights [*lumières*] and work. Whoever participates in this
society, has a Right acquired by this communication, for he had contributed
to rendering it general through his own work. From this, one must conclude
the constitutive Rights of society exclude every idea of an exclusive privilege."
Anyone who attempts to get an exclusive privilege "lands a direct blow, the
most dangerous of all, on society," for he steals his portion from the common

"while he conserves for his own profit the growth of forces that result from" the communication among society. He "breaks so far as is possible in him from all the constituent ties of society."[87] Autonomy in invention rested upon a false account of how the communication of technique sustained all further developments; its legal enshrinement in patents and privileges was a violation of a social compact about the circulation and communication of knowledge and technique. Abeille made no distinction between microinventions, produced through a long evolutionary process, and macroinventiveness, produced only by geniuses breaking existing ways of working.

Debates about genius and collective invention were supervenient on the broader politics of ownership in knowledge.[88] A hundred years later, the battle over patent abolition in Britain rested on two opposed schools on the nature of mind and the nature of creativity. The great engineer Isambard Kingdom Brunel captured the tenor of the intellectual program underlying abolition clearly: "The most useful and novel inventions and improvements of the present day are mere progressive steps in a highly wrought and highly advanced system, suggested by, and dependent on, other previous steps. . . . [I]n most cases they result from a demand which circumstances happen to create."[89] A mid-nineteenth-century French critic of patents as a form of antiquated privilege put it more bluntly: "Who discovered the motive force of steam? . . . It is Watt? Is it Fulton? No; rather it is the eighteenth century; just as the nineteenth century has discovered railroads and the electric telegraph."[90] Reflection on the broader cultural sources of creativity undermined claims of individual property.

Humility

In a draft of his never-completed logic text, Stanhope celebrated the humbling implications of mechanizing reason:

> The \Mechanical Instrument/ ~~Machine~~ above described for drawing Consequences, will shew to Man in what consists his boasted faculties! It will cause him to perceive the Weakness of his Nature. . . . \That inanimate thing,/ ~~Machine~~, by leading Man to acquire Knowledge, will make him discover his Deficiency. And by learning, he will only learn that he is ignorant. The Consequence of which \it will be/ ~~must~~ be, to \weaken/ self-conceit, and to lessen Pride that baneful CANKER of the human mind.[91]

Like his predecessors in the seventeenth century, Stanhope used his machine to prompt reflection on human capacity. The self-reflection occasioned by the machine ought to lead us to recognize the infinite gulf between humanity and the creator: "One piece of knowledge, however, of the first importance, by learning to <u>reason</u>, he will acquire; namely, that there is a DISTANCE, literally INFINITE, between HIS CREATOR and himself. Man will thus feel his <u>Nothingness</u>, when \his thoughts ascend towards/ that BEING OMNISCIENT, and OMNIPOTENT, the AUTHOR OF FELICITY."[92]

The difference and analytical engines likewise bore powerful lessons about the nature of the creator. In 1837, Charles Babbage explained that we come to glimpse the nature of the divine creator by extrapolating: "We take the highest and best of human faculties, and, exalting them in our imagination to an unlimited extent, endeavour to attain an imperfect conception of that Infinite Power which created every thing around us."[93] We grasp the capacity of a creator by studying the product: "The estimate we form of the intellectual capacity of our race, is founded on an examination of those productions which have resulted from the loftiest heights of individual genius, or from the accumulated labours of generations of men." The same holds no less for the divine: "The estimate we form of the Creator of the visible world rests ultimately on the same foundation."[94] Given this inductive process, our conception of divine creativity must change as technology and science develop: "All those discoveries which arm human reason with new power, and all additions to our acquaintance with the material world, must from time to time render a revision of that notion necessary."[95] Babbage's calculating engines demanded just such a revision. The engines "greatly engaged" his "own views respecting the extent of the laws of Nature."[96]

Using a detailed reflection on his engine, Babbage turned to the old question that separated Clarke and Newton from Leibniz as well as earlier theologians. Are miracles exceptions to the ordinary laws of creation or expressions of the laws? Does the creator violate his own laws or did he, as an expression of his skill and foresight, build the apparent exceptions into an order of laws higher than those we have hitherto perceived? Babbage used his engine to defend the latter position. Imagine an engine that counts from 1 to 100,000,000 one by one. Watching it from the outside, we could induce that its law was to add one each. Our induction would be false, for the next number produced would be 100,010,002, then 100,030,003, then 100,060,004: "The law which seemed at first to govern this series fails at the hundred million and second term."[97] Babbage adduced further changes in the law; in each case, we can induce a law for the behavior of the machine, but we cannot predict what the next higher-level law might be. While the "consecutive introduction" of new law-regulated behavior "at various definite intervals is a necessary consequence of the mechanical structure of the engine, our knowledge of analysis does not enable us to predict the periods themselves at which the more distant laws will be introduced."[98]

Consider two machines: one in which an operator must change a setting for each new form of the law and the other that requires no such intervention: "Which of these two engines would, in the reader's opinion, give the higher proof of skill in the contriver? He cannot for a moment hesitate in pronouncing that that on which, after its original adjustment, no superintendance was required, displayed far greater ingenuity than that which demanded, at every change in its law, the intervention of its contriver."[99] The engine thus serves to motivate a hermeneutics for looking at the geological and animal record as the product of a most skilled creator: "The minutest changes, as well as those transitions apparently the most abrupt, have throughout all time been the necessary, the inevitable consequences of some more comprehensive law impressed on matter at the dawn of its existence."[100] To deny this would be to deny God the highest skillfulness in behavior.

Babbage used the engine to set up a hierarchy of machines crafted by ever more skilled creators:

> A machine that alters its behavior in some particular way at a given time;
> A machine that alters its behavior in different ways at different times;
> A machine that alters its behavior when adjusted by someone and will act in a manner entirely foreknown to the creator.

His analytical engine was this third sort of machine: "The engine . . . may be set, so as to obey any given law; and, at any periods, however remote, to make one or more seeming exceptions to that law." The machine was capable of tricking human inductive reasoning: "The apparent law which the spectator arrived at, by an almost unlimited induction, is not the full expression of the law by which the machine acts."[101] Far from suggesting that the mind was ultimately a material or mechanical contrivance, Babbage's Difference Engines taught a lesson in humility about the nature of creation. The more a machine can do, the more we understand the possible extent of skill and foresight in human beings and, a fortiori, in the divine being. The more we can construct a machine that replicates reason, Babbage suggests, the more we ought to expand our estimation of the genuine creator and the possibilities of creative activity itself.

Like his collaborator Lovelace, Babbage resisted the idea that the human mind itself might be mechanical: a machine might anticipate, but it never originated. One commentator has noted, "Indeed, one might argue that the powers of his Engines served to distinguish precisely between those aspects of the human mind that were thought to be merely mechanical and those that were regarded as being truly original, creative, rational, and ultimately divine in character."[102] In contrast with Stanhope, Babbage understood his machine as expanding the domain of the mechanical, not by eliminating the mind, but by defending its nonmechanical distinctness.

Unmachinelike Machines: Mechanizing Association

By the middle of the nineteenth century, to be machinelike was precisely not to be creative or original.[103] Imitation had become the antithesis of the creative and free. In his autobiography, John Stuart Mill described his nature after his upbringing following principles of strict rational utilitarianism: "The description so often given of a Benthamite, as a mere reasoning machine, though extremely inapplicable to most of those who have been designated by that title, was during two or three years of my life not altogether untrue of me."[104] Machinelike behavior was habitual and algorithmic, imitative without freedom to act otherwise, to improve, or to transform.

Eighteenth-century associationist psychology fit poorly with any reduction of thinking to a calculation or an algorithmic mechanism: while reasoning could be understood in material terms, it was hardly a mechanical procedure. In a remarkable piece of juvenilia from 1870 called "Ye Machine," Alfred Marshall, not yet the pathbreaking economist, sketched the design principles of a machine to explore anew how much of mind could plausibly be replaced with machinery. Marshall's envisioned machine materializes the processes of association so central to eighteenth- and nineteenth-century psychology and philosophy of mind. Marshall posited that machines might not only have "machinelike," algorithmic behavior; they could conceivably be far more flexible. As he described it, his machine was not machinelike in following predetermined steps: the machine had to be taught; it had to be allowed to develop the full range of associations human beings do through their education. Rather than building a mechanism for doing addition or multiplication, the machine needed to have mechanisms of association, sensation and memory, and pleasure and pain, through which it would learn to add.

Marshall declared that mechanisms of association could well be capable of originality; "a machine of very great power—by means of the enormous numbers of associations which it would have ever present with in—might in [geology] as in other sciences, discover laws that we have not yet attained to." The same held in aesthetic domains. The machine might do better than human beings: "Music, of course, it could easily construct for itself. . . . Nay, even, if it had come in contact with men, it might thus exhibit the lives and actions of

men, ordinary and straordinary [sic], worked out all with complete originality, and yet with systematic subordination to truth."[105] Marshall envisioned a
machine capable of imitation—one capable of free creation, not only aping.

Cultural production, then, was entirely within the machine's ambit, so
human distinctiveness must lie elsewhere. Its limitations came instead in its
complete inability to grasp both self-consciousness and the consciousness of
others: "Of the secret springs of action it could say nothing"—for "nothing
corresponding to them would have ever entered the machine."[106]

From Machines That Imitate to Machines That Emulate

In his 1947 lecture on an envisioned "Automatic Computing Engine," Alan
Turing reflected on the claim that "computing machines can only carry out
the processes that they are instructed to do." True, by design: "the intention
in constructing these machines in the first instance is to treat them as slaves,
giving them only jobs that have been thought out in detail." They need not
always be so treated: their programs ("instruction tables") could be written so
that they "might on occasion, if good reason arose, modify" themselves. Turing offered an analogy with learning through apprenticeship and the surpass-
ing of one's teacher: "In such a case one would have to admit that the progress
of the machine had not been foreseen when its original instructions were put
into it. It would be like a pupil who had learnt much from his master, but had
added much more by his own work."[107] The machine would improve as it absorbed past knowledge and technique; imitation would be the precondition
for improvement. A machine so devised could do far more than anticipate.

Like Marshall, though entirely independent of him, Turing insisted that
machines need not have machinelike behavior. Machines need not be exclusively tied to predefined algorithmic behaviors: "It is certainly true, that
'acting like a machine' has become synonymous with lack of adaptability. But
the reason for this is not obvious." Memory—not processing ability—was the
problem. So long as we demand a machine be infallible, Turing argued, "it
cannot also be intelligent." Rather than failing to give an answer, "we could
arrange that it gives occasional wrong answers."[108] If we want a computer to
develop intelligence, we need not to treat it like a mere imitator. It must un-
dergo the practical training of building skills and knowledge that cannot be
reduced merely to procedure. A machine might know *how*: it need not exclusively follow encoded patterns of knowing *that*.

In his earlier paper introducing his famous abstract machine, Turing had
coupled Romantic unconscious influence and Gödel's incompleteness theo-
rem. He wrote in 1940, "In pre-Gödel times it was thought by some that it

would be possible to carry" the program of formal logic "to such a point that all the intuitive judgements of mathematics could be replaced by a finite number of these rules. The necessity for intuition would then be entirely eliminated."[109] With his constructive proof of propositions that are formally undecidable, Gödel had explored the dream of reducing mathematical inquiry to formal logic. "The activity of the intuition," Turing explained, "consists in making spontaneous judgements which are not the result of conscious trains of reasoning."[110] Mathematics could never be reduced to mechanism.

In describing the possibility of a machine capable of thinking, Turing insisted on the collective and social nature of knowledge production, and of originality, even in mathematics: "A human mathematician has always undergone an extensive training. . . . One must not expect a machine to do a very great deal of building up of instruction tables [programs] on its own." The contributions of individuals are always cumulative and minor, but decidedly not exclusively imitative: "No man adds very much to the body of knowledge, [so] why should we expect more of a machine?"[111] Building on the work of others, machines and human beings alike could add a bit to the body of knowledge. We ought neither to overestimate our novelty nor to disregard it entirely. Given the lessons of humility central to his writing about the collective nature of knowledge, Turing's place as a transformative genius in the idealist history of computing becomes just a bit ironic.

111111　Final Carry

Epilogue

Anybody, with a little bit of study, can think up a calculating machine; it is no trouble at all. There are any number of ways to go about it, and devise something in the air or on paper; but it is a very difficult proposition to make one that will be simple and produce accurate results in use.

D. E. FELT, 1916[1]

Practical calculating machines, effortlessly keeping it digital, are ubiquitous. Materialized solutions to the problem of carry surround us. Nearly all electronic digital computers since the Second World War have a dedicated arithmetical unit, where the solution to carrying is a central defining feature of the architecture. Choosing the mechanism for carrying binary numbers is a major design decision in constructing a microprocessor and, despite all our advances in engineering, still subject to trial and error.

Within a history of computing focused on the logical foundations of the information revolution, the details of how a computer adds are far less important than the generality of the stored-program computer. The abstract, extraordinarily simple "Turing machine" reduces addition to more basic operations. For all its logical and philosophical import, the austerity of Turing's imagined machine bears little relationship to the architecture of most working machines and little connection to actually materialized computational devices. Turing knew this very well. In his 1947 lecture on the Automatic Computing Engine, Turing outlined the functions a dedicated addition unit—called an "adder"—must have. Like his predecessors for centuries before him, Turing left a blank box for the carry mechanism (see figure E.1).

Leaving the box empty is no option. Drawing upon the results of the field of "Computer Arithmetic," microprocessor designers look to a dizzying array of trade-offs in speed, size, complexity of circuitry, power consumption, and heat production in deciding among various ways of materializing algorithms

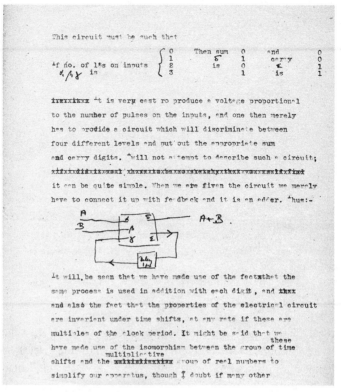

FIGURE E.1. Alan Turing, carry diagram, "Lecture to L. M. S. Feb. 20 1947," on Automatic Computing Engine. Turing Archives, King's College, Cambridge, AMT/B/1/14. Used by permission of the Turing Estate, King's College, and MIT Press.

for carrying. One size does not fit all; logical elegance inspires attempts to implement with extant materials and processes; the constraints of building at the nanoscale, in turn, have inspired new logics of design.

While the importance of logical reflections around Turing's abstract machine for the subsequent history of computing is often overstated, formal logic has been essential to the design of superior carry mechanisms. Straightforward logical manipulation reveals that carries can be computed independently of and in parallel to the addition of individual digits. So a circuit could, in the same unit of time, compute all the additions of digits and the carries involved—just as Babbage sought to do with the design of his "anticipating carry."[2] Performing carries in parallel to additions involves a recursive algorithm of considerable logical elegance. The recursive quality of the logical relationship requires that the carry following the addition of the nth digit have $n + 1$ inputs. Materialization of this recursive algorithm takes up considerably

more area on a microprocessor, with much longer and highly irregular connections and greater power consumption and heat dissipation than a simpler
adding mechanism would require. The complexity grows very quickly with
the number of digits (see figure E.2).[3] The cost of increased speed is a high
degree of "fan in" of the digital circuitry.[4]

The undeniable speed gain comes with a tremendous increase in the complexity of the mechanism—as Babbage realized in designing his anticipating
carry. Values of speed and security must compete with the realities of materialization, as heat, cost, and complexity each must be considered.

The advent of the integrated processor did not make the questions of
materiality braided through the preceding chapters any less pressing. David
Brock and Christophe Lécuyer have stressed, "far from being 'dematerialized,'
microelectronics is an intensely material technology—one that relies on complex materials and intricate ways of manipulating them."[5] Materializing carry,
now as in the past, involves the subtle interplay of plausible designs and logical forms with the properties of materials produced through dense networks
of skill, expertise, and formal reasoning. Neither an idealist, nor a materialist,
nor a social history alone suffices. *Plus ça change . . .*

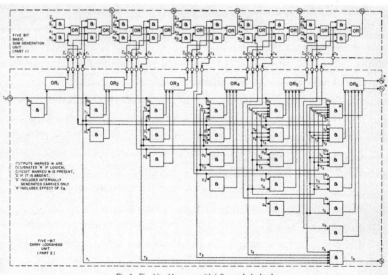

Fig. 2—Five-bit adder group with full carry look-ahead.

FIGURE E.2. Five-bit adder with full carry look-ahead. Note that each additional digit requires adding
a circuit with an ever greater number of logic gates; these are the embodiment of the logical recursion
of computing the carries independent of the main additions. O. L. Macsorley, "High-Speed Arithmetic
in Binary Computers," *Proceedings of the IRE* 49, no. 1 (1961):67–91, at p. 69. Used by permission of IEEE.

To Market, to Market

A 1925 advertising triptych for a Burroughs adding machine promised "quality," "accuracy," and "durability" (see figure E.3). The advertising copy contained a folk sociology of the market success of a machine: "Burroughs Machines were never built simply to sell but, first of all, to do their work perfectly; then they were made to last indefinitely—then of course their sale could not be stopped." Calculating machines did not gain a large market because they had already been perfected; and no preexisting market was just waiting for such machines. Starting in the middle of the nineteenth century, advocates of calculating machines managed to convince an initially small group of potential users, primarily those in actuarial work, that their machines functioned *well enough* for contemporaries to integrate them into that work.[6] The creation of a market for the machines spurred the many competitive firms to produce a series of improvements in contrivance, design, and manufacture. The manufacture and sale of calculating machines exploded in the 1890s.[7] Reliability was not the precondition of the emergence of a market: superior accuracy and durability resulted from the processes made possible through the emergence of that market.

The first commercially successful digital calculating machine was Thomas de Colmar's Arithmometer, a product of a wealthy inventor working within the culture of emulation. In 1822, the journal of the foremost French society for emulation—the *Société d'Encouragment pour l'Industrie Nationale*—carried a detailed description of Colmar's new calculating machine. The description stated that the efforts of a century and a half of development had suggested that functioning mechanisms for calculating machines might never be found: "If one could assign limits to our intellectual faculties, it would seem that the numerous means already found for calculating mechanically would have exhausted research in this domain and that nothing remains to be done, given the efforts of famous savants of so many countries." Colmar, the author claims, finally "defeated all the difficulties" in the machine he presented to the Society of Encouragement.[8] It is "certain," the report opines, that Colmar "had no knowledge of those when he imagined his own, and that he *could not have helped himself* with the works of his predecessors."[9] The claims of novelty were misplaced. A German commentator noted that the French appeared rather ignorant about Babbage's efforts.[10] No less than three eighteenth-century machines—those of Stanhope, Müller, and Hahn—performed all four of the standard arithmetical operations with a high degree of success, probably equivalently to most early Colmar machines. In England, Stanhope's instruments were regularly singled out in the late nineteenth

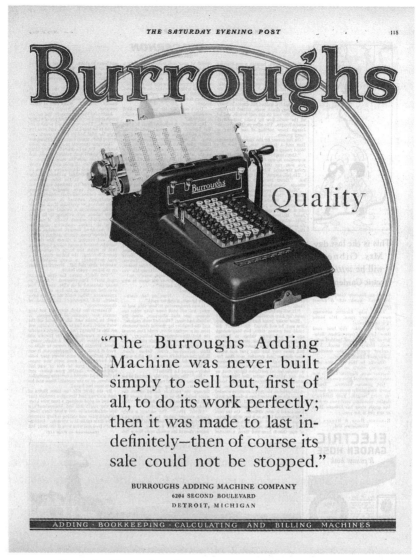

FIGURE E.3. Burroughs adding machine advertisement, 1925. Author's collection.

century as the first to perform well.[11] Only Hahn and his former assistants produced more than a few machines; but Müller, Hahn, and Stanhope each created machines whose design was intended for larger manufacture. Stanhope had gone so far as to cast fundamental parts of his machines in iron. None was manufactured in any number.

Neither were any devices of Colmar's—for some time. The Colmar Arithmometer was indeed the first calculating machine to enter widespread production. A wealthy insurance magnate, Colmar paid for many years of development of the machine. Around fifty years after Colmar had received his initial five-year patent, his machines began to be produced in larger numbers, around one hundred a year in the 1870s.[12] Stephan Johnston has shown that Colmar, like every other inventor of a calculating machine, focused above all on the carry mechanisms and produced at least five distinct mechanisms up to the 1860s "in response to publicly voiced concerns about the delicacy of the tens carry." Like other inventors, he sought to lessen the dependence on springs.[13] A series of clever contrivances provided much of the felt superiority of the Colmar machine, as a report to the English Society of Arts testified.[14]

Yet the Colmar machine succeeded not because its carry was perfected—the machines were still quite prone to error—but because advocates for the machine fostered multiple markets for a machine with its blend of perfections and imperfections. The trade-offs of working with the fallible machines were proclaimed to be favorable; more and more users came to agree. Increasing numbers of people in the insurance and actuarial industries deemed that the machine was sufficiently secure and robust, and they transformed their working processes to make use of it.[15] Unlike nearly all the machines covered in this book, Colmar's machine has a rich history of users who transformed its use and intervened in the further development of the machine.[16] At a meeting of the Society of Arts in 1886, for example, the delightfully named Professor Cathorn Unwin explained: "About twelve years ago he obtained one of the French calculating machines and had used it occasionally a good deal. Recently Mr Tate had been kind enough to add some of his improvements to it and he should probably use it for some time to come."[17] Tate, an English improver of Colmar's machines, was happy to modify Colmar machines for other users. That maven of tech transfer, the locksmith-turned-philosopher Philippe Vayringe, would likely have approved.

Nearly every serious contemporary discussion of the Colmar machines mentions their problematic qualities: "With hard wear the French machine soon gets out of order, owing, no doubt, to the parts being made of soft yellow brass and not being cut out of the solid."[18] Colmar machines remained works of handcraft, not commodity products of standardized manufacture, and were priced accordingly.[19] The Colmar machines had moved from the craftsmanship of risk toward the craftsmanship of certainty but not to standardized manufacture—yet. The success of the machines did not hinge upon the availability of new machine tools for harder metals; indeed, the machines made of soft brass succeeded despite not being made of hard metals through new precision machining techniques and tools. Many among the dizzying array of competitors that soon emerged drew upon just such precision machining in harder metals to create superior machines. They did so after, not before, the market for machines was large and profitable.

Subsequent inventors worked from the failings and inconveniences they discerned in the series of Colmar machines. The many international societies of emulation duly evaluated the ramifying number of these calculating machines, often reflecting on their durability in heavy use and suggesting new ways of using the machines. The English Society of Arts, to take one example, noted in 1885, "In some of the early foreign machines the workmanship was rather faulty and unnecessarily complex, so that when the machine was worn, it was difficult to repair." Tate, for example, "has simplified the construction, and has employed the finest workmanship, although it may be doubted if the abolition of the locking expedients in the original [Colmar] machine . . . is altogether advisable."[20] Observers did not agree about which trade-offs of complexity, simplicity, and materials would produce optimal results—ultimately it depended upon the purpose of the machine. Manufactures developed machines tailored for a wide variety of uses and niches, users modified and adapted them, and firms responded in kind. Soon keyboard input, printing, and other features became prominent, permanent aspects of calculating devices up through the 1970s.

Led by actuarial work, sales of calculating machines boomed in the 1890s through the First World War and beyond. By the turn of the twentieth century, makers of calculating machines, especially adding machines, were profitable and growing quickly. Debates about priority raged; American makers in particular were prone to litigate about patent violations. The Felt and Tarrant Corporation, maker of the Comptometer, sued the American Arithmometer Corporation, the future Burroughs, for copying its patented metal case design. This litigation came amid a long series of lawsuits filed by the two competitors over infringement of their respective patents.[21] In the course of their ongoing litigation, associates of Felt and Burroughs wrote competing histories of the development of calculating machines, focused on the small contrivances that reduce them to practice.[22] Indeed, much of the early historiography on calculating machines arose from patent litigation within jurisdictions and rivalries among different nations. Competition produced both machines and the early narratives about them.

Users put these machines to many novel applications; manufacturers responded to the needs of their clients, including scientists.[23] Digital computation using mechanical and electromechanical machines had grown to impressive proportions in accounting, life insurance, and other business domains, as well as in government work. Up through the 1940s, it remained far from obvious that these primarily business-oriented machines performing elementary mathematics would dominate scientific computation.[24] With the support of a small number of early advocates, digital calculating machines had moved from novelties to regular tools of laboratory and observational work in the decades around the turn of the twentieth century.[25] The British mathematician Leslie Comrie pioneered and advocated the use of commercial adding machines and Hollerith punch card machines for large-scale scientific computation.[26] In a retrospective written just after World War II, Comrie recounted the history of the machines from Pascal to Stanhope and then through their multinational commercialization in the late nineteenth century: "Although these machines were developed primarily for the wide fields of commercial and accounting application, they lend themselves to mathematical and scientific computing, which consists, after all, of the four fundamental processes of addition, subtraction, multiplication and division— just the processes that any calculating machine is designed to perform." Comrie encouraged modifying existing machines rather than creating new ones: "As far as possible commercial machines should be used in their standard form. . . . The computer's art lies mainly in his ability to manipulate the problem and to apply ingenuity and low cunning in developing techniques that take advantage of the mechanical features of the machine."[27] In this statement about gradual invention through modification of existing machinery, Comrie described how he made good on Babbage's dreams of automatic computation for scientific purposes. The early electronic and electromechanical computers of wartime work stemmed from just such adaptations of mechanical calculating machines to large-scale numerical efforts. No royal road led calculating machines to become general-purpose computers: users, both individual and institutional, put digital arithmetic at the center of new general-purpose machines.

Comrie's retrospective appeared in an issue of *Mathematical Tables and other Aids to Computation*. This important journal tracked the gradual emergence of the digital age of automatic computation in the years just after World War II.[28] The previous issue of the journal included a seminal article on one of the greatest of such wartime machines, the University of Pennsylvania's Electronic Numerical Integrator and Computer (ENIAC). The further development of this specialized machine foretold a future of automatic computation, programming, and storage, always with digital carry as a central feature.[29] This future was not as obvious in 1946.

Before, during, and just after the Second World War, indeed, special-purpose *analog* machines, not general-purpose digital ones, seemed to many the future of computing machinery.[30] The differential analyzer, envisioned by Lord Kelvin and brought into successful practice under Vannevar Bush, dominated scientific and engineering computing before World War II and well into the 1950s. Differential analyzers used physical analogs to model the phenomena. They directly performed high-level integrations through approximative physical analogs rather than approximative numerical solutions. High-level machines with specialized functions, they had a directness absent from rather more abstract digital calculating machines capable of basic arithmetical operation. These highly specialized analog machines largely dropped from historical sight from the mid-1950s as the obviousness of the digital revolution, and then a binary digital revolution, took hold.

A fundamental facet of electronic digital computers is their "generality." Historians working in an idealist mode have long attributed this generality to Turing and Von Neumann, who conceived of a general machine as one capable of rewriting its own instructions, with its instructions and data in common storage. Before digital computers were understood (and became) general in this Turing–Von Neumann sense, however, they were "general" in that they could perform *any* sort of calculation.[31] Comrie celebrated the lowly desktop calculating machine as just such a general machine to be tried before building a special-purpose machine. Performing arithmetic—and logical reflection on performing arithmetic—turns out to be rather more central to the development of computing than is told in much journalistic and more philosophical historiographies celebrating the primacy of logic for the development of the digital computer.[32]

Material Foundations of a Digital Age

The fundamental architectural move from analog to digital computers in the middle of the last century entailed a transformation in the appropriate expertise around the performing of computation. The heirs of the natural philosophers of this book—mathematicians, logicians, and physicists—became crucial paths of access to the new computational machines.[33] Crucial, but never sufficient: making and using the new machines demanded far more than competence in mathematics and formal logic. Programming electronic digital computers was expected to be an easy application of logical principles and a clerical task, a task low in the hierarchy. Programming proved to be none of these things. In the decade following the Second World War, mathematicians and engineers alike came quickly to realize the difficulty and the challenge of programming, a practice autonomous from symbolic logic, however dependent upon it.[34] Programming remains a craft, its best practitioners liberal artists: free to innovate in their activity, free to use well-trod paths, free to improve on them while learning from them, free to produce collectively and to invent new forms of collective action.[35] Programmers are among the greatest heirs to the noble history of emulation.

Writing code is an intellectual craft; so too is making hardware. In reviewing Intel's latest processor architecture, the technology journalist Anand Shimpi describes the process of design: "Microprocessor design is one giant balancing act. You model application performance and build the best architecture you can in a given die area for those applications. Tradeoffs are inevitably made as designers are bound by power, area and schedule constraints. You do the best you can this generation and try to get the low hanging fruit next time. . . . Obviously you can't predict everything that will happen, so you continue to model and test as new applications and workloads emerge."[36] Grounded in sophisticated electrical engineering and chemical manufacture, chip design is still a journey through different materializations, not simply a projection of a design known from first principles to be optimal.[37] The challenge remains no less true in materializing arithmetic. The challenge of the trade-offs among speed, area, power, and complexity inherent in materializing algorithms for binary addition has motivated the search for new algorithms devised with implementation in silicon in mind. Logical manipulation and chemical and electrical engineering are creatively bound together. As materials, fabrication processes, and user demands have changed, so have the priorities in the design of adders for microprocessors. However much hylomorphism remains a desideratum in chip design, chip designers still cannot accurately predict many of the qualities of the chip in operation, including speed and heat. The computer arithmetic community cannot accurately predict the speed and heat of various designs. In Very Large Scale Integration (VLSI) processor design, "the performance of a chosen [adder] topology will be known only after the design is finished. Therefore a lingering question remains: could we have achieved a higher performance, or could we have had a better VLSI adder topology? The answers to those questions are generally not known. There is no consistent and realistic speed estimation method employed today by the computer arithmetic community."[38] Purely "rational" design remains a goal, not a reality.

Charles Babbage never produced his Baconian philosophy of systematic discovery and invention. Nor did Leibniz. The hubristic savants populating this book learned quickly the consequence of trivializing material making and the social relations and political configurations that make it possible. Enabled by symbolic logic, contemporary programmers and chip designers cannot afford to think that they work in a dematerialized world of perfect design; neither can we, in reflecting upon the potential of interconnected digital technologies. Technologies provide no panacea for social ills, but neither do disembodied ideals. Neglecting the network of knowledge, labor relations, and political economies of invention precludes the work necessary to materialize our aspirations for a decent digital future. Only well-chosen material embodiments, under iterated emulation, can realize the autonomy, interconnection, and access to knowledge our information technologies can make possible. Realizing a more just sociotechnical order means reckoning with matter.

Acknowledgments

"The instruments necessary for culture are not the work of the cultivator," Louis-Paul Abeille wrote in 1765. "In no art," he explained, "is the effort and exercise of a single man sufficient. The existence of society depends, then, on the communications of forces, of insights and work." Every chapter of this book took root at conferences and colloquia where I presented at the invitation of colleagues. My thanks to Nicholas Dew, James Dalgorno, Mary Terrall, Alex Marr, Deborah Harkness, Tony Grafton, Ken Alder, John Tresch, Nathan Ensmenger, Robert Kohler, Massimo Mazzotti, Cathryn Carson, Mimi Kim, Tony Lavolpa, Hall Bjornsted, Heidi Voskuhl, Lissa Roberts, David Kaiser, and David Mindell, who all offered inspiration, direction, and much-needed criticism. I was fortunate to get Mike Mahoney's advice on the first fruits of this work before his untimely death. Michael Gordin, Debbie Coen, Daniel Magoscy, and Jean-François Gauvin generously worked through the entire manuscript. Mario Biagioli, Margaret Schotte, and Lorraine Daston greatly sharpened key arguments. Two incisive anonymous reviewers improved every facet of the book. Florin Morar shared his excellent thesis work and knowledge of Leibniz's calculating machines. The project began many years ago when I was fortunate to study with Simon Schaffer, and every good page bears the imprint of his insight. At Columbia, intellectual companionship came from Carol Gluck, Samuel Moyn, Debbie Coen, Pamela Smith, Pierre Force, Philip Kitcher, Carl Wennerlind, Roosevelt Montás, Joel Kaye, George Saliba, Martha Howell, Will Deringer, Abram Kaplan, Mike Neuss, and Christia Mercer. Eben Moglen and David Isenberg invited me into the networks attempting to materialize our higher aspirations for the digital age. Patrick McMorrow and Patricia Morel ensured open waterways to project resources. George Lee Jr. offered crucial emendations. My editor Karen Merikangas Darling supported this project with equal parts insight and patience. Mihaela Bacou facilitated my work in Paris over many years. Reid Hall in Paris and the Max Planck Institute in Berlin provided the rare commodities of time and space to think, converse, and write.

Numerous curators, librarians, and other scholars made the empirical foundation of this project possible. The interlibrary loan staffs at Columbia and the Max Planck Institute in Berlin found troves of arcane antiquarian articles with cheer and alacrity. Jane Wess, then at the Science Museum in London, enabled me to inspect the Stanhope machines at short notice. Tony Simcock at the Museum of the History of Science, Oxford, guided me in using the Morland and Babbage manuscripts there. The excellent staff at the Kent History and Library Centre in Maidstone welcomed me warmly and allowed me to pore over the third Earl Stanhope's papers on several visits. Birgit Zimny aided me in using Leibniz's papers in Hanover. Expert manuscript librarians guided me in collections across Europe and the United States: the British Library, the Houghton Library at Harvard, the Leibniz Archiv at the Gottfried-Wilhelm-Leibniz Bibliothek, the Bibliothèque nationale de France, the Académie des sciences in Paris, the Bibliothèque de Genève, the Lambeth Palace Library, the Royal Society Archives, CNAM in Paris, and the special collections at Stanford and Columbia. Alan Gabbey, Michael R. Williams, Benjamin Elman, and Simone Rieger generously shared their knowledge of, and often copies of, manuscripts and rare secondary sources.

Numerous talented students aided me throughout. At key stages, Sarah Petrak and Jenne O'Brien saved me from countless infelicities, repetitions, and errors. Stacey Van Vleet and Daniel Asen translated Chinese materials; Adam Bronson consulted a key Babbage manuscript in Japan; Hannah Elmer worked through a swath of dense German correspondence; Mike Neuss helped with manuscripts in Philadelphia. Marcus Anderssen, Brian Mackus, and Lily Cutrono provided research support.

This material is based on work supported by the National Science Foundation under Grant No. 0551849 and Grant No. GER-9452875. Additional research funds came from the James R. Barker Chair in Contemporary Civilization, the Lenfest Distinguished Faculty Award, and Columbia College. My work on eighteenth-century philosophy of mathematics was supported by the Defining Wisdom Project Grant, Arete Initiative, University of Chicago. The Leonard Hastings Schoff fund of the Columbia University Seminars supported the cost of reproductions and permissions. The Leibniz Fund underwrote final research and production costs.

An earlier version of chapter 3 appeared as "Improvement for Profit: Calculating Machines and the Prehistory of Intellectual Property," in *Nature Engaged: Science in Practice from the Renaissance to the Present*, edited by

Mario Biagioli and Jessica Riskin (New York: Palgrave Macmillan, 2012); sections of chapter 6 have appeared in different forms as "Space, Evidence, and the Authority of Mathematics in the Eighteenth Century," in *The Routledge Companion to Eighteenth-Century Philosophy*, edited by A. Garrett (London: Routledge, 2014), and "Reason, Calculating Machines and Efficient Causation," in *Efficient Causation: The History of a Concept*, edited by Tad Schmaltz (Oxford: Oxford University Press, 2014). Substantially altered sections from chapter 2 and 6 are to appear in the *Oxford Handbook of Leibniz*.

The epigraph to the fifth carry is from B. H. Fairchild, "The Art of the Lathe" from *The Art of the Lathe*. Copyright © 1998 by B. H. Fairchild. Reprinted with the permission of The Permissions Company, Inc., on behalf of Alice James Books, www.alicejamesbooks.org.

I dedicate *Reckoning with Matter* to my family in Nevada: my parents and sister nurtured the interests that animate this book, provided the material and early computational resources and/or toys that made it possible, and set the example of generosity the book attempts to emulate.

This book was envisioned just before Sophie came into our lives, first drafted as Athena did, and completed as Arete joined the merry band. Their intense love of books suggests they may read it someday—and, without doubt, uncover the remaining errors and infelicities, all which of remain my fault alone. Elizabeth Lee has lived with this project for nearly a decade. She stressed, in the words of John M. Ford, "Say what you mean. Bear witness. Iterate." She perfected every iteration of the manuscript and perfects every day together. 我愛妳們!

Conventions

In transcriptions of manuscripts, words added above a line are enclosed in slashes: "the \green/ cat" represents "green" added above the line between "the" and "cat."

Dates or other information in square brackets are conjectural; a question mark indicates a lower degree of certainty. Dates are given in the calendar form (Julian or Gregorian) used by the author.

Abbreviations

A   Leibniz, Gottfried Wilhelm. *Sämtliche Schriften und Briefe*. Edited by
           Deutsche Akademie der Wissenschaften, etc. Berlin, Munich, etc. 1923–.
           Cited A[series], [volume]:[page].
BL   British Library, Department of Manuscripts, London.
BNF   Bibliothèque nationale de France, Department of Manuscripts, Site
           Richelieu, Paris.
BPU   Bibliothèque de Genève, Department of Manuscripts, Geneva.
CKS   Kent History and Library Centre, Maidstone, Kent.
GP   Leibniz, Gottfried Wilhelm. *Die philosophischen Schriften*. Edited by
           K. Gerhardt. Berlin: Weidmann. 1875.
JM   Pascal, Blaise. *Oeuvres complètes*. Edited by Jean Mesnard. 4 vols. to date.
           Paris: Desclée de Brouwer, 1964–.
JS   Journal des Savants
LBr   Leibniz Briefwechseln, Gottfried-Wilhelm-Leibniz Bibliothek, Hanover,
           Germany
LH   Leibniz Handschriften, Gottfried-Wilhelm-Leibniz Bibliothek, Hanover,
           Germany
RS   Royal Society Archives, London
RS HS   J. F. W. Herschel Papers, Royal Society Archives, London
RT   Transcriptions of Babbage papers by C. J. D. Roberts, M.A., now available at
           https://web.archive.org/web/20071022212318/http://babbage.bravehost.com/
           (with * indicating that quotation not checked against original
           manuscript)

Notes

Introduction

1. Hoart 1870.

2. Charles Babbage, Buxton MS 7, Museum of the History of Science, Oxford. 7.11.1822, ff. 12–13; edition in Buxton and Hyman 1988, p. 60.

3. Preface to the *Memoirs of the Analytical Society*, 1813, p. xxi, in Babbage 1989, vol. 1, p. 59. For the significance of Bacon to Babbage, see Durand-Richard 1996, pp. 468–69.

4. For the importance of the study of failures, see the now classic discussion in Pinch and Bijker 1987, pp. 22–24.

5. Adam 1832, p. 400, punctuation modified. The Oughtred Society serves as the foremost organization dedicated to the study of these instruments.

6. For business machines from the late nineteenth to mid-twentieth century, see Cortada 2000; Yates 2000; Heide 2009; Warwick 1995.

7. For the history of early calculating machines, see especially Williams 1985 and Marguin 1994. Major works include Martin 1992 (now available in an expanded and illustrated form at "Rechenmaschinen Illustrated," http://www.rechenmaschinen-illustrated.com/); Archibald 1943; Bischoff 1990; Chase 1980; Goldstine 1972; Goldstine 1977; Ocagne 1986; Swartzlander 1995; and Stein and Kopp 2010. A fine source for continuing research is Stephan Weiss, "Beiträge zur Geschichte des mechanischen Rechnens," at http://www.mechrech.de/. For calculation before these machines, see Barnard 1916; Bennett 1987; and Pullan 1969.

8. Grade-school children in the United States are unlikely still to learn addition in such a manner. For an earlier generation of the new math, see Phillips 2014.

9. Morland 1673, pp. 13, 14.

10. Numeracy has been far less studied than literacy. See Thomas 1987; Netz 2002; and Bullnynck 2008.

11. The foremost study remains Cohen 1982; see also the sophisticated Emigh 2002. For evidence using "age-heaping" of a significant expansion of numeracy before the expansion of schools in Western Europe by 1600, see A'Hearn, Baten, and Crayen 2009. This technique has spawned something of a cottage industry of using numeracy as a proxy for "human capital" in economic history.

12. Thomas 1987, pp. 116–17.

13. Jones 2006, pp. 32–38, 277n49.

14. Berkeley 1948–57, 4:86.

15. Hobbes 1981, I, 1, 2; for the Latin, see Hobbes 1999.

16. See Erickson et al. 2013, ch. 1. For anxieties about symbolic reasoning in the eighteenth century, see Jones 2014.

17. For failures and successes in crossing the "savant-fabricant" divide, see Mokyr 2005a; and Jones 2008, ch. 4. Compare Lécuyer 2007 on the mix of manufacturing capacities and diverse management competencies in the history of Silicon Valley.

18. For discussions about the "overly socialized" history of technology, insufficiently attentive to materials, see Lécuyer and Brock 2006 in dialogue with Ceruzzi 2005; Constant 1999; Scranton 2000; and Williams 2000. From alternate theoretical traditions, see, notably, Ingold 2007 and Bennett 2010.

19. For healthy skepticism and a historical perspective on "innovation," see Godin 2008, alongside his other works.

20. See the formulations in Edgerton 1999, quotation at p. 126.

21. Hughes 1971 provides the classic example of a study of invention avoiding these pitfalls; see also Carlson 1999.

22. Arguments against patents from the mid-nineteenth century in Britain anteceded much of the more recent discussion, but they have long been less popular than accounts of heroic invention.

23. See Kelty 2008; Nuvolari 2005, 2004; Coleman 2012; and Sennett 2008, pp. 24–27. See also Nielsen 2012.

24. For collaboration among knowledge communities, see the discussion of "public proprietary knowledge" in Johnson 2009, ch. 8. For the gap between computer programming as science and craft, see Ensmenger 2010.

25. For the relation of models of the history of invention and accounts of mind, see the remarkable study McGee 1995.

26. See Sennett 2008.

27. For discussions of authorship and its problems in science studies, see Biagioli and Galison 2003.

28. Compare the case study and analysis in Jones 2008.

29. See Roberts, Schaffer, and Dear 2007, as well as Roberts and Inkster 2009, with a more global purview.

30. Ingold 2010, p. 92; Ingold 2007.

31. For personae, see Daston and Sibum 2003 and Condren and Hunter 2008.

32. For the relationship of natural philosophers and artisans, see the classic Rossi 1970 and Smith 2004a. For maker's knowledge, see Pérez-Ramos 1988.

33. For a recent survey of technical expertise in the early-modern period, see Ash 2010 and the accompanying papers.

34. Mahoney 1988, pp. 116–17.

35. For a powerful corrective to the idealist history of computers, see Priestley 2011; see also Agar 2003. For a strong articulation of the idealist view, see Davis 2001, pp. 186–87.

Chapter One

1. Pierre Petit to Jacques Buot, 23.9.1646, in Buot 1647, pp. 191–92; partial edition in JMII:345. Petit was no disinterested observer, as he had invented his own machine for aiding multiplication.

2. JMI:464.

3. Sir Balthasar Gerbier to Samuel Hartlib, 4.10.1648, Hartlib Correspondence (CD-ROM), 10/2/13A.

4. Pye 1968, pp. 20-24.

5. Most of the existing machines were for operations on nondecimal currencies; one is explicitly for computing with nondecimal lengths; and three are for "abstract" amounts. See Mourlevat 1988.

6. Morland 1673, p. 12.

7. JMII:341.

8. For Pascal's "marketing" campaign, see Descotes 1989.

9. The vocabulary of proxies and tools comes from Collins and Kusch 1998, p. 124.

10. On these issues, see especially Schaffer 1994 and Schaffer 1999.

11. Classic studies include Polanyi 1962; then Collins 1985, 1990, 2001; Collins and Kusch 1998. Important recent contributions are Sibum 1994; Jackson 2000; Iliffe 1995; Smith 2004a; and Mukerji 2006. For the case of Descartes and his relations with his artisan Ferrier, see Burnett 2005 and Gauvin 2006a. For the longer-term history, see Long 1991, 2001. See also the citations from economic history cited in chapters 3 and 4.

12. For another useful taxonomy, see Gordon 1988. Compare the distinction of "somatic" and "collective" tacit knowledge in Collins 2010, chs. 5-6.

13. Grabiner 1998 and Sibum 1994.

14. See Sennett 2008, ch. 5.

15. For invisible and visible technicians, see Shapin 1994, ch. 8. For artisanal spaces as places of scientific innovation and discovery, see, e.g., Jackson 2000; Pantalony 2004; and Smith 2004a.

16. Samuel Morland to Charles Stuart, sixth duke of Lennox and third duke of Richmond, 2.4.1669, BL Add. MS 21947, f. 217r. Compare Hilaire-Pérez 2013, ch. 6, for a rich study of the social and epistemic relations revealed in eighteenth-century London account books.

17. See Crinò 1955, 1957; discussed in Ratcliff 2007, pp. 175-78, and Miniati 1993.

18. For a fine exposition of calculating instruments using the principle of Napier's bones, see Williams 1983, pp. 276-88.

19. 8.10.1668, BL Add. MS 21951, ff. 7v-8r.

20. For Morland's calculating machines, see Ratcliff 2007 and Dickinson 1970.

21. Samuel Morland to Charles Stuart, sixth duke of Lennox and third duke of Richmond, 2.4.1669, BL Add. MS 21947, f. 217v.

22. For Blondeau, see Webster 1975, pp. 405-11; see Pepys's Diary for 24.11.1662.

23. Blondeau 1653, sig. A1v.

24. See Eagleton and Jardine 2005; Daumas 1953, p. 95; Willmoth 1993, p. 149.

25. See Pell to Collins, BL Add. MS 4278, ff. 117v, 118r.

26. Derham 1700, p. 88; Taylor 1954, #292.

27. See BL Add. MS 21947, ff. 203r, 204r, 206r, 207v, 208r-v.

28. For these processes of technological transfer, see the literature cited in chapters 3 and 4. The exemplar of Morland's machine presented to the Medici duke is signed by Henry Sutton and Samuel Knibb. Like Fromantle, Knibb was a well-known maker of luxurious clocks.

29. See Morland and Pell's correspondence concerning the quadrature of the circle: BL Add. MS 4417, ff. 47, 52.

30. Morland to [Tenison], from Hammersmith, 25.10.1694; 26.5.1695; 11.6.1695, Lambeth Palace MS 931, items 4-6, e.g., item 4, f. 1r.

31. See Marguin 1993 for an overview.

32. Leibniz to Johann Sebastian Haes, 5–6.1695, A3, 6:383.

33. See, e.g., Thomas 1987.

34. Moore 1681, p. 20; see also Willmoth 1993, pp. 148–49.

35. Moore's then famous arithmetic is slightly more descriptive about the process of carrying but just gives an example, no general rules. See Moore 1650, p. 20.

36. Morland 1673, p. 13. On books as advertisement for services, see Margócsy 2014, p. 112.

37. Morland 1673, p. 15.

38. For Pascal's machine, see Mourlevat, Jean-Baptiste, and Formento 1981; Mourlevat 1988.

39. JMII:337. He first described problems with using the counting board, an extremely widespread way of performing arithmetic using counters.

40. JMII:337.

41. As they are introduced only in the twentieth century, the terms "analog" and "digital" are anachronistic, but they clarify the issue. For the development of this distinction, see Mindell 2002, pp. 318–19.

42. The machine is described in Schickard to Johannes Kepler, 25.2.1624, in Schickard 2002, vol. 1, pp. 140–42. For these machines and their makers, see von Freytag Löringhoff 1978; Kistermann 1995; Kistermann 2001; Williams 1983, pp. 282–96; for the machine's carrying mechanism, see Stein and Kopp 2010, pp. 22–27. The rediscovery of this letter in the early twentieth century led to a predictably uninformative priority dispute between German and French scholars.

43. See Williams 1983, p. 285.

44. For the sautoir, see Fréchet 2004, pp. 357–61; Marguin 1994, pp. 54–55. The term likely refers to part of a harness serving as a stirrup.

45. JMII:336.

46. In his manuscript design of a calculating machine, Gilles Personne de Roberval, responsible for displaying and selling Pascal's machine, noted that Pascal's machine worked only in one direction: "1000. rangs irioent aussi facilemt que deux, et s'arreteroient aussi justemt, chacun en sa propre place. [. . .] il ne sçut trouuer le moien faire aller la Machine de gauche à droite, et de droit à gauche, sur les memes nombres." Roberval, "Pour la Machine d'Arithmetique à leues ou leviers, et Chappelets," [after 21.8.1669], Archives, Académie des sciences, Paris, Fonds Roberval (Suppl.) 116, f. 16r. I thank Alan Gabbey for pointing me to this manuscript and to this passage in particular. See also Carcavy to Leibniz, 5.12.1671 A2, 1:307.

47. For the importance of such corrections, see Collins and Kusch 1998, pp. 121–24. More advanced calculators and computers do not make this simple error in most cases but can be made to produce errors of this sort. See Collins 1990, pp. 62–70.

48. Marguin 1994, p. 46; Marguin 1993.

49. How and if Schickard solved this problem can only be discerned from reconstructions, which use a spring mechanism to keep the gears in digital positions.

50. Roberval, "Pour la Machine d'Arithmetique à leues ou leviers, et Chappelets," [after 21.8.1669], Archives, Académie des sciences, Paris, Fonds Roberval (Suppl.) 116, f. 16r. For subtraction on Pascal's machine, see Kistermann 1998.

51. A manuscript on the use of the machine notes, "Pour se server de cette machine il est necessaire de la poser a plat." Descotes 1986, p. 9.

52. Turner 2008, p. 280n50. Diderot's diagram suggests the pins were inserted.

53. For example, the elasticity of the springs must fit within a narrow range of values, and they must not change that elasticity too quickly over time; the weight of the parts must be set and not too variable. When Pascal was working on his machine, Hooke's law—stating that the force of a spring is proportional to its displacement—was unknown, and so were the constants

applicable to different springs made of different metals. Compare Leibniz's explicit reflections on this subject in chapter 2.

54. JMII:335–36 (italics mine). On this document, see Nagase 1998 and Meurillon 1982.

55. The known examples postdate Pascal's complaint; we do not know whom in particular Pascal was targeting. Jean-François Gauvin's dissertation presents the most thorough account of such arguments in early-modern France. See Gauvin 2008, as well as Gauvin 2006a.

56. JMII:340.

57. See, e.g., Roberval, "Pour la Machine d'Arithmetique à leues ou leviers, et Chappelets," [after 21.8.1669], Archives, Académie des sciences, Paris, Fonds Roberval (Suppl.) 166, f. 16r; Huygens to Leibniz, 27.12.1694, A3, 6:261.

58. For the ideology of engineers in the period, see Vérin 1993.

59. Sonenscher 1989, pp. 41–42, 60–64.

60. JMII:338–39.

61. Gauvin 2008, ch. 3; see also Marguin 1994, p. 53.

62. "Fake" scientific instruments intended for display and not use, especially astrolabes, were extremely common both in the Islamic world and in early-modern Europe.

63. JMII:338–39.

64. Here I'm glossing Michaux 2001, pp. 211–12.

65. JMII:338.

66. Compare Collins and Kusch 1998, p. 184.

67. JMII:339.

68. Compare Descartes on good and bad forms of habituation in mathematics in Jones 2006, ch. 1.

69. Domat 1989, p. 238.

70. Alder 1997, p. 132, offers an important reminder of how often engineers and managers remain in a position of weakness relative to producers and artisans, especially outside of conditions of strong monopoly capitalism.

71. JMII:340.

72. "Privilège de la machine arithmétique," 22.5.1649, JMII:712–15.

73. Roberval, BNF NAF 5175, ff. 5–8; Gabbey Catalogue # MEb 2: "Ce mémoire n'est que pour moy seul."

74. Roberval, "Pour la Machine d'Arithmetique à leues ou leviers, et Chappelets," [after 21.8.1669], Archives, Académie des sciences, Paris, Fonds Roberval (Suppl.) 116, f. 16r.

75. Arnauld to Perrier, 5.9.1673, in Arnauld 1775, pp. 714–15.

76. Grillet 1673, p. 4. For Grillet's machine, see Williams 1983, pp. 291–93. Williams was kind enough to share Grillet's manuscript with me; it appears to be a version of the *Suite de la nouvelle machine D'arithmetique de novveav rectifiee de l'invention dv sievr Grillet* in Grillet 1673.

77. Grillet 1673, p. 4.

78. For ubiquitous skills, see Collins and Kusch 1998, esp. pp. 144, 162, 185–87.

79. Desnoyers to Roberval, 26.6.1646.

80. Targosz 1982, p. 164. For Burattini, see Targosz 1977; Taton 1982. For the effort of the new monarchs to create a more absolutist court, au courant with the latest Western European developments, see Targosz 1995; see also the introduction to Favaro 1896.

81. See the references in chapter 4.

82. JS (5.1733), p. 867. See also Anonymous 1732, p. 116.

83. See Neher-Bernheim 1983.

84. Gallon 1735–77, esp. vol. 1.

85. JS (8.1751), pp. 510–11; this is an edition of the Procès-Verbaux of the Académie of the demonstration.

86. Neher-Bernheim 1983, pp. 393–95.

87. Ronfort 1989, p. 57.

88. Chaulnes 1768, p. 10.

89. JS (8.1751), p. 508.

90. Compare Bertucci 2006.

91. JS (8.1751), p. 511.

92. Among a vast literature on this topic, see, recently, Stalnaker 2010, ch. 3; Pannabecker 1998; Bender and Marrinan 2010.

93. Diderot, s.v. "Prospectus," Diderot and Alembert 1751–72, vol. 1, p. 4; see Stalnaker 2010, p. 100.

94. For the state of technical drawing in France, see Lavoisy 2004.

95. Diderot, s.v. "Arithmétique (machine)," Diderot and Alembert 1751–72, vol. 1, p. 681 (June 1751).

96. Diderot, s.v. "Prospectus," Diderot and Alembert 1751–72, vol. 1, p. 4.

97. JMII:338–39.

98. For an overview of the lowering of "access costs" to knowledge in the eighteenth century, see Mokyr 2005a; see also Jones 2008.

99. Morland 1695, pp. 31–32.

100. Morland 1695, pp. 32–33.

101. JMII:692.

102. Mdm. Périer, JMI:576–77; 608.

103. Pascal 2000, S617.

104. Pascal 2000, S617n402, following McKenna 1990. For these arguments, see Gouhier 1978, pp. 124–25; Orcibal 1950; Lesaulnier 1992.

105. Walton 1655, pp. 200–201.

First Carry

1. Babbage 1989, vol. 11, p. 86.

2. For the long-term history of tables, see Campbell-Kelly 2003. For the dangers of too narrowly conceiving Babbage's focus on producing tables, see Swade 2003, esp. ch. 3; Swade 2010. On the dangers of anachronism in proclaiming Babbage's importance for the history of computing, see Cohen 1990.

3. By convention, Difference Engine written with capital letters means one of Babbage's machines, and with lower letters, those of other inventors.

4. Daston 1994; Grier 2005; Grattan-Guinness 1990. For Babbage and contemporary table making, see Swade 2003, pt. I.

5. For the parallelism, see Bromley 1987.

6. Babbage planned mechanisms for nonconstant differences very early in the process. See Babbage to Bromhead, 9.2.1822 [RT*] and Babbage to Herschel, 9.4.1822, RS HS 2.171 [RT]. See also Collier 1990, pp. 107–16; Roberts 1987; Tee 1994, p. 135; Swade 2003, pp. 142–43; and Swade 2010.

7. Though the machine is often said to produce tables, it more precisely performs only subtabulation between given values, a point Maurice Wilkes insisted upon. See Swade 2003, pt. I for subtabulation.

8. On Difference Engine 0, see Taylor 1992; Tee 1994.

9. Babbage to Herschel, RS HS 2.173, received 12.6.1822 [RT].

10. Babbage to Bromhead, 24.3.1823 [RT*].

11. Swade 1995, 2001, 2005.

12. These remain understudied. For the serious study of the engineering drawings, see Bromley 1998; Bromley 2000; Bromley 1983.

13. See http://plan28.org/.

14. 7.11.1822, Buxton MS 7, Oxford Museum of the History of Science, p. 7; Buxton and Hyman 1988, p. 55. My thanks to Tony Simcock for providing his guidance and draft catalog notes concerning this complicated and misunderstood manuscript.

15. 7.11.1822, Buxton MS 7, Oxford Museum of the History of Science, pp. 8–9; Buxton and Hyman 1988, p. 58.

16. For technical descriptions of DE 1, see Roberts 1990a; Collier 1990, appendix; Lindgren 1990, pp. 239–42.

17. Lardner, in Babbage 1989, vol. 2, p. 155.

18. Lardner, in Babbage 1989, vol. 2, p. 158.

19. "On the Mathematical Powers of the Calculating Engine," 26.12.1837, in Babbage 1989, vol. 3, p. 38.

20. On these mechanisms in DE1, see Lindgren 1990, pp. 242–46; cf. Lardner, in Babbage 1989, vol. 2, p. 159.

21. Lardner, in Babbage 1989, vol. 2, p. 159.

22. Babbage to Bromhead, 10.8.1823 [RT*]

23. Babbage 1989, vol. 8, §319.

24. Babbage 1989, vol. 8, §320.

25. Babbage 1989, vol. 11, p. 85.

26. Herschel to Babbage, 10–17.4.1828, RS HS 2.225 [RT].

27. Herschel to Babbage, 12.2.1828, RS HS 2.219 [RT].

28. Babbage to Herschel, 9.5.1828, RS HS 2.226 [RT]. See Second Carry for a more detailed discussion.

29. For these letters, see Williams 1992; see also the notes scattered throughout RT.

30. For this contrast, see Pye 1968; for the importance of implementers with "competence" in the first industrial revolution, see Meisenzahl and Mokyr 2011.

31. Babbage to Herschel, 9.5.1828, RS HS 2.226 [RT].

32. Herschel to Babbage, 18.22.1827, RS HS 2.218 [RT].

33. Herschel to Babbage, 12.2.1828, RS HS 2.219 [RT].

34. Babbage to Herschel, 9.5.1828, RS HS 2.226 [RT].

35. For the DE2 design, see Swade 1995, pp. 68–69, with implications discussed in Swade 2010.

36. Babbage to Lord Ashley, 25.11.1829, BL Add. MS 37184, f. 432 [RT].

37. Duke of Somerset to Babbage, 7.11.1830, BL Add. MS 37185, f. 336 [RT].

38. Babbage 1989, vol. 11, p. 86. For the development of the anticipating carry, see Collier 1990, pp. 127–29.

39. Babbage 1989, vol. 11, p. 87.

40. Babbage 1989, vol. 3, p. 31.

41. See Bromley 1980, pp. 13–14, 33–34.

42. See, for example, Science Museum Archives [BAB] M11, 12.

43. Babbage 1989, vol. 3, p. 58; see Collier 1990, pp. 129–30.

44. Babbage 1989, vol. 3, p. 31.

45. Leibniz 1966, pp. 307–8.

Chapter Two

1. "Dissertatio exoterica de usu geometriae," [8–9.1676], A7, 6:488.

2. Leibniz to Johann Sebastian Haes, 29.3.1695, A3, 6:332. See also A3, 4:407.

3. See Leibniz to Gilbert Burnet, 10.9.1701, A1, 20:449. For Leibniz and China, see Perkins 2004; on his exchanges with the Jesuits about technical questions, see pp. 125–26.

4. These machines have not yet had sufficient historical and technical investigation. See the museum catalog Liu 1998, pp. 96–102; Li Di, Bai Shangshu, and Williams 1992; Bai Shangshu and Li Di 1980, esp. pp. 81–82; Gingerich 1986, and Graf 1994. My thanks to Daniel Asen and Stacey Van Vleet for translations from Mandarin.

5. Graf 1994. There were also multiplying and dividing machines based on Napier's bones.

6. In 1736, Jean-Baptist du Halde described native Chinese clockmakers in disparaging terms; see Pagani 2001, p. 76.

7. Pagani 2001.

8. Hilaire-Pérez and Verna 2006, p. 544. For "mental capital" in the eighteenth century, see Inkster 1990.

9. Leibniz to Arnauld, [early Nov. 1671], A2, 1*:286.

10. The remarkable manuscripts on the geometric, algebraic and even analytical machines have not to my knowledge been studied closely. See LH 35,3A,20, ff. 1–4 (a.k.a. Cc 816); LH 35,12,1, f. 13; A6, 3:412–13; LH 35,13,1, f. 408 (Cc 1069); LH 35,13,1, f. 444–45. For Leibniz's interest in cryptographic machines, see Rescher 2014.

11. For artisanal knowledge in early-modern Europe, see the citations in chapter 1.

12. Sonenscher 1989, p. 45; see likewise Alder 1997, pp. 129–32.

13. Recent critics of the social history of technology have suggested that such history is too socialized and not recognizant of the material; see, e.g., Lécuyer and Brock 2006.

14. For his trip to London, see "Observata philosophica in itinere Anglicano sub initium" [3.1673], A8, 1:3–19.

15. For the development of Leibniz's machine, see Morar 2014 as well as von Mackensen 1968a, which unfortunately was never published. Portions of the latter are summarized in von Mackensen 1969, 1968b; see also Wilberg 1977, pp. 11–28. For a more technical survey, see Stein and Kopp 2010, 2014; Badur and Rottstedt 2004 contains especially beautiful and illuminating diagrams, esp. pp. 129–35. The primary study of the development of the machine drawing upon the manuscripts is now the learned Walsdorf 2014, which came to my attention just as this book went to press.

16. Birch 1756–57, vol. 3, p. 73.

17. Draft Contract, [early 1673], LH 42,5, 61r, in von Mackensen 1968a, p. 163.

18. For Leibniz's discussion of the requirements of the "multiplying wheels," see especially "Project de la machine," [before 24.5.1673], LH 42,5, 54v; edition in von Mackensen 1968a, pp. 145–46. For the cogwheel, see also the undated LH 42,5, f. 29; Lehmann 1987; and Stein and Kopp 2010, pp. 37–40. In his earliest papers on a calculating machine, he considers performing multiplication directly.

19. For the stepped drum, see LH 42,5, f. 14r, edition in von Mackensen 1968a, pp. 168–69, which he dates to c. 1672; "Memoire pour Monsieur Ollivier touchant la machine arithmetique

perfectionée," 15.7.1677, LH 42,4, f. 7r, 8v; and the illustrated "Machine d'Arithmetiq[ue]," 8.5.1682, LH 42,4, f. 40; "Ma Machine Arithmetiq[ue] de la maniere que je l'ay fait faire à Paris, l'an 1674," 7.1685, LH 42,5, f. 11r; cf. dating of Stein and Kopp 2010, pp. 34, 41–42. For the influence of the stepped drum, see Kistermann 1999; compare, however, Johnston 1997, note 9. The first known published description of the stepped drum appears in Pütter 1765, pp. 243–46; see ch. 4n16 below.

20. See the remarkable thinking on paper in "Project de la machine," [before 24.5.1673], LH 42,5, 57r–v; edition in von Mackensen 1968a, p. 153.

21. "Project de la machine," [before 24.5.1673], LH 42,5, 57v; edition in von Mackensen 1968a, p. 154.

22. "Project de la machine," [before 24.5.1673], LH 42,5, 56v; edition in von Mackensen 1968a, p. 150. For Leibniz's solutions to the challenges of carry in more detail, see Stein and Kopp 2010, esp. pp. 43–49.

23. Leibniz to Oldenburg, 26.2.1672/3, in Oldenburg 1965–86, vol. 9, pp. 489, 493; see the discussion in Iliffe 1992, p. 38.

24. 5.3.1672/3, in Birch 1756–57, vol. 3, p. 77. Hooke's manuscript notes concerning the meeting do not contain any information not included in Birch; see RS MS 847.

25. 5.3.1672/3, in Hooke 1935.

26. 5.3.1672/3, in Birch 1756–57, vol. 3, p. 77.

27. For the role of such models, see Popplow 2002.

28. Hooke's papers in the Royal Society Archive contain no traces of the technical details of the machine. RS Classified Papers 20/54, f. 117r, records the existence of "Severall arithmetick [illeg deletion] engines." Some records, perhaps manuscript materials or models, still appear to have existed in the early eighteenth century. Waller, "Life of Hooke," in Hooke and Waller 1705, p. xix: "In 1674. he shew'd an Engine or Instrument to perform any Arithmetical Operation, but the more particular account of this and other Instruments not describ'd in this Volume, I shall reserve for another opportunity."

29. See Iliffe 1992, pp. 37–39.

30. Leibniz to Oldenburg, 26.2.1672/3, in Oldenburg 1965–86, vol. 9, pp. 489, 493.

31. Iliffe 1995, p. 314.

32. Willmoth 1993.

33. Daumas 1953, pp. 93–94. For their collaboration, see Iliffe 1995, pp. 289–90, 310–11, 313; Simpson 1989, pp. 47–48, 52–54.

34. 5.2.1672/3, in Birch 1756–57, vol. 3, p. 75.

35. See Clifton 1996, p. 250.

36. Hooke 1935, 5.3.1672/3.

37. John Pell noted Hooke's mention of his machine, and inquired after it: "I asked him What this way was. He answered. An Engine for multiplying and dividing." BL Add. MS 4422, f. 78r.

38. Hooke claimed to be holding back on revealing his design until Leibniz had revealed his. Hooke to Th. Haak for Leibniz, 4(14).7.1678, A3, 2:472; Robert Hooke to Th. Haak for Leibniz, [22.7.1680], A3, 3:235–36.

39. For Hooke, see Bennett 2006; Bennett 2003; Chapman 2005.

40. Iliffe 1995, pp. 286–87.

41. For another good example, see Leopold 1980.

42. "Saw Sir S. Morelands Arithmetick engines very silly." Hooke 1935, Entry for 31.1.

43. "Concerning Arithmetick Instruments," brought in by Robert Hooke, 7.5.1673, RS RBO/4/52, p. 198; printed version in Birch 1756–57, vol. 3, pp. 86–87.

44. On models, disclosure and early-modern privileges and patents, see Biagioli 2006a, esp. pp. 154–55.

45. Birch 1756–57, vol. 3, p. 386, discussed in Iliffe 1995, p. 298.

46. Leibniz to Jakob Bernoulli, 4.1703, in Bernoulli 1993, 109; discussed in Davillé 1920–22, part III, p. 38.

47. Leibniz to Christian Habbeus, 5.5.1673, A1, 1:417.

48. Leibniz to Christian Habbeus, 5.5.1673, A1, 1:416–17. On shopping for instruments in Paris, see Bennett 2002; see also Turner 1998 and Daumas 1953. For a roughly contemporary guide, see Blegny 1878, originally published in 1692.

49. Leibniz to Christian Habbeus, 5.5.1673, A1, 1:417.

50. In fact, Leibniz made off not just with artisanal skills but with information about Colbert's plan and means for acquiring artisanal skill. Leibniz to Christian Habbeus, 5.5.1673, A1, 1:417. Leibniz was as avid a collector of legal secrets as he was of natural philosophical ones—see chapter 3.

51. Leibniz to Christian Habbeus, 5.5.1673, A1, 1:416–17.

52. LH 38, 98r.

53. See, e.g., the signed contract of A3, 2:473–74.

54. Hansen to Leibniz, 5.6.1679, A1, 2:484–85; Hansen to Leibniz, 21.8.1679, A1, 2:511.

55. For a recent overview of the literature, see Hilaire-Pérez and Verna 2006; for the foreign skills essential to Britain's industrial revolution, see MacLeod 2004; Harris 1998; and Inkster 1990. For the close connection of patents and technology transfer, see Biagioli 2006a and MacLeod 1988. For Leibniz on the need for skilled workers in German industry, see Robinet 1994, p. 267, as well as Elster 1975.

56. Leibniz to Louis Ferrand, [5.1672], A1, 1:452; cf. Iliffe 1995, p. 315. For Leibniz and Ollivier, see now Walsdorf 2014, pp. 66–73.

57. Leibniz for Académie des sciences [early 1675], LH 42,5, f. 33v; von Mackensen 1968a, p. 177.

58. For Parisian instrument makers, see the classic Daumas 1953, esp. pp. 97–103, as well as the more recent Bennett 2002 and Turner 1998; Ollivier does not appear in any of the major guides to early-modern instrument makers or clockmakers, though Leibniz's correspondence makes clear he was known to various savants as a skilled maker of clocks and instruments.

59. Leibniz sets out Ollivier's terms in detail in Leibniz for the Académie des sciences [?] [1674–75], in von Mackensen 1968a, pp. 169–73.

60. Leibniz quickly got himself into trouble with the Royal Society by rashly claiming that his machine was finished. Leibniz to Oldenburg, 15.7.1674, in Oldenburg 1965–86, vol. 9, p. 44; Oldenburg to Leibniz, 8.12.74, in Oldenburg 1965–86, vol. 9, p. 141.

61. Leibniz to Oldenburg, 26.2.1672/3, in Oldenburg 1965–86, vol. 9, pp. 489, 493.

62. Ollivier to Leibniz, 24.5.1677, A3, 2:148.

63. Ollivier to Leibniz, 19.6.1677, A3, 2:164.

64. Wilberg 1977, pp. 33–37, appears to take the draft contract discussed below, [probably from early 1679], LBr 119, ff. 27r–29r; A3, 2:598–602, to be the memoire.

65. For a recent survey of early-modern technical drawing, see the essays in Lefèvre 2004.

66. See, for example, Lehmann 1993; Badur and Rottstedt 2004; Stein et al. 2006; and Stein and Kopp 2010; Badur and Rottstedt 2004; Stein and Kopp 2014; see the critique in Morar 2009, 2014.

67. See, e.g., Elster 1975, p. 78.

68. Leibniz, "Memoire pour Monsieur Ollivier touchant la machine arithmetique perfectionée," 15.7.1677, LH 42,4, f. 7v (italics mine).

69. "Project de la machine arithmetique," [early 1673, before 24.5.1673], LH 42,5, 56v; von Mackensen 1968a, p. 150.

70. "Memoire pour Monsieur Ollivier touchant la machine arithmetique perfectionée," 15.7.1677, LH 42,4, f. 7r.

71. See chapter 5 and the nuanced discussions in Alder 1997, pp. 129, 136–53; Lubar 1995, esp. pp. S70–74; Lavoisy 2004.

72. LH 38,277r. Published in *Journal de Trévoux*, 1718; edition in Leibniz 1906, 131.

73. Leibniz for [Académie?], [before 15.12.1676], LH 42,5, f. 37v, 38r; von Mackensen 1968a, pp. 172, 173.

74. Pye 1968, pp. 20–24.

75. Guiffrey 1881, c. 781.

76. Leibniz, "Memoire pour Monsieur Ollivier touchant la machine arithmetique perfectionée," 15.7.1677, LH 42,4, f. 8v.

77. Ollivier to Leibniz, 11.4.[1678], LH 42,4, 20r; noted as lost in A3, 2:384.

78. Ollivier to Leibniz, 15.11.1678, A3, 2:536.

79. See discussion of difficulties with calculating machine: 22.8.1679, LH 42,4, 1r.

80. Ollivier to Leibniz, 13.11.1678, A3, 2:536.

81. Leibniz's papers contain only a heavily edited draft of the contract, at LBr 119, ff. 27r–29r.

82. Leibniz contract for Ollivier, [mid 1.1679], A3, 2:599. Compare the analysis in Wilberg 1977, pp. 33–37.

83. Leibniz contract for Ollivier, [mid 1.1679], A3, 2:601–2.

84. Leibniz contract for Ollivier, [mid 1.1679], A3, 2:602.

85. Hansen (now in Oxford) to Leibniz, 28.9.1680, A1, 2:431. See the similar approach in Walsdorf 2014, pp. 75–76.

86. In one letter, for example, Hansen explained that he had sought Ollivier at his boutique ten times, only to find it closed. Hansen to Leibniz, 4.10.1677, A1, 2:294; A3, 2:242.

87. Hansen to Leibniz, 31.10.1678, A1, 2:375; Hansen to Leibniz, 14.11.1678, A1, 2:382.

88. Leibniz's concern with secrecy echoes throughout the correspondence with Hansen. Mariotte promised to serve Leibniz "sub fide silentij." Compare Hansen to Leibniz, 1/11.4.1678, A1, 2:331 (where Hansen promises not to say anything to Mariotte for the time being) with Hansen to Leibniz, 13.6.1678, A1, 2:342 (the promise is said to be a quotation from Mariotte).

89. Hansen to Leibniz, 7.11.1678, A1, 2:380.

90. Hansen to Leibniz, 7.11.1678, A1, 2:380.

91. The contract exists only as a heavily revised draft; there is no fair copy; and the letters never appear to refer to it as signed.

92. Hansen to Leibniz, 30.7.1677, A1, 2:283.

93. Hansen to Leibniz, 11.10.1677, A1, 2:297.

94. Ollivier to Leibniz, 24.5.1677, A3, 2:148; Ollivier to Leibniz, 19.6.1677, A3, 2:164.

95. Hansen to Leibniz, 9.8.1677, A1, 2:288.

96. Hansen to Leibniz, 4.10.1677, A1, 2:242:

97. Hansen to Leibniz, 30.7.1677, A1, 2:283.

98. Leibniz for the Académie des sciences [?], [1674–75], in von Mackensen 1968a, p. 170.

99. In a companion document, Leibniz indicates that the subsistence came to forty sous (two livres) a day for Ollivier alone, and an ecu (sixty sous, three livres) if his "boy" was also

engaged in the project. Leibniz for the Académie des sciences [early 1675], LH 42,5, f. 33r–v; von Mackensen 1968a, p. 176.

100. Leibniz for the Académie des sciences [?],[1674–75], in von Mackensen 1968a, p. 170.

101. Leibniz for the Académie des sciences [?], [1674–75], in von Mackensen 1968a, p. 171.

102. See the remarkable description of Leibniz's back and forth with Ollivier at Leibniz for the Académie des sciences [?],[1674–75], in von Mackensen 1968a, pp. 174–75.

103. Ollivier to Leibniz 24.5.1677, A3, 2:148.

104. Ollivier to Leibniz 24.5.1677, A3, 2:148 (italics mine). Again, the next year: Ollivier to Leibniz, 11.4.[1678], LH 42,4, 20r.

105. For artisans and the ownership of labor from this period to the nineteenth century, see Sonenscher 1989 and Rule 1987; see also Linebaugh 2003.

106. Hansen to Leibniz, 20.12.[1677], A3, 2:296. So difficult was the production of machines in Germany, however, that years later Leibniz suggested to l'Hôpital that it might be better to send a completed machine for France to be copied and reproduced in greater numbers. See the letter of 21.10.1697, A3, 7:393.

107. LH 38,277v (1714–15?).

108. Leibniz, draft contract for Ollivier, [mid 1.1679], A3, 2:602.

109. See, for example, Christian Philipp [Hamburg] to Leibniz, 1/11.10.1679, A1, 2:520; Tschirnhaus to Leibniz, 5.12.1679, A3, 2:906; Johann Jakob Ferguson to Leibniz, 10.9.1682, A3, 3:735.

110. Hansen to Leibniz, 21.8.79, A1, 2:511.

111. See A3, 2:806. Leibniz to Johann Friedrich, 5.1678, A1, 2:175; von Mackensen 1968a, pp. 93, 117; cf. Wilberg 1977, p. 42.

112. On invisible technicians, see Shapin 1994, ch. 8.

113. Morar 2009, pp. 21–22, citing LBr 489 and A1, 17:11.

114. On the problematic nature of the nineteenth-century reconstructions, see Morar 2014.

115. Version of the verb "to perfect" (perfecter, perfectieren, etc.) appears throughout the correspondence in describing the nature of this activity.

116. Leibniz to Landgrave Karl of Hessen-Kassel, 7.1.1701, A1, 19:334. The correspondence with and around Wagner, Buchta, and Teuber is available in preliminary transcriptions in series A1 and A3. See Transkriptionen für die Leibniz-Akademieausgabe der Leibniz-Forschungsstelle Hannover, available at http://www.gwlb.de/Leibniz/Leibnizarchiv/Veroeffentlichungen/Transkriptionen.htm; as of this writing, these transcriptions do not yet include the crucial diagrams and charts. On the Helmstedt stage, see Stein 1888; Wilberg 1977; Scheel 2001; Morar 2014; and Walsdorf 2014, pp. 90–98. For analysis of Wagner's trials, with illustrations, see Badur and Rottstedt 2004, pp. 139–43.

117. See Leibniz's letters of recommendation to the chancellor and duke, A1, 19:108–11.

118. Wagner to Leibniz, 15.2.1701; Wagner to Leibniz, 22.10.1706, in Transkriptionen.

119. See Schmiedecke 1969 and Walsdorf 2014, pp. 99–101. For portions of Leibniz's correspondence with Teuber, see Leibniz 1845, to be used in conjunction with Transkriptionen and the manuscript letters in LBr 916.

120. Leibniz to Teuber, 8.4.1712, in Leibniz 1845, p. 23.

121. Leibniz to Teuber, 3.1.1712, in Leibniz 1845 and see Leibniz to Teuber, 23.9.1714 and 7.10.1714, in Transkriptionen. Morar 2009 discusses important aspects of this stage of the construction and the nature of the work relationship (esp. at p. 46).

122. See Leibniz to Teuber, 20.2.1711, in Leibniz 1845.

123. Leibniz to Buchta, 15.11.1713. in Transkriptionen.

124. Leibniz to Buchta, 21.3.1714. in Transkriptionen.

125. For the state of the machine at Leibniz's death, see Morar 2009, pp. 23, 29, drawing upon the letters of Teuber.

126. For a sophisticated account of the problems of claims about whether the machine "works," see Morar 2009 and Morar 2014.

127. A1, 11:168; discussed in Robinet 1994, p. 267. Leibniz contradicts this elsewhere with praise for mere implementors; see Leibniz to Johann Friedrich, [8?].1679, A1, 2:188.

128. Leibniz to Samuel Clarke, [1716], GP7:357.

129. For Leibniz's account of the organic as a perfectly organized machine to infinity, see Smith 2011 and Duchesneau 2003.

130. [*Monadologie*], 1714, GP6:618.

131. See Benziger 1951.

132. [*Monadologie*], 1714, GP6:611.

133. *Nouveaux essais sur l'entendement humaine*, [1703–4] A6, 6:50–51.

134. See Jones 2006, ch. 6.

135. "Recommandation pour instituer la science générale," [4–10.1686(?)], A6, 4:713, 712.

136. Among others, see the remarkable "Discours touchant la Méthode de la Certitude" [8.1688–10.1690?], A6, 4:959–61.

137. A6, 4:961.

138. Leibniz's detailed studies on elasticity, during the period of the invention of the calculating machine and the early steps of the calculus, comprise a major part of A8, 1. For the importance of elasticity for Leibniz, see Bertoloni Meli 1993, pp. 50–55.

139. A6, 4:586; Leibniz 1969, p. 291, trans. modified. See Summers 1987, pp. 182–83.

140. For anxieties around skill in assaying within natural philosophy, see Schaffer 2002; see also Pastorino 2009.

141. A6, 6:262. For a fine case study about mathematicians and the traditions of skilled gauging of barrels, see Grabiner 1998; cf. Alder 1997, p. 150.

142. A6, 6:262.

143. "Recommandation pour instituer la science générale," [4–10.1686(?)], A6, 4:710–11.

144. "Recommandation pour instituer la science générale," [4–10.1686(?)], A6, 4:711.

145. "Recommandation pour instituer la science générale," [4–10.1686(?)], A6, 4:712.

146. Leibniz to Malebranche, 22.6.1679, A2, 1:720.

147. This was likely part of an attempt to gain membership in the French Académie des sciences.

148. Leibniz to Johann Friedrich, 21.1.1675, A1, 1:493.

149. Compare Wakefield 2010 and, more generally, Antognazza 2009.

Second Carry

1. Draft of Statement, 16.12.1828, BL Add. MS 37184, f. 163 [RT]. For the initial effort to gain government support, see Collier 1990, pp. 32–43; Roberts 1990b; Swade 2001, pp. 32–38. The manuscripts transcribed in [RT], especially the documents from the Treasury, fill out these accounts. I have not independently checked the Treasury manuscripts.

2. For the fortunes of the reputation of the inventor in Britain, see MacLeod 2007.

3. These are Babbage's memories of the meeting written up seven years later; Babbage to Lord Goderich, 10.1.1830, BL Add. MS 37185, f. 10 [RT].

4. Edward Bromhead to Babbage, undated [c. 3–5.1823], BL Add. MS 37182, ff. 433–34; see Collier 1990, 40.

5. For these figures, see the detailed analysis of Roberts 1990c; same amounts in Swade 2001, p. 67.

6. For his demonstrations, see Schaffer 1994.

7. Babbage to Bromhead, 14.9.1822 [RT*].

8. Babbage 1822, pp. 11, 12. See Swade 2003, pp. 119–21.

9. Babbage to Bromhead, 28.8.1822 [RT*].

10. See Fox 2009, p. 463, on Babbage and Brewster on the need for money allowing disinterest.

11. Babbage 1822, p. 11. On the thematic of the realization of bits of Laputa, see Fox 2009, p. 465.

12. Peel to Croker, 6.3.1823, in Jennings 1885, vol. 1, p. 263.

13. Peel to Croker, 6.3.1823, in Jennings 1885, vol. 1, p. 263.

14. Peel to Sir Humphry Davy, 13.12.1824, in Parker 1891, p. 364; MacLeod 1971, p. 83.

15. Peel to Croker, 6.3.1823, in Jennings 1885, vol. 1, p. 263.

16. Croker to Peel, 21.3.1823, in Jennings 1885, vol. 1, p. 264; see the discussion in Collier 1990, pp. 37–38.

17. Minute of the Board of the Treasury, 25.3.1823, Treasury Office, Whitehall, T29/219, f. 427 [RT*].

18. Francis Baily to Babbage, 16.4.1823, BL Add. MS 37182, f. 12.

19. Secretary of the Royal Society to the Treasury, 1.5.1823; Treasury 1823–9621 [RT*]. For Babbage's intelligence of—and attempts to control—the deliberations, see Charles Babbage, Manuscript Journal, 1820–25; Waseda University Library Manuscript Collection, Tokyo, pp. 56–57. See George Airy's comment quoted in Swade 2001, p. 37.

20. Babbage became well known for his criticism of the Royal Society; see Babbage 1989, vol. 7, ch. 4, esp. pp. 33–52.

21. Johns 2009, pp. 259–64. For the range of alternatives to the Royal Society, see Stewart 1999; for a nuanced view of the relationship of the Royal Society and practical invention in the eighteenth century, see Miller 1999.

22. See MacLeod 2007 for the program of inventive genius, pp. 145–52; for the radical critique, see pp. 161–70. The latter prefigures recent literature on collective invention, such as Allen 1983; Nuvolari 2004; as well as Basalla 1988.

23. Davies Gilbert to Babbage, 28.5.1823, BL Add. MS 37172, f. 32 [RT].

24. Davies Gilbert and other members of the Royal Society, 3.7.1823, Treasury 1823–15012 [RT*].

25. Address of Henry Thomas Colebrook . . . on Presenting the Gold Medal to Charles Babbage, 18.7.1823, in Babbage 1889, p. 224.

26. For the importance of this wording, see Roberts 1990b.

27. Anonymous 1828, copy in RS HS 27.54.

28. Warrant for Babbage, 3.8.1823, T52/111, f. 1 [RT*].

29. See Turvey 1991; the theme is stressed in Collier 1990.

30. Draft of Statement, 16.12.1828, BL Add. MS 37184, f. 163 [RT]; Swade 2001; Collier 1990; Roberts 1990b document the subsequent story in detail.

31. Chancellor of Exchequer H. Goulburn to Lord Ashley [copy], 13.12.1829, BL Add. MS 37184, f. 451–52. [RT]; discussed in Collier 1990, p. 67.

32. J. Stewart to President of Royal Society, 24.12.1828, T27/89, f. 27 [RT*]; RS DM/4/133.

33. Babbage had suggested Herschel. Babbage to Duke of Wellington, 11.12.1828; Treasury 1828–22672 [RT*]; also BL Add. MS 37184, f. 156.

34. "Babbage Engine Committee," 26.1.1829, RS DM/4/135.

35. Indeed, Herschel's drafts indicate that he considered adding more detail about Babbage's mechanical notation, but such theoretical innovation is not focused upon in the finished report. Herschel, "Notes for the Report," n.d. [late 1828–early 1829], RS HS 27.54.

36. "Report of the Committed Appointed . . . to consider the subject referred in Mr. Stewert's Letter Relative to Mr. Babbage's Calculating Engine," 2.1829, in Babbage 1889, p. 233.

37. "Report of the Committed Appointed . . . to consider the subject referred in Mr. Stewert's Letter Relative to Mr. Babbage's Calculating Engine," 2.1829, in Babbage 1889, p. 234.

38. This is a late addition to the draft; "the parts of the wheelwork on which the calculations are registered" is canceled. Manuscript of "Report of the Committed Appointed . . . to consider the subject referred in Mr. Stewert's Letter Relative to Mr. Babbage's Calculating Engine," 2.1829, RS DM/4/135; the final version is in RS CMB/1/26.

39. Herschel, "Actual State of the Machine and Drawings," n.d. [late 1828–early 1829], RS HS 27.54. All canceled with two vertical lines; not in final report.

40. Manuscript of "Report of the Committed Appointed . . . to consider the subject referred in Mr. Stewert's Letter Relative to Mr. Babbage's Calculating Engine," 2.1829, RS DM/4/135.

41. "Minutes of the Council of the Royal Society relating to the Report of the Committee on Mr. Babbage's Calculating Machine," 12.2.1829, in Babbage 1889, p. 232.

42. Extract from a Treasury Minute no. 2964, 28.4.1829, RS DM/4/138.

43. See the reprints of the report and the council approval in *Gleanings in Science*, 2(1830), pp. 96–98.

44. Babbage, Memorandum of Interview of Lord Ashley with Chancellor, 24.2.1830, BL Add. MS 37185, f. 69 [RT]; see Collier 1990, p. 70.

45. Peel to William Buckland, BL Add. MS 40,514, f. 223, quoted in Hyman 1982, pp. 190–91.

46. See Turvey 1991 for a reconstruction of the events.

47. Babbage, Notes for Interview with Sir R. Peel, 11.1842, BL Add. MS 37192, f. 182 [RT; emendations from RT]

48. Babbage, Recollections of an Interview with Sir R. Peel, 11.11.1842, BL Add. MS 37192, f. 189 [RT].

49. Babbage, Recollections of an Interview with Sir R. Peel, 11.11.1842, BL Add. MS 37192, f. 189 [RT].

Chapter Three

1. Leibniz for Académie des sciences [early 1675], LH 42,5, f. 33r–v; edition in von Mackensen 1968a, pp. 175–76.

2. See the considerable recent literature on early-modern capitalism, state contracting and new natural knowledge, including Smith 1994; Nummedal 2007; Cook 2007; Margócsy 2014 and the essays in Smith and Findlen 2002; For the growing niches for technical expertise in the Enlightened cameralist state, see Heilbron 1993 and Heilbron 2011.

3. For entrepreneurship in early-modern France, see, e.g., Mukerji 2002, 2006.

4. A1, 2:154; see Grua 1956, p. 329.

5. See Woodmansee 1994, esp. p. 36; for the ubiquity of the romantic author in modern American intellectual property jurisprudence, see Boyle 1996.

6. For the importance of collective invention in the development of steam engines after Watt, see, for example, Nuvolari 2004; for the collective production of the Canal de Midi, see Mukerji 2009.

7. Pottage and Sherman 2010.

8. "Lettre dédicatoire a Monseigneur le Chancelier . . . Avec un avis nécessaire," 1645, JMII:340.

9. "Privilège de la machine arithmétique," 22.5.1649, JMII:714.

10. JMII:712. See Biagioli 2006a, p. 172n39 and see the study of the privilege in Gauvin 2008.

11. JMII:713.

12. The most extended treatment of the machine is Mourlevat 1988.

13. Elsewhere the privilege suggests that it covers all instruments for aiding arithmetic; the discussion of "principal invention and essential movement" here seems to limit it to machines that are supposed to be able to perform carries automatically.

14. Prager 1964, pp. 281–82.

15. Breaking with older teleological accounts of the development of patent systems, recent scholarship has emphasized the gulf between the privileges of early-modern states and the patent systems developed from the revolutionary period of the late eighteenth century onward. The revisionist literature includes Biagioli 2006c; Biagioli 2006a; Bracha 2004; Bracha 2005 and Johns 2009. These works build on Belfanti 2004; MacLeod 1988; Hilaire-Pérez 2006; Long 1991, 2001. For France before 1700, see Isoré 1937; Prager 1964; Frumkin 1947–48; Silberstein 1961, pp. 209–52.

16. Belfanti 2006, pp. 335–36; for artisans, see especially the key study Long 1991. Long stresses a lengthy gestation of ideas about the sense of ownership of skills and procedures by craftspeople.

17. A privilege for royal manufacture in France, for example, often included allowances for nobility to be involved, exemptions from taxes, the naturalization of foreign workers, a monopoly on manufacture and sale, and a requirement that manufacturers produce at a given level. Cole 1939, vol. 2, p. 136; many examples given in Boissonnade 1932, e.g., 328–32. See also the privilege for artisans, including mathematicians, to live in the Louvre in Colbert, Clément, and Brotonne 1861, vol. 5, pp. 526–27. Leibniz remarked on these aspects of Colbert's privileges: see Leibniz 1906, p. 141.

18. Compare, however, the arguments against such a sharp division in Isoré 1937, pp. 102–7, and the subsequent examples.

19. For the English case, which is very clear, see Bracha 2004, pp. 12–14; Bracha 2005, pp. 17–20. For the French case, where working clauses were largely implied, see Isoré 1937, pp. 108–9, but see pp. 115–16; Hilaire-Pérez 1991, p. 914. Privileges that did not result in a working manufactory seem to have been simply treated as nullified. Germany seems largely to be an exception to the requirement for reduction to practice; see Popplow 1998, p. 107, drawing on Silberstein 1961.

20. JMII:340.

21. JMII:713.

22. JMII:713–14. The use of the past and present tense should not deceive one into thinking an early privilege existed; such use of tenses was boilerplate in privileges. The granting of patrimonial rights to successors was standard in French privileges through the mid-eighteenth century. See Isoré 1937, pp. 109–11.

23. Noted by Prager 1964, p. 281.

24. For the details of the manuscripts of Pascal's privilege, see JMII:711–12. Whether the original exists is unknown.

25. For the practice of using the two types of wax, see Isoré 1937, p. 117. For Pascal's privilege, see JMII:711; the original is lost, but the copyist of the Guerrier version noted "Copié sur l'original en parchemin, scellé du grand sceau de cire jaune" (JMII:715, note *).

26. Biagioli 2006c, p. 1146.

27. For counterfeiters, see Isoré 1937, pp. 114–15, and esp. Gauvin 2008, pp. 228–34.

28. JMII:713; my italics.

29. JMII:713. For this distinction in modern patent jurisprudence, see Pottage and Sherman 2010, pp. 30–32.

30. JMII:338.

31. Compare the case of Babbage in Schaffer 1994.

32. Belfanti 2006, pp. 335–36; Long 1991.

33. Biagioli 2006a, p. 143.

34. Strongly emphasized in Bracha 2004 and Biagioli 2006c.

35. For the patronage relationship, see Meurillon 2001, pp. 96–97; Richelieu called upon Etienne Pascal to help investigate a scheme to determine longitude. See JMII:82–6; Prager 1964, pp. 273–77.

36. According to some accounts, Pascal initially named the machine the "Séguier."

37. See Meurillon 2001, pp. 96–97, which notes that the dedicatory letter stresses the honor Séguier did to the entire Pascal family.

38. Solon 1905, p. 117; the privilege is in Colbert, Clément, and Brotonne 1861, vol. 5, pp. 548–49.

39. Although the effort collapsed a few years later, the system of regularly departing carriages with standardized routes is the best known and most successful of the inventive efforts of Pascal and the duc de Roannez, who shared religious views and a certain entrepreneurial spirit. See Mesnard 1965, p. 766.

40. The king referred their petition to his council, headed by Colbert. "Arrêt pour accorder à Messieurs les ducs de Roannez, marquis de Sourches et de Crenan, la permission d'établir des carroisses publics dans la ville et faubourgs de Paris," 19.1.1662, JMIV:1397.

41. See Mesnard 1985; Sturdy 1995, p. 56. Not enough is known about privileges in mid-century France to state how important Séguier was. Ismael Boulliau to Huygens, 21.6.1658, in Huygens 1888, vol. 2, p. 186, noted that the "grace" of a privilege depends "absolument de M. le Chancelier."

42. Compare the case of the intellectualization of seeing in the creation of the academies for art, discussed in Heinich 1993.

43. See Hamscher 1987, pp. 16–17; Church 1972, p. 338; Kerviler 1874, 97–102.

44. Mellot 1984.

45. See JMI:464; Meurillon 2001.

46. For the Pascals in Normandy, see Beaurepaire 1900–1901; Cléro 2001; Pouzet 2001, pp. 75–85, as well as Gauvin 2008. For this revolt, see Ménard 2005; Verthamont and Floquet 1842; Foisil 1970.

47. On the limits to absolutism, especially fiscal limits, see Collins 1988.

48. See, amid a large literature, Bonney 1978, p. 26; Parker 1989, pp. 50–51; and Church 1972, pp. 269–72.

49. "Lettre à la sérénissime reine de Suède," 6.1652, JMII:924 (italics mine).

50. "Lettre à la sérénissime reine de Suède," 6.1652, JMII:925.

51. "Lettre à la sérénissime reine de Suède," 6.1652, JMII:924.

52. "Lettre à la sérénissime reine de Suède," 6.1652, JMII:924.

53. For England, see the treatment in Bracha 2004, pp. 17–18.

54. Calendar of State Papers, Domestic, Charles II, 16–28.2.1664. See the discussions in MacLeod 1988, pp. 16, 32; Hulme 1917, pp. 65–67.

55. See Isoré 1937, p. 101.

56. Receuil Le Nain, Conseil, vol. 49, pt. 2, f. 334v, quoted in Isoré 1937, p. 112; see also p. 115.

57. Arrêt de *Parlement* de Paris, 7.2.1662, JMIV:1401.

58. Ismaël Boulliau to Huygens, 21.6.1658, in Huygens 1888, vol. 2, pp. 185–86; discussed in Gauvin 2008, p. 238, n112.

59. Biagioli 2006c, p. 1143; for an earlier incisive articulation of the shift engendered by speci-fication, see Gomme 1946, pp. 26–27.

60. "The language of privilege, of royal sanctioned rights, often served as the unacknowl-edged partner to *liberté*." Smith 1995, p. 238.

61. See the fine example in Hanley 1997.

62. Leibniz to Johann Sebastian Haes, [5–6.1695], A3, 6:383–84. For Leibniz on Morland, see [1674], LH 42,5, f. 10v.

63. Warrant of Charles II to Samuel Morland, 8.8.1667, Museum of the History of Science, Oxford, MS Gunther 49, discussed in Ratcliff 2007.

64. London *Gazette*, 16.4.1668.

65. London *Gazette*, 18.6.1666; quoted in Walker 1973, p. 113.

66. Walker 1973, p. 113.

67. MacLeod 1988, p. 34.

68. See BL Add. MS 21947, ff. 204, 208, 217.

69. See, e.g., Morland to Her Grace the duchess of Laderdal [i.e., Lauderdale], in Windsor, 20.12.1681, BL Add. MS 35125, ff. 333r–4v.

70. For "the conventions of commercial practitioners who used the designing of suppos-edly unique instruments as a means of marketing their skills and publications," see Willmoth 1993, p. 144; compare Biagioli 2006b, pp. 7–13; Biagioli 2006a, p. 165; and Margócsy 2014.

71. Letter from Sir Samuel Morland to Tenison, [1689], containing an account of his life, Lambeth Palace MS 931, item 1; edition in Dickinson 1970, pp. 117, 118.

72. For the development of specification, see, e.g., Dutton 1984, p. 75.

73. For Morland's pumps, see Dickinson 1970, pp. 56–73.

74. Patent #175, quoted in Dickinson 1970, p. 61.

75. See Biagioli 2006a, pp. 154–57.

76. See 29.1.1674, *Journal of the House of Commons*, 9: 1667–87 (1802), pp. 300–301; 24.2.1674, *Journal of the House of Commons*. 9: 1667–87 (1802), p. 314.

77. "REASONS Offered against the Passing of Sir Samuel Morland's Bill TOUCHING WATER-ENGINES," [n.d.], BL Shelfmark 816.m.10.(94).

78. Magalotti 1980, p. 62.

79. See Thirsk 1978, p. 11.

80. "*Sir* SAMUEL MORLAND'*s ANSWER to several Papers of* Reason *against his Bill for his New Water-Engins*," [n.d., 3–4.1677], BL Shelfmark 816.m.10.(93), recto. Another copy is at BL Cup.645.b.11.(23.). This document is summarized in CSP [April 2?].1677, pp. 73–74.

81. "*Sir* SAMUEL MORLAND'*s ANSWER to several Papers of* Reason *against his Bill for his New Water-Engins*," [n.d., 3–4.1677], BL Shelfmark 816.m.10.(93), recto.

82. BL Shelfmark 816.m.10.91.

83. "De Machina ad usum transferenda," LH 45,5, 19v. Compare the strategy of Huygens and his partners for a new form of carriage in Jansen 1951, pp. 173–74; see also Huygens's strategies around the pendulum clock, discussed in Howard 2008.

84. Louis Ferrand to Leibniz, 11.2.1672, A1, 1:183.

85. Compare the protocols for priority at the Royal Society and elsewhere in Iliffe 1992.

86. Pierre de Carcavy to Leibniz, 12.1671, A2, 1*:307, 308.

87. For Carcavy's job of running "the tandem collections" of the royal library "as a machine of public administration," see Soll 2009, pp. 99–101.

88. See the fine example in Leibniz 1906, p. 141; Leibniz to Christian Habbeus, 5.5.1673, A1, 1:416.

89. Leibniz to Christian Habbeus, 5.5.1673, A1, 1:416.

90. Leibniz to Louis Ferrand, [5.1672], A1, 1:452.

91. Pierre de Carcavy to Leibniz, 5.12.1671, A2, 1:307.

92. See Oldenburg to Leibniz, 30.1.1672/3, in Oldenburg 1965–86, vol. 9, p. 431.

93. [1674], LH 42,5, f. 10v.

94. [1674], LH 42,5, f. 10v; compare the similar discussion at LH 42,4, f. 67v.

95. Leibniz to Elisabeth of Bohemia (?), [11.1678], A2, 1*:661.

96. Salomon-Bayet 1978, p. 165.

97. For Dalencé's activities as a broker of inventions and privileges, see BNF NAF 560, ff. 197, 198, 200. Leibniz called upon him later to obtain a privilege for Martinus Lipenius, but Dalencé accidentally had it issued to Leibniz (A1, 3:522).

98. Leibniz for [Académie?], [before 15.12.1676], LH 42,5, f. 37v, 38r; von Mackensen 1968a, pp. 172, 173.

99. A6, 4:712.

100. Guiffrey 1881, c. 781.

101. Leibniz to Johann Friedrich von Linsingen, 4.1675, A1, 1:495.

102. Leibniz to Johann Friedrich, 21.1.1675, A1, 1:492. For Leibniz and the move to Hanover, see Rescher 1992.

103. Joachim Dalencé to Leibniz, [29.10.1675], A3, 1:303 (emphasis in original).

104. Leibniz to Gallois, 2.11.1675, A3, 1:306.

105. Salomon-Bayet 1978, p. 156.

106. For Leibniz and the Académie, see Salomon-Bayet 1978.

107. Leibniz to Colbert, 11.1.1676, A1, 1:457.

108. The history of this effort remains to be written. See, however, Elster 1975 and especially Wakefield 2010.

109. A1, 2:126.

110. Leibniz to Johann Friedrich, [Fall 1678], A1, 2:85.

111. See Leibniz 1906, p. 142.

112. Leibniz to Johann Friedrich, [2?.1679], A1, 2:129.

Third Carry

1. Babbage to Duke of Wellington, 23.12.1834, BL Add. MS 40611, f. 183v; reprinted in Babbage 1989, vol. 3, p. 7.

2. Brewster 1830, p. 333; see Johns 2009, pp. 250–58.

3. Brewster 1830, p. 333.

4. See Bracha 2004, pp. 33–35.

5. Brewster 1830, p. 333. This anonymous review is sometimes erroneously credited to Babbage, e.g., Dutton 1984, p. 70; Mokyr 2009, p. 351.

6. Babbage to Duke of Wellington, 23.12.1834, BL Add. MS 40611, f. 183v (emphasis in original); in Babbage 1989, vol. 3, p. 7.

7. Babbage to Ashley, BL Add. MS 37184, ff. 459–60; in Collier 1990, p. 68.

8. Babbage to Herschel, 27.6.1823, RS HS 2.184 [RT].

9. Babbage to Herschel, 9.5.1828, RS HS 2.226 [RT].

10. Babbage to Herschel, 9.5.1828, RS HS 2.226 [RT].

11. Babbage to Bryan Donkin and George Rennie, 11.4.1829, BL Add. MS 37184, ff. 254–55, quoted in Collier 1990, pp. 53–54.

12. On debates over ownership, I build upon Schaffer 1994 and Ginn 1991.

13. Babbage to Lord Ashley, 25.11.1829, BL Add. MS 37184, f. 432 [RT].

14. Report of Messrs Rennie and Donkin, 22.4.1829, BL Add. MS 37184, f. 266 [RT].

15. Babbage, "Report on the Calculating Machine," BL Add. MS 37185, f. 264. See Schaffer 1994; for Babbage's estimation of Clement's work, see Ginn 1991, pp. 172–75.

16. For the evidence of Clement's innovations, see above and Williams 1992, p. 74.

17. Babbage 1989, vol. 8, §325, p. 185.

18. Babbage 1989, vol. 8, §79, p. 47.

19. Schaffer 1994.

20. See Pottage and Sherman 2010, ch. 2, drawing on Berg 2002.

21. For artisanal ownership in the works of skill, see Rule 1987, esp. at pp. 104–5.

22. See, among a large literature, Sunder 2007, esp. p. 109.

Chapter Four

1. Hume 1987, pp. 328–29.

2. Merck 1784, p. 270.

3. Merck 1784, p. 271. The journal is now easily accessible through "Retrospektive Digitalisierung wissenschaftlicher Rezensionsorgane und Literaturzeitschriften des 18. und 19. Jahrhunderts aus dem deutschen Sprachraum," http://www.ub.uni-bielefeld.de/diglib/aufklaerung/index.htm.

4. Hahn 1779.

5. Merck 1784, p. 273.

6. Merck 1784, p. 273.

7. Diderot and Alembert 1751–72, s.v. "Emulation."

8. For the history of Prometheus, see Raggio 1958 and MacLeod 2007, pp. 47–58.

9. Merck 1784, p. 269.

10. Merck 1784, p. 275.

11. Or perhaps all centuries: see the classic account of Basalla 1988.

12. Merck 1784, p. 275.

13. Compare Pancaldi 2003 on "imitation-competition." See more broadly Inkster 1990; Berg 2007; Harris 1998; and Bertucci 2006.

14. See now Reinert 2011.

15. Bischoff 1990, p. 124.

16. JS (12.1776), p. 871. A description of the stepped drum, without a diagram, was printed in Pütter 1765, pp. 243–46; this obscure publication appears to have been known to no other makers of calculating machines. If anyone knew of it, it would likely have been Müller, who had strong connections with Göttingen. For the Göttingen "Modellkammer," see Müller 1904, p. 112n1. See Walsdorf 2014, pp. 103–5.

17. Hahn 1779, p. 139. Early in his inventive process, Hahn knew that Leibniz's remaining machine was in Göttingen. Hahn to Lavater, 4.10.1773, in Paulus 1975, p. 64.

18. Compare Vera Keller's work on the lists of lost inventions from antiquity: Keller 2008, 2015.

19. The Göttingen mathematician Abraham Kästner, who studied the machine closely, reflected on the level of Leibniz's success, to discern "daß Leibnitz von Erreichung seiner Absichten nicht so weit entfernt gewesen, sondern die Maschine würklich zu Stande gebracht." Pütter 1765, p. 243. Kästner elsewhere referred to the machine as a "relic." See Morar 2009, p. 29n36.

20. Poleni 1709, p. 27.

21. Anonymous 1751, p. 669.

22. De Fouchy 1763, pp. 152–53.

23. Gersten 1735, p. 81.

24. For mechanics and the republic of letters in another context, see Sibum 2003.

25. Gersten 1735, p. 82.

26. Kratzenstein to Le Sage, 2.2.1772, in Prévost 1805, p. 403; the letter is at BPU MS Suppl. 513, ff. 189–201. See Splinter 2005 and Académie impériale des sciences 1897–1911, vol. 2, p. 540, meeting of 16.5.1765.

27. See Kratzenstein 1779, p. 142.

28. See Mokyr 2009 on access costs.

29. Compare Biagioli 2006b.

30. Berg 2004, pp. 126–27.

31. See, e.g., MacLeod 2004; Rosenband 2000; Inkster 1990; Jones 2008.

32. Rosenband 2007, p. 390; Berg 2004.

33. For a survey of practical improvement in the empire, see Umbach 2000, ch. 4. For the challenge of judging technical expertise in the empire, see, for example, Smith 1994; Nummedal 2007; Fors 2014, ch. 3.

34. Young 1759, p. 41.

35. For control of artisanal spaces, see, for example, Schaffer 1994 and Jackson 2000.

36. See Vayringe's autobiography in dom Calmet 1751, cc. 987–1000; an English translation may be found in Manning 1860, ch. 16. For the creation of the machine, see B*** 1847, p. 321. For Vayringe, see Bedini 1995, p. 163ff; Lepage 1854, including his ducal privilege.

37. dom Calmet 1751, c. 994.

38. Voltaire to Nicolas Claude Thieriot, 15.5.1735, D870.

39. dom Calmet 1751, c. 994; Manning 1860, p. 136, trans. modified.

40. dom Calmet 1751, c. 994; Manning 1860, pp. 136, 137, trans. modified.

41. dom Calmet 1751, c. 994; Manning 1860, pp. 137.

42. Hilaire-Pérez and Verna 2006, p. 545, following MacLeod 2004.

43. dom Calmet 1751, c. 990; Manning 1860, p. 130.

44. Hilaire-Pérez 2007.

45. Keyssler 1751, 24.6.1731, p. 1488 (Vayringe); 1.8.1730, p. 1223 (Rowley). For Rowley's spheres, see King and Millburn 1978, pp. 154–57, and for important corrections, see Bedini 1994, esp. pp. 63–64; http://www.mhs.ox.ac.uk/sphaera/index.htm?issue3/articl5.

46. A machine in Vienna is signed "Ph. Vayringe Fecit A Luneville." See King and Millburn 1978, p. 153.

47. dom Calmet 1751, c. 992; Manning 1860, p. 133. For Desaguliers' status between craftsman and engineer, see Poni 1993.

48. Vayringe 1732.

49. Voltaire to Jean Baptiste Nicolas Formont, 25.6.1735, D882; see Gauvin 2006b.

50. Janetschek 1988, p. 174.

51. Nagel 1960.

52. Janetschek 1988, p. 174.

53. Reynolds 1997, p. 107.

54. *Reflexion* 778 (on Baumgarten); Ak 15:340–41; trans. modified from Gammon 1997, 578.

55. Sulzer 1773–75, s.v. "Nachahmung," pt. 2, vol. 1, p. 283, transcription at http://www.text log.de/2817.html.

56. Young 1759, pp. 66–67.

57. For the aesthetic background, see Weinbrot 1985; Hont 2005, p. 117; and Pigman 1980.

58. "Emulation" demands combining concrete economic history with the history of economic reflection; compare the recent debate in Sewell 2010; Sonenscher 2012; and Sewell 2012.

59. Hont 2005, p. 121; Reinert 2011. For societies of emulation, see Reinert 2011; Shovlin 2003; Fox 2009, with full reference to the literature.

60. Anonymous 1778, pp. 3–4.

61. Hamann to Kant, 27.7.1759, Ak 10:12–13; trans. Kant 1986, pp. 40–41; see Gammon 1997, 567.

62. Kant, *Critique of Judgment*, §47, Ak 5:308; Kant 2000, p. 187.

63. Kant, *Critique of Judgment*, §46, Ak 5:308; Kant 2000, pp. 186–87 (his emphasis).

64. Kant, *Critique of Judgment*, §47, Ak 5:309; Kant 2000, p. 188.

65. Kant, *Critique of Judgment*, §47, Ak 5:310; Kant 2000, p. 189.

66. Merck 1784, pp. 271–72.

67. For Leupold, see briefly Ferguson 1971; Matschoss 1939, p. 113.

68. Leupold 1727, p. 38.

69. Lavater to Hamann, 26.12.1777, in Hamann 1894, p. 3.

70. For Hahn's machines, the essential reference is Anthes 1989. Hahn's notebooks (Hahn 1987) unfortunately contain relatively few references to his calculating machine project, each analyzed in Anthes.

71. Lavater, quoted in Hahn 1979, pp. 270–71n116.

72. See, e.g., letters of Hahn to Lavater, in Paulus 1975, pp. 68, 71, 73.

73. Hahn 1779, p. 140. For the effort to get Hahn to write this article, see Christoph Martin Wieland to Merck, 28.5.1778, in Merck 2007, vol. 2, p. 93.

74. For nonhylomorphic design, see Ingold 2010, p. 92; Ingold 2007; and Pye 1968. In a different idiom, see Scranton 2006.

75. Hahn 1779, p. 140.

76. Hahn 1779, p. 143.

77. See the journal entries for 7–8.4.1777, in Hahn 1979.

78. Hahn 1779, p. 144.

79. Sander 1784, p. 68.

80. See Hahn to Lavater, 4.10.1773, in Paulus 1975, p. 64.

81. Hahn 1779, p. 146.

82. Müller to Lichtenberg, 4.1.1785, in Lichtenberg 1983–2004, vol. 3, p. 5. For Müller and his work, see Lindgren 1990.

83. Müller to Lichtenberg, 22.5.1783, in Lichtenberg 1983–2004, vol. 2, p. 616; a version of the letter appeared as Müller 1783.

84. Müller explained the process in a later autobiography as well as in letters written as he was in the process of having the machine made. Diehl 1930, p. 5.

85. Müller to Lichtenberg, 22.5.1783, in Lichtenberg 1983–2004, vol. 2, pp. 616–17.

86. Diehl 1930, p. 5.

87. Diehl 1930, p. 5.

88. Lange notes that Müller's machine involves "wesentliche Verbesserungen." Lange 1981, p. 2.

89. Müller to Lichtenberg, 22.5.1783, in Lichtenberg 1983–2004, vol. 2, p. 616.

90. Hahn 1785, p. 93.

91. Hahn 1785, p. 87.

92. Hahn speculated further that he had not kept his machine very secret. Outside the realm of print, Hahn made it known that he believed Merck, the editor of the *Teutsche Merkur*, had transmitted details of the mechanism of his machines to his relative Müller. Hahn 1785, p. 94; cf. Lichtenberg 1983–2004, vol. 3, p. 5.

93. Hahn 1785, p. 87.

94. Hahn 1785, p. 87.

95. Müller and Klipstein 1786, p. 3–4.

96. Müller and Klipstein 1786, p. 4, note *.

97. A. G. Kästner to the members of the Akademie, 10.10.1784, MS, Akademie der Wissenschaften, Göttingen, quoted in Weber 1980, p. 20; compare Kästner 1784.

98. Kästner 1784, p. 1205.

99. Pye 1968.

100. Müller to Lichtenberg, 22.5.1783, in Lichtenberg 1983–2004, vol. 2, p. 618.

101. Müller to Lichtenberg, 9.9.1784, in Lichtenberg 1983–2004, vol. 2, p. 903.

102. Müller to Lichtenberg, 9.9.1784, in Lichtenberg 1983–2004, vol. 2, p. 903.

103. Diehl 1930, p. 6; c. 1786; see Lichtenberg 1983–2004, vol. 5.1, p. 96; Baldinger 1787, pp. 93–94.

104. Baldinger 1787, p. 94.

105. Now published as Bischoff 1990.

106. 11.12.1784, in Hahn 1983, p. 159.

107. 16.12.1784, in Hahn 1983, pp. 160–61.

108. 14.12.1784, in Hahn 1983, p. 159. See Matthews 2011, p. 58.

109. For the centrality of machines in nineteenth-century Romanticism, see Tresch 2012.

110. For Herder's account of Spinoza and its significance, see Beiser 1987, pp. 158–64, and Herder 1940.

111. Herder 1940, p. 133; Herder 1881–1913, vol. 16, p. 492.

112. Herder 1940, p. 103; Herder 1881–1913, vol. 16, p. 450.

113. Herder 1940, p. 103; Herder 1881–1913, vol. 16, p. 451.

114. Hahn to Johann Ludwig Ewald, 16.7.1778, in Brecht 1981, 370–71.

115. See Stäbler 1992 for the most robust account of Hahn's theological thinking.

116. 7.3.1788, in Hahn 1983, p. 301.

117. [31.3.1788], in Hahn 1983, p. 308.

118. 21.3.1788, in Hahn 1983, p. 305.

119. 3.8.1788, in Hahn 1983, p. 347.

120. Such a view violated Lutheran strictures that grace must utterly alter the sinner; rather, for Hahn, "salvation results through a process of sanctification, as the divine seed grows and subdues the flesh." Hayden-Roy 1994, p. 60; see introduction to Hahn 1979, p. 27.

121. Introduction to Hahn 1979, p. 27.

122. Hahn 1779, p. 140.

123. Young 1759, p. 12.

124. Here I follow Abrams 1971, pp. 160–67; see Sambrook 2005; Biagioli 2009.

125. 21.3.1788, in Hahn 1983, p. 305.

Fourth Carry

1. 7.11.1822, Buxton MS 7, Oxford Museum of the History of Science, p. 14.

2. Dan Moore to Babbage, 23.8.1822, BL Add. MS 37182, 437r [RT].

3. Babbage to Moore, 28.8.1822, BL Add. MS 37182, 439r [RT].

4. Francis Baily to Babbage, 26.5.1823, BL Add. MS 37183, f. 30 [RT]. See the discussion in Swade 2003, pp. 43–45.

5. Versailles 10.7.1789, BL Add. MS 37183, ff. 76r–77v.

6. Williams 1981, pp. 238–39. It is not clear when he acquired these machines and books.

7. The *Extractor*, 1.11.1828, copy in RS HS 27.54.

8. See *Oxford Dictionary of National Biography*, s.v. "Kollmann."

9. Kollman 1828.

10. Lindgren 1990 presents a compelling but circumstantial case for the importance of Müller for Babbage; most Babbage scholars appear to bracket this question.

11. Müller to Lichtenberg, 9.9.1784, in Lichtenberg 1983–2004, vol. 2, p. 90; trans. Lindgren 1990, p. 67.

12. J. W. F. Herschel, translation and paraphrase of *Beschreibung seiner neu erfundenen Rechenmaschine: nach ihrer Gestalt, ihrem Gebrauch & Nutzen*, BL Add. MS 37198, f. 190v; cf. Müller and Klipstein 1786, pp. 48–50. See Lindgren 1990, p. 244–46.

13. Letter from Stephen Lee (RS) to [fourth] Earl Stanhope, 28.4.1822, CKS U1590/C83/2.

14. Fourth Earl Stanhope to Stephen Lee, 27.7.1822, CKS U1590/C83/2. Cf. Lee to Babbage, 22.8.1822. BL Add. MS 37182, 435r.

15. Charles Babbage, Manuscript Journal, 1820–25; Waseda University Library Manuscript Collection, Tokyo, p. 55 [RT]. My thanks to Adam Bronson for consulting this manuscript on my behalf.

16. Reported in Harley 1890, p. 124; compare Edmondson 1885.

17. Lardner, in Babbage 1989, vol. 2, p. 158.

18. Harley 1890, p. 122. I have been unable to trace the provenance of the machines, including Babbage's purchase of them. Babbage offered to loan the two Morland machines and three Stanhope machines, all now in the Science Museum, for an exhibition in 1863. See Babbage 1989, vol. 11, pp. 116–17.

19. Jarvis to Babbage, n.d. [1830/1831?], BL Add. MS 37185 f. 419 [RT].

20. Babbage to Herschel, 9.4.1822, RS HS 2.171 [RT].

21. Brewster, Scott, and Stark 1832, p. 296.

Chapter Five

1. For simplicity's sake, I refer to Charles Mahon as Stanhope; he was the Lord (or Viscount) Mahon from the death of his older brother in 1763 until the death of his father in 1786, at which point he became the third Earl Stanhope, the name used in almost of all the literature on his inventions.

2. CKS U1590/C83/1/1/16v; see also CKS U1590/C83/1/1/24. The manuscript pages are not numbered; I use my own foliation for reference.

3. CKS U1590/C83/1/1/16r.

4. CKS U1590/C83/1/2/16r.

5. This chapter, following Tim Ingold, "assigns primacy to the processes of formation as against their final products." Ingold 2010, p. 92. See also Ingold 2007 and the discussion papers that follow.

6. For a survey of relations of mind and hand, see Roberts and Inkster 2009; Roberts, Schaffer, and Dear 2007.

7. Draft treatise on arithmetical machines, c. 1775, CKS U1590/C83/1/21.

8. For contemporaneous Birmingham and other such environments, see Jones 2008, ch. 4.

9. See Bennet 1992 (copy at CKS). For Stanhope's life, see DNB; Stanhope and Gooch 1914; and Newman 1969.

10. The theory remains important to fringe scientific work: see Edwards 2002 for recent perspectives.

11. Stanhope to Le Sage, 18.3.1774, BPU MS Suppl. 515, f. 84r.

12. Stanhope and Gooch 1914, p. 11. Cf. Le Sage to Stanhope, 1.2.1776, BPU MS Suppl. 518, f. 497v.

13. Du Pan to Fredenrich, BPU MS 1545, quoted in Engel 1946, p. 489; see also Bennet 1992, p. 38.

14. Stanhope to du Roveray, 15.1.1777, CKS U1590/C62, p. 29.

15. Du Pan to Fredenrich, BPU MS 1545, quoted in Engel 1946, p. 489; discussed in Bennet 1992, p. 38.

16. See Ivernois 1782, vol. 1, pp. 54–57, for Stanhope's activities and popularity.

17. Stanhope to Le Sage, undated but [4–mid 6.1774], BPU MS Suppl. 515, f. 89v.

18. For these movements, see Starobinski 1989; Magnin and Marcacci 2001; Candaux and Sigrist 2001.

19. Saussure 1774, 130–31, quoted in Magnin and Marcacci 2001, 423.

20. Diderot and Alembert 1751–72, vol. 4, s.v. "Dent" (mécanique).

21. Le Cerf and Mahon 1778, pp. 950, 951; a more developed example of this appears in Preud'homme 1778, pp. 81ff. It is unclear why Stanhope published it under the name of the clockmaker Le Cerf rather than Preud'homme. See the report of the misattribution, H. 1780.

22. A review of the paper questions its claims to certainty while praising the goal. Anonymous 1780, p. 46.

23. See Séris 1987, p. 181.

24. Preud'homme 1778, p. 100.

25. Preud'homme 1778, p. 100.

26. See Candaux and Sigrist 2001; Crosnier 1910.

27. Anonymous 1778, pp. 3–4.

28. For a powerful study of artisans, machines, and affect, see Voskuhl 2013. For industrial espionage, see Jones 2008, pp. 149–60; Harris 1998.

29. Anonymous 1778, p. 4.

30. Compare the "industrial enlightenment" in Mokyr 2005a, p. 322, and the careful study Jones 2008. For a powerful restatement of the attack of historians of technology on the linear model, see Scranton 2006.

31. See Smith 2004b, and the classic if dated Landes 1979. More generally, see Harris 1998.

32. Stanhope to [?], 24.4.1777, CKS U1590/C66; for these efforts, contemporaneous with the calculating machine work, CKS U1590/C711/8.

33. CKS U1590/C111; see Mahon 1778; BPU MS Suppl. 515, ff. 108v–9r. I have been unable to find the timber in the Royal Society Archives.

34. CKS U1590/C711/5: "Proposed Exp. to prove what is exactly the influence of the air in a vibrating pendulum"; discussed in Stanhope to [?], 24.4.1777, CKS U1590/C66.

35. Stanhope to Le Sage, 24.7.1774, BPU MS Suppl. 515, f. 107r.

36. CKS U1590/C85/8. Cf. Carlson and Gorman 1990.

37. CKS U1590/C83/1/2/36bis.

38. CKS U1590/C83/1/5.

39. Note that Stanhope's cylinders are toothed, unlike any of the other versions. For the apparent lack of knowledge concerning Leibniz's mechanisms into the nineteenth century, see chapter 4.

40. CKS U1590/C83/2/29.

41. CKS U1590/C85/8 (all emphases his).

42. Stanhope credited his approach to invention to his old Genevan mentor, Le Sage. CKS U1590/C86/4 [prob. after 1800]; see BPU MS 2002/2; Prévost and Le Sage 1804, p. 324.

43. Stanhope to North, 11.10.1801, CKS U1590/C86/3.

44. For a strong defense of the analogical nature of almost all invention, see Basalla 1988, e.g., p. 45.

45. Carlson and Gorman 1990, p. 394. Compare Pancaldi 2003, esp. chs. 3, 6.

46. Hilaire-Pérez 2007, p. 137.

47. Mokyr 2007, p. 187.

48. Mokyr 2002. See likewise Jacob and Stewart 2004; Jacob 2007; Stewart 2007; Mokyr 2007; Jones 2008. All these views challenge a historiography seeing the inventive activity of the industrial revolution as the product of tinkerers, culminating in Mathias 1969.

49. Sulzer 1773–75, s.v. "Nachahmung" pt. 2, vol. 1, p. 283, transcription at http://www.text log.de/2817.html.

50. Compare Schuster 1984.

51. CKS U1590/C83/1/2/36bis.

52. Compare Swade's notes on Babbage's diagrams of his own spiral wheel in Swade 1995, pp. 71–74.

53. CKS U1590/C83/1/36bis.

54. CKS U1590/C83/1/3; CKS U1590/C83/1/2.

55. CKS U1590/C83/2/24.

56. Ferguson 1977, p. 828.

57. Lubar 1995, p. S64; Ferguson's book-length account is less susceptible to these forms of criticism.

58. Alder 1997, p. 129.

59. McGee 2004, p. 83.

60. Compare Kemp 2004, pp. 171–73, for the autonomy of sketching within Leonardo's design and natural philosophical practice.

61. For a sophisticated take on the role of drawing in invention, see Carlson 1999 and Popplow 2004.

62. For the wastefulness of the movement from a craftsman tradition done mostly without drawing toward a mechanical tradition of draftsmanship, see McGee 1999, pp. 219–20.

63. Wright 1994, p. 8.

64. For pinned barrels more broadly, see Bedini 1964; Voskuhl 2013, pp. 132–34n11.

65. Stanhope to [Le Sage?], 24.4.1777, CKS U1590/C66.

66. Mahon to second Earl Stanhope, 3.6.1770, CKS U1590/C53/6,7; also in Stanhope and Gooch 1914, p. 10.

67. See the engravings in CKS U1590/C711/2–3, dated 1773 and 1774. He never published his memoire.

68. Stanhope to Le Sage, 18.3.1774, from Paris, BPU MS Suppl. 515, f. 84r.

69. Stanhope to Le Sage, undated but [4–mid 6.1774], from Paris, BPU MS Suppl. 515, ff. 89r, 90v.

70. William Pitt, Lord Chatham to James Grenville, 28.11.1774, quoted in Newman 1969, p. 134.

71. Stanhope to Le Sage, undated but [c. 7–8.1774], BPU MS Suppl. 515, f. 98r.

72. For Ramsden, see McConnell 2007.

73. See BPU MS Suppl. 515, f. 110r; 117r.

74. Stanhope to Le Sage, 5.8.1776, BPU MS Suppl. 515, f. 119v. For a remarkable study of Vulliamy, see Hilaire-Pérez 2013, pp. 353–64, 409–15.

75. Stanhope to Le Sage, 24.4.1777, BPU MS Suppl. 515, f. 121r.

76. Stanhope to Le Sage, [mid or late 5.1778], BPU MS Suppl. 515, f. 129v.

77. The three extant Stanhope machines, dated 1775 and 1777, bear little resemblance to the cylinder A machine. Was any part of "cylinder A" machine actually produced? Almost certainly substantial parts were made, which he retasked. CKS U1590/C83/1/1.

78. Fourth Earl Stanhope to Stephen Lee, 22.7.1822, CKS U1590/C83/2; copy in BL Add. MS 37182, ff. 436r–v.

79. Cornelius Varley, Biographical Note on Samuel Varley, American Philosophical Society, Philadelphia, Penn. Manuscripts, 509.078 M582. Thanks to Michael Neuss for consulting the Varley manuscripts.

80. Varley 1820, p. 32.

81. Stanhope and Gooch 1914, p. 275.

82. See, for example, Varley 1825; discussed in Brooks 1989, p. 32.

83. Gill 1822, p. 416.

84. Mokyr 2005b, p. 1159.

85. Stanhope to Le Sage, 24.4.1777, BPU MS Suppl. 515, f. 121r.

86. Wright 1994, pp. 8–9.

87. Murphy 1966, p. 186; Jaffee 2010, pp. 152, 174–79.

88. North to Stanhope, 27.8.1812, CKS U1590/C88.

89. Stanhope to North, 3.9.1812, CKS U1590/C88.

90. North to Stanhope, 14.10.12, CKS U1590/C88.

91. For his refusal to patent his innovations in printing, see Newman 1969, p. 183.

92. Stanhope to North, 11.10.1801, CKS U1590/C86/3.

93. Stanhope to North, 11.10.1801, CKS U1590/C86/3.

94. North to Stanhope, 6.2.1802, CKS U1590/C86/3.

95. CKS U1590/C85/8. Compare Priestley's account of Newton, discussed in Schaffer 1990, pp. 89–90.

96. CKS U1590/C85/8.

97. Pictet 1802, p. 297.

98. Anonymous 1801, p. 106.

99. On these, see Harley 1879, 1890; Gardner 1982; Wess 1997.

100. S.P. 1804, pp. 149–50.

101. S.P. 1804, p. 150.

102. CKS U1590/C85/8 (emphases his; cancellations omitted).

103. CKS U1590/C85/8.

Fifth Carry

1. Varley 1832, p. 157.

2. Varley 1825; Varley 1820. See the list in Cornelius Varley, "Life of C. Varley, written by himself," n.d., American Philosophical Society, Philadelphia, Penn. Manuscripts, 509.078 M582, f. 7r.

3. Smiles 1864, p. 304.

4. Herschel to Babbage, 12.2.1828, RS HS 2.219 [RT].

5. Gill 1822, p. 416.

6. Smiles 1864, p. 298.

7. H. R. Palmer, Address, 2.1.1818; quoted in *Civil Engineer and Architect's Journal* 1(1838):138.

8. Herschel to Babbage, 12.2.1828, RS HS 2.219 [RT].

9. Herschel to Babbage, 10–17.4.1828, RS HS 2.225 [RT].

10. Herschel to Babbage, 10–17.4.1828, RS HS 2.225 [RT]. Compare Herschel's notes for this letter, RS HS 27.42.

11. Herschel to Babbage, 10–17.4.1828, RS HS 2.225 [RT].

12. Herschel, "Actual State of the Machine and Drawings," n.d. [late 1828–early 1829], RS HS 27.54; draft for RS report.

13. The brief was whether the progress of the machine will "prove adequate to the important object which it was intended to attain." J. Stewart to President of Royal Society, 24.12.1828, T27/89, f. 27 [RT*]; RS DM/4/133.

14. Manuscript of "Report of the Committed Appointed . . . to consider the subject referred in Mr. Stewert's Letter Relative to Mr. Babbage's Calculating Engine," 2.1829, RS DM/4/135. Canceled and marked "Not to be."

15. "Report of the Committed Appointed . . . to consider the subject referred in Mr. Stewert's Letter Relative to Mr. Babbage's Calculating Engine," 2.1829, in Babbage 1889, p. 233.

16. Clement to Babbage, Statement of Account as at 9th May 1829, BL Add. Ms 37184, f. 291 [RT].

17. Clement 1825, pp. 141–42; see Williams 1992.

18. Exposition of 1851, ch. 13, "Calculating Engines," p. 173.

19. Herschel, "Notes for the Report," n.d. [late 1828–early 1829], RS HS 27.54 [not all cancelations included]. Cf. "Report of the Committed Appointed . . . to consider the subject referred in Mr. Stewert's Letter Relative to Mr. Babbage's Calculating Engine," 2.1829, in Babbage 1889, p. 233.

20. Swade 2003, pp. 149–55.

21. R. Wright to Babbage, 18.6.1834, BL Add. MS 37188, f. 390.

22. Quoted in Musson 1975, p. 115; discussed in Schaffer 1994.

23. Brooks 1992; Brooks 1989.

24. BL Add. MS 37196, f. 251, quoted in Collier 1990, p. 221–22; I follow Collier here.

25. See BL Add. MS 37195, ff. 130–31, quoted in Collier 1990, p. 220.

26. Contract of Employment between Babbage and Jarvis, 24.11.1835, BL Add. MS 37189, f. 203 [RT].

27. Jarvis to Babbage, n.d. [1830/1831?], BL Add. MS 37185, f. 419 [RT].

28. Jarvis to Babbage, c. 15.2.1831[?], BL Add. MS 37185, f. 476. The context here is Jarvis's dispute over his salary and status under Clement.

29. Jarvis to Babbage, 25.8.1833, BL Add. MS 37188, f. 39 [RT]. See Hyman 1982, pp. 131–32.

30. See Schaffer 1994, p. 215.

31. Jarvis to Babbage, 25.8.1833, BL Add. MS 37188, f. 39 [RT].

32. Swade 2005, p. 76. See also the lively retelling of the history of the rebuilding effort in Swade 2001 and the technical description of the engine in Swade 1995.

33. Swade 2005, p. 75.

34. Swade 2005, p. 76.

35. For an insightful and highly critical appraisal of similar claims for Leibniz's machine, see Morar 2009, 2014.

36. For these difference engines, I follow closely Lindgren 1990; see the additional material in Swade 2003, ch. 5.

37. Lindgren 1990, p. 114.

38. Lindgren 1990, p. 168.

39. Lindgren 1990, quotation on p. 172.

40. Merzbach 1977, pp. 23–25.

41. Gravatt 1857.

42. Jarvis to Babbage, 25.8.1833, BL Add. MS 37188, f. 39 [RT].

43. Jarvis to Babbage, 25.8.1833, BL Add. MS 37188, f. 39 [RT].

44. Martin Archer Shee, *Rhymes on Art*, 1805, quoted in Fox 2009, p. 448.

45. See especially Woodmansee 1994.

46. See McGee 1995 and MacLeod 2007.

Chapter Six

1. http://www.aaai.org/AITopics/pmwiki/pmwiki.php/AITopics/AIEffect#kearns.

2. Babbage 1844.

3. For a powerful revisionist account of machines in their nineteenth-century Romantic context, see Tresch 2012.

4. Note G, http://www.fourmilab.ch/babbage/sketch.html, reprinted in Babbage 1889, p. 44. For an informed view of authorship of the notes, see Fuegi and Francis 2003; much the relevant correspondence is in Huskey and Huskey 1980.

5. Turing 1950, p. 450.

6. Note G, http://www.fourmilab.ch/babbage/sketch.html, reprinted in Babbage 1889, p. 44.

7. McGee 1995.

8. CKS U1590/C85/8. Compare Priestley's account of Newton, well known to Stanhope: Priestley 1775, vol. 2, p. 167; see the discussion in Schaffer 1990, pp. 89–90.

9. CKS U1590/C85/8.

10. Compare Mahoney's reminder to keep the histories of logic and of mechanical computation appropriately separated. Mahoney 1988, pp. 116–17.

11. See Voskuhl 2013 for a similar case of the mismatch between our modern expectations concerning the significance of automata and Enlightenment views of their import during the eighteenth century.

12. Du Marsais 1743, p. 174.

13. Despite a modish return of a Whiggish narrative that connects materialism and atheism, it is very clear that materialist accounts of human beings came just as often from nonorthodox Christians, not only members of a putative "radical enlightenment." See Thomson 2008.

14. For these debates, see especially Yolton 1991; Thomson 2008.

15. See the similar claims in Erickson et al. 2013, ch. 1.

16. For a sophisticated analysis of forms of nonmechanical materialism, see Wolfe and Terada 2008; Wolfe 2010, 2014.

17. See Hartley 1791; Glassman and Buckingham 2007; Smith 1987.

18. Hobbes 1981, I, 1, 2.

19. Fries 1803, p. 205; see Schubring 2005, p. 494; Gregory 1983. Fries's essay concerns Reinhold, who briefly became a major exponent of Bardili.

20. Bardili 1800, pp. 1–2. For Bardili's logical realism and its opponents, see the introduction to Hegel 1986, pp. 15–18; Zahn 1965a, 1965b and Zöller 2000.

21. For Hegel's attack on logic as formal identity in Reinhold and Bardili, see Hegel 1977, 179–83, 186–88; for context, see the introduction to Hegel 1986 and the preface in Kant 1992a, p. 525.

22. Lichtenberg to Georg Friedrich Werner, 29.11.1788, in Lichtenberg 1983–2004, vol. 3, p. 599. See Clark 2002, p. 78.

23. Condorcet 1847–49, vol. 3, p. 27.

24. See Jones 2014 for a fuller version of the following argument.

25. Buffon 1749, pp. 53–54; see Tonelli 1959, pp. 46–48; Richards 2006, pp. 703–4.

26. *De l'interpretation de la nature* [1754], §3, in Diderot 1975–, 9:29.

27. Playfair and Maskelyne 1778, p. 321.

28. "Of Infinities," Berkeley 1948–57, vol. 4, p. 235.

29. "The Analyst," Berkeley 1948–57, vol. 4, p. 66.

30. For nuanced accounts of these developments, see Easton 1997 and Owen 2002.

31. Kant 1996, A717/B745.

32. Playfair and Maskelyne 1778, p. 320.

33. For this distinction, see Sherry 1991, p. 46.

34. Berkeley 1948–57, vol. 4, p. 235.

35. Berkeley 1948–57, vol. 4, p. 86.

36. Kant 1996, A727/B755.

37. CKS U1590/C85/8.

38. Mdm. Périer, JMI:576–77; 608.

39. Pascal 2000, S617.

40. LH 42,4, f. 33r.

41. Bayle 1820, vol. 13, p. 237, s.v. "Sennert" rem. C., trans. in Des Chene 2006, pp. 225–26.

42. Bayle denied the possibility of the production of order without consciousness of it as a goal; Leibniz responded in numerous places, as at GP 4:356; 3:374.

43. Hunter and Davis 1999–2000, vol. 5, p. 354.

44. Leibniz to Lady Masham, GP 3:374.

45. For Leibniz and "plastic natures," see Smith 2011, pp. 127–35.

46. Leibniz to Lady Masham, GP 3:374.

47. Coward 1702, pp. 123–24. See Thomson 2010, pp. 28–33.

48. Bentley 1838, vol. 3, p. 45.

49. Coward 1702, p. 101.

50. Lichtenberg 1966, B380.

51. For Mayer's Newtonianism, see Forbes 1970, p. 18.

52. Lichtenberg 1966, J1416 (and numerous remarks in that J notebook); and the considerable edition of Lichtenberg's manuscript notes on Le Sage in Lichtenberg 2003.

53. For Lichtenberg's reflections over many years, see Lichtenberg 2003; more briefly, Lichtenberg 1983–2004, vol. 3, p. 617.

54. Lichtenberg 1966, J1521; trans. in Stern 1959, p. 297.

55. Lichtenberg 1966, E32; trans. in Stern 1959, p. 88.

56. Lichtenberg 1966, F324; trans. in Stern 1959, p. 291.

57. Lichtenberg 1966, J393.

58. Young 1759, p. 12.

59. Kant 1992b, p. 164; Ak 2:122–23 (translation revised); cf. Kant 1996, A627–8/B655–56.

60. For imitation in the Renaissance, see Pigman 1980 and Kemp 1977.

61. Shaftesbury 2001, vol. 1, p. 208.

62. For the long reach of Leibniz's organicism, see Benziger 1951.

63. McGee 1995.

64. Here I follow closely the classic account of Abrams 1971, pp. 160–67; see also the more recent Sambrook 2005. For a nuanced comparison of mid-eighteenth-century materialisms, see Thomson 2008, ch. 6 and Wolfe 2014.

65. Hartley 1791, 1, prop. 11, p. 72; see Garrett 2013.

66. Duff 1767, pp. 6–7.

67. For example, Batteux 1747.

68. Gérard 1774, p. 84.

69. Gérard 1774, p. 63.

70. Gérard 1774, p. 63.

71. Abrams 1971, pp. 202–3.

72. See Biagioli 2009.

73. The "best poet" claim is attributed to Gauss.

74. For the relationship of Cartesian standards of truth and classical rhetorical ideas, see Gaukroger 1997; Jones 2006, ch. 2. For Kästner's embedding in Cartesian aesthetic reflection, see Kästner 1751, 1756.

75. Kästner 1783, p. 340; Horace, *De Ars Poetica*, 148–50, trans. from http://data.perseus.org /citations/urn:cts:latinLit:phi0893.phi006.perseus-eng1:125.

76. Kant 2000, §43, Ak 5:304, p. 183.

77. Kant 2005, No. 829, Ak 15:370, p. 507.

78. Kant 2005, No. 829, Ak 15:370, p. 508.

79. Kant 2005, No. 829, Ak 15:370, p. 508.

80. Kant 2005, No. 829, Ak 15:370, p. 508.

81. Kant 2000, §43, Ak 5:304, p. 183.

82. Kant 2000, §22 General Remark, Ak 5:241 (his italics); trans. Guyer 1996, p. 275.

83. Kant 2005, No. 943, Ak 15:419, p. 516.

84. Kant 2005, No. 945, Ak 15:419, p. 516.

85. Kant 2000, §43, Ak 5:303, p. 183.

86. Kant 2005, No. 943, Ak 15:419, p. 516. Bold in original.

87. Abeille 1765, pp. 22–24.

88. See MacLeod 2007, pp. 267–71.

89. Quoted in MacLeod 2007, p. 267.

90. Arthur Legrand, quoted in MacLeod 2007, p. 267.

91. CKS U1590/C85/8.

92. CKS U1590/C85/8.

93. Babbage 1989, vol. 9, p. 1.

94. Babbage 1989, vol. 9, p. 4.

95. Babbage 1989, vol. 9, p. 1.

96. Babbage 1989, vol. 9, p. 5.

97. Babbage 1989, vol. 9, p. 5.

98. Babbage 1989, vol. 9, pp. 7–8.

99. Babbage 1989, vol. 9, p. 8.

100. Babbage 1989, vol. 9, p. 11.

101. Babbage 1989, vol. 9, p. 30.

102. Green 2005, p. 44.

103. Though see Tresch 2012.

104. Mill 1981, p. 111.

105. Marshall, "Ye Machine," in Raffaelli 1994, 129. For this machine, see Cook 2005.

106. Marshall, "Ye Machine," in Raffaelli 1994, 129.

107. Turing 2004a, pp. 392–93.

108. Turing 2004a, pp. 393–94.

109. Turing 2004b, pp. 192–93.

110. Turing 2004b, p. 192.

111. Turing 2004a, p. 394.

Final Carry

1. Felt 1916, p. 17.

2. Discussed in Bromley 1998.

3. Macsorley 1961.

4. So complicated are the choices involved that new forms of representation have been devised to visualize the choices; see Choi and Swartzlander 2005.

5. Lécuyer and Brock 2006, p. 317; see also Brock and Lécuyer 2012.

6. Compare Warwick 1995 and the classic manifesto for such a conceptualization of success and failure, Pinch and Bijker 1987.

7. Cortada 2000, pp. 31–43.

8. Hoyau 1822, p. 356.

9. Francoeur 1822, p. 34 (my emphasis).

10. Anonymous 1823, p. 121.

11. Boys 1886, p. 377.

12. See Johnston 1997; and the updated version at http://www.mhs.ox.ac.uk/staff/saj/arithmometer/; for patents and numerous original documents, see http://www.arithmometre.org/.

13. Johnston 1997, p. 14.

14. Boys 1886, p. 377.

15. See, for example, the remarkable and influential papers Hannyngton 1871; Gray 1873; Hannyngton 1865, discussed in Warwick 1995.

16. Johnston 1997; for users' history, see the studies in Oudshoorn and Pinch 2003, among a large literature; for users and democratic invention, see von Hippel 2005; for a combining of these two traditions, see Mody 2011.

17. Boys 1886, p. 388.

18. West 1889, p. 159.

19. Compare Gordon 1988 on the limits of standardized manufacture in the nineteenth century.

20. M'Leod 1885, p. 943.

21. Excerpts of the documents are available at http://www.home.ix.netcom.com/~hancockm/felt_%26_tarrant.htm.

22. For example, see Felt 1916, and, with detailed studies of patents, Turck 1921 versus Karpinski 1933.

23. For an overview, see Jones 2016.

24. For business and administrative uses, see Yates 2000, 2005; Agar 2003; Cortada 1996; Chandler and Cortada 2000; Cortada 2000; among others.

25. Kidwell 1990; Warwick 1995.

26. For Comrie, see Priestley 2011, pp. 60–63.

27. Comrie 1946, p. 149.

28. The best way to get a sense of this transition moment is to survey the journal *Mathematical Tables and Other Aids to Computation.*

29. Goldstine and Goldstine 1946.

30. For these machines, see especially Owens 1986, 1996, and Mindell 2002.

31. Priestley 2011, ch. 6, esp. at pp. 124–25, 153–55. For a sophisticated account of the interplay of engineering and logic in Turing's work, see Copeland 2006.

32. For a powerful reconceptualization of the history of precomputing focused on users, see Swade 2011; most serious histories of computing involve nuanced appraisals of the role of Turing among other factors; see, among a steadily growing literature, Grier 2005; Pelaez 1999; Agar 2001; Campbell-Kelly and Aspray 1996; Agar 2003; Priestley 2011; Mindell 2002; Rojas and Hashagen 2000.

33. Owens 1996, p. 38.

34. See especially now Ensmenger 2010, pp. 35–39 (and *passim*).

35. On the latter, see Kelty 2008; Nuvolari 2005.

36. Shimpi 2012; cf. Scranton 2006; Brock and Lécuyer 2012.

37. For a recent consideration, Zeydel, Baran, and Oklobdzija 2010; the earlier literature is provided in Swartzlander 1990.

38. Oklobdzija et al. 2003, p. 272.

References

Abeille, Louis-Paul. 1765. *Effets d'un privilége exclusif en matiére de commerce, sur les droits de la propriété, &c.* Paris.

Abrams, M. H. 1971. *The Mirror and the Lamp: Romantic Theory and the Critical Tradition.* London: Oxford University Press.

Académie impériale des sciences. 1897–1911. *Procès-verbaux des séances de l'Académie impériale des sciences depuis sa fondation jusqu'à 1803.* 4 vols. Saint Petersburg: Tip. Imp. Akademia nauk.

Adam, Anderson. 1832. Arithmetic. In *The Edinburgh Encyclopaedia Conducted by David Brewster, with the Assistance of Gentlemen Eminent in Science and Literature,* edited by D. Brewster. Philadelphia: J. and E. Parker, 2:345–400.

Agar, Jon. 2001. *Turing and the Universal Machine: The Making of the Modern Computer.* Cambridge: Icon.

———. 2003. *The Government Machine: A Revolutionary History of the Computer.* Cambridge, MA: MIT Press.

A'Hearn, Brian, Jörg Baten, and Dorothee Crayen. 2009. Quantifying Quantitative Literacy: Age Heaping and the History of Human Capital. *Journal of Economic History* 69 (3):783–808.

Alder, Ken. 1997. *Engineering the Revolution: Arms and Enlightenment in France, 1763–1815.* Princeton: Princeton University Press.

Allen, Robert C. 1983. Collective Invention. *Journal of Economic Behavior & Organization* 4 (1): 1–24.

Anonymous. 1732. Machines ou inventions approuvées par l'académie en M. DCCXXX. *Mémoires de l'Académie des sciences (Paris)*:115–116.

———. 1751. Review of *Abhandlung vom Grundlegen der Flächen . . .* Wien, 1751. *Zuverlaessige nachrichten von dem gegenwaertigen zustande: Veraenderung und wachsthum der wissenschaften* 12:665–72.

———. 1778. Précis sur l'origine, le but, et le régime de la Société établie à Genève pour l'encouragement des Arts & de l'Agriculture. *Mémoires de la Société établie à Genève pour l'encouragement des Arts & de l'Agriculture* 1:1–6.

———. 1780. Review of "Account of the Advantages of a newly invented Machine [. . .] By Mr. Le Cerf [. . .]." *Monthly Review* 62 (January):44–48.

———. 1801. Earl Stanhope. *The Public Characters* 3:65–107.

————. 1823. Ueber die neueste Rechnen=Maschine oder über das Arithmometer des Chevalier Thomas zu Colmar. *Dingler's Polytechnisches Journal* 11 (1–3):121–22.

————. 1828. Calculating Machinery. *The Record* (no pagination). Copy in RS HS 27.54.

Anthes, Erhard. 1989. Die Rechenmaschinen von Philipp Matthäus Hahn. In *Philipp Matthäus Hahn, 1739–1790: Ausstellungen des Württembergischen Landesmuseums Stuttgart und der Städte Ostfildern, Albstadt, Kornwestheim, Leinfelden-Echterdingen.* Stuttgart: Württembergisches Landesmuseum Stuttgart, 456–78.

Antognazza, Maria Rosa. 2009. *Leibniz: An Intellectual Biography.* Cambridge: Cambridge University Press.

Archibald, Raymond Clare. 1943. Seventeenth Century Calculating Machines. *Mathematical Tables and Other Aids to Computation* 1 (1):27–28.

Arnauld, Antoine. 1775. *Lettres.* Paris: Chez Sigismond d'Arnay & Compagnie.

Ash, Eric H. 2010. Introduction: Expertise and the Early Modern State. *Osiris* 25 (1):1–24.

B***. 1847. *Chroniques barroises du IVe au XIXe siécle.* Bar-le-Duc: Numa Rolin.

Babbage, Charles. 1822. *Letter to Sir Humphry Davy, Bart, President of the Royal Society, &c.&c. On the Application of Machinery to the Purpose of Calculating and Printing Mathematical Tables.* London: J. Booth and Baldwin, Crodock and Joy.

————. 1989. *The Works of Charles Babbage.* Edited by Martin Campbell-Kelly. 11 vols. London: W. Pickering.

Babbage, Henry Prevost. 1889. *Babbage's Calculating Engines, Being a Collection of Papers Relating to Them, Their History and Construction.* London: E. and F. N. Spon.

Babbage, J. (pseudonym). 1844. The New Patent Novel Writer. *Punch* 7:268.

Badur, Klaus, and Wolfgang Rottstedt. 2004. Und sie rechnet doch richtig!: Erfahrungen beim Nachbau einer Leibniz-Rechenmaschine. *Studia Leibnitiana* 36:129–146.

Bai Shangshu and Li Di. 1980. Hand Calculating Machine [*sic*] of Ancient Times in the Collection of the Palace Museum (in Chinese). *Journal of the Palace Museum,* no. 1:76–82.

Baldinger, Ernst Gottfried. 1787. Herrn Ingenieur Capitain Müllers in Gießen Rechenmaschine. *Medicinisches Journal* 4:93–94.

Bardili, Christoph Gottfried. 1800. *Grundriss der ersten Logik, gereinigt von den Irrthümmern bisheriger Logiken Oberhaupt, der Kantischen insbesondere.* Stuttgart: F. Ch. Löflund.

Barnard, Francis Pierrepont. 1916. *The Casting-Counter and the Counting-Board: A Chapter in the History of Numismatics and Early Arithmetic.* Oxford: Clarendon Press.

Basalla, George. 1988. *The Evolution of Technology.* Cambridge: Cambridge University Press.

Batteux, C. 1747. *Les beaux arts reduits à un même principe.* Paris: Chez Durand.

Bayle, Pierre. 1820. *Dictionnaire historique et critique.* Vol. 13. Paris: Desoer Libraire.

Beaurepaire, Charles de. 1900–1901. Le séjour de Pascal à Rouen. *Précis Analytique des Travaux de l'Académie des Sciences, Belles-Lettres et Arts de Rouen*:211–311.

Bedini, Silvio. 1964. The Role of Automata in the History of Technology. *Technology and Culture* 5 (1):9–42.

————. 1994. In Pursuit of Provenance: The George Graham Proto-Orreries. In *Learning, Language, and Invention: Essays Presented to Francis Middleton,* edited by W. D. Hackmann and A. J. Turner. Aldershot: Variorium/Ashgate, 54–77.

————. 1995. The Fate of the Medici-Lorraine Scientific Instruments. *Journal of the History of Collections* 7 (2):159–70.

Beiser, Frederick C. 1987. *The Fate of Reason: German Philosophy from Kant to Fichte.* Cambridge, MA: Harvard University Press.

Belfanti, Carlo Marco. 2004. Guilds, Patents and the Circulation of Technical Knowledge: Northern Italy during the Early Modern Age. *Technology and Culture* 45:569–89.

———. 2006. Between Mercantilism and Market: Privileges for Invention in Early Modern Europe. *Journal of Institutional Economics* 2:319–38.

Bender, J., and M. Marrinan. 2010. *The Culture of Diagram*. Palo Alto, CA: Stanford University Press.

Bennet, Angela C. 1992. The Stanhopes in Geneva: A Study of an English Noble Family in Genevan Politics and Society. MA Thesis, University of Kent.

Bennett, James A. 1987. *The Divided Circle: A History of Instruments for Astronomy, Navigation, and Surveying*. Oxford: Phaidon.

———. 2002. Shopping for Instruments in Paris and London. In *Merchants and Marvels: Commerce, Science and Art in Early Modern Europe*, edited by P. H. Smith and P. Findlen. New York: Routledge, 370–95.

———. 2003. Hooke's Instruments. In *London's Leonardo: The Life and Work of Robert Hooke*, edited by J. Bennett, M. Cooper, M. Hunter, and L. Jardine. Oxford: Oxford University Press, 63–104.

———. 2006. Instruments and Ingenuity. In *Robert Hooke: Tercentennial Studies*, edited by M. Cooper and M. Hunter. Aldershot: Ashgate, 65–76.

Bennett, Jane. 2010. *Vibrant Matter: A Political Ecology of Things*. Durham: Duke University Press.

Bentley, Richard. 1838. *The Works of Richard Bentley*. Edited by A. Dyce. London: F. Macpherson.

Benziger, James. 1951. Organic Unity: Leibniz to Coleridge. *PMLA* 66 (2):24–48.

Berg, Maxine. 2002. From Imitation to Invention: Creating Commodities in Eighteenth-Century Britain. *The Economic History Review* 55 (1):1–30.

———. 2004. In Pursuit of Luxury: Global History and British Consumer Goods in the Eighteenth Century. *Past & Present* (182):85–142.

———. 2007. The Genesis of "Useful Knowledge." *History of Science* 65:123–33.

Berkeley, George. 1948–57. *The Works of George Berkeley, Bishop of Cloyne*. Edited by A. A. Luce and T. E. Jessop. 9 vols. London: Nelson.

Bernoulli, Jacob. 1993. *Der Briefwechsel von Jacob Bernoulli*. Edited by A. Weil, C. Truesdell, and F. Nagel. Basel: Birkhäuser.

Bertoloni Meli, Domenico. 1993. *Equivalence and Priority: Newton versus Leibniz*. Oxford: Clarendon Press.

Bertucci, Paola. 2006. Public Utility and Spectacular Display: The Physics Cabinet of the Royal Museum in Florence. *Nuncius* 21:323–36.

Biagioli, Mario. 2006a. From Prints to Patents: Living on Instruments in Early Modern Europe. *History of Science* 44:139–86.

———. 2006b. *Galileo's Instruments of Credit: Telescopes, Images, Secrecy*. Chicago: University of Chicago Press.

———. 2006c. Patent Republic: Representing Inventions, Constructing Rights and Authors. *Social Research* 74:1129–72.

———. 2009. Nature and the Commons: The Vegetable Roots of Intellectual Property. In *Living Properties: Making Knowledge and Controlling Ownership in the History of Biology*, edited by J.-P. Gaudillière, D. J. Kevles, and H.-J. Rheinberger. Max Planck Institute for the History of Science Preprint 382, 241–50.

Biagioli, Mario, and Peter L. Galison, eds. 2003. *Scientific Authorship: Credit and Intellectual Property in Science*. New York: Routledge.

Birch, Thomas. 1756–57. *The History of the Royal Society of London for Improving of Natural Knowledge, from Its First Rise*. London.

Bischoff, Johann Paul. 1990. *Versuch einer Geschichte der Rechenmaschine: Ansbach 1804*. Edited by S. Weiss. München: Systhema Verlag.

Blegny, Nicolas de. 1878. *Livre commode des adresses de Paris pour 1692*. Edited by E. Fournier. 2 vols. Paris: Daffis.

Blondeau, Peter. 1653. *A most humble mem[o]randum from Peter Blondeau, concerning the offers made to him by this Commonwealth, for the coyning of the monie, by a new invention, not yet practised in any state of the world, the which will prevent counterfeiting, casting, washing, and clipping of the same: Which coyn shall be marked on both the flat sides, and about the thickness or the edge; of a like bigness and largness, as the ordinarie coyn is: And will cost no more than the ordinarie unequal coyn, which is used now*. London: n.p.

Boissonnade, P. 1932. *Colbert, le triomphe de l'étatisme; la fondation de la suprématie industrielle de la France, la dictature du travail (1661–1683)*. Paris: M. Rivière.

Bonney, Richard. 1978. *Political Change in France under Richelieu and Mazarin, 1624–1661*. Oxford: Oxford University Press.

Boyle, James. 1996. *Shamans, Software, and Spleens: Law and the Construction of the Information Society*. Cambridge, MA: Harvard University Press.

Boys, C. V. 1886. Calculating Machines. *Journal of the Society of Arts* 34:376–88.

Bracha, Oren. 2004. The Commodification of Patents 1600–1836: How Patents Became Rights and Why We Should Care. *Loyola of Los Angeles Law Review* 38:177–244.

———. 2005. Owning Ideas: History of Intellectual Property in the United States. S.J.D. Dissertation, Harvard Law School.

Brecht, Martin. 1981. Hahn und Herder. *Zeitschrift für Württembergische Landesgeschichte* 41: 364–87.

Brewster, D., W. Scott, and J. Stark. 1832. *Letters on Natural Magic: Addressed to Sir Walter Scott, Bart*: London: John Murray, Albemarle Street.

Brewster, David. 1830. Review of Charles Babbage's *Reflections on the Decline of Science in England, and on Some of Its Causes. The Quarterly Review* 43:305–42.

Brock, David C., and Christophe Lécuyer. 2012. Digital Foundations: The Making of Silicon-Gate Manufacturing Technology. *Technology and Culture* 53 (3):561–97.

Bromley, Allan G. 1980. *The Mechanism of Charles Babbage's Analytical Engine circa 1838*, Technical Report 166. Sydney: Basser Department of Computer Science, University of Sydney.

———. 1983. *Babbage's Calculating Engines*, Technical Report 215. Sydney: Basser Department of Computer Science, University of Sydney.

———. 1987. The Evolution of Babbage Calculating Engines. *Annals of the History of Computing* 9 (2):113–36.

———. 1998. Charles Babbage's Analytical Engine, 1838. *Annals of the History of Computing* 20 (4):29–45.

———. 2000. Babbage's Analytical Engine Plans 28 and 28a—the Programmer's Interface. *Annals of the History of Computing* 22 (4):5–19.

Brooks, John. 1992. The Circular Dividing Engine: Development in England 1739–1843. *Annals of Science* 49 (2):101–35.

Brooks, Randall Chapman. 1989. The Precision Screw in Scientific Instruments on the 17th–19th Centuries: With Particular Reference to Astronomical, Nautical and Surveying Instruments. PhD diss., Department of Astronomy, University of Leicester.

Buffon, Georges-Louis Leclerc. 1749. *Histoire naturelle générale et particulière: Avec la description du Cabinet du Roy.* Vol. 1. Paris.

Bullnynck, Maarten. 2008. The Transmission of Numeracy: Integrating Reckoning in Protestant North-German Elementary Education (1770–1810). *Pedagogica Historica: International Journal of the History of Education* 44 (5):563–85.

Buot, Jacques. 1647. *Usage de la roue de proportion . . .* Paris: Chez Melchior Mondiere.

Burnett, D. Graham. 2005. *Descartes and the Hyperbolic Quest: Lens Making Machines and Their Significance in the Seventeenth Century.* Vol. 95, pt. 3, of *Transactions of the American Philosophical Society.* Philadelphia: American Philosophical Society.

Buxton, H. W., and Anthony Hyman. 1988. *Memoir of the Life and Labours of the Late Charles Babbage Esq., F.R.S.* Cambridge, MA: MIT Press.

Campbell-Kelly, Martin. 2003. *The History of Mathematical Tables: From Sumer to Spreadsheets.* Oxford: Oxford University Press.

Campbell-Kelly, Martin, and William Aspray. 1996. *Computer: A History of the Information Machine.* 1st ed. New York: Basic Books.

Candaux, Jean-Daniel, and René Sigrist. 2001. Saussure et la Société des Arts. In *H.B. De Saussure (1740–1799): Un régard sur la terre,* edited by J.-D. Candaux and R. Sigrist. Geneva: Georg Editeur, 431–51.

Carlson, W. Bernard. 1999. *Banishing Prometheus? Edison, Invention, and Representation.* Available from http://www.stanford.edu/dept/HPS/WritingScience/etexts/Carlson/Prometheus.html.

Carlson, W. Bernard, and Michael E. Gorman. 1990. Understanding Invention as a Cognitive Process: The Case of Thomas Edison and Early Motion Pictures, 1888–1891. *Social Studies of Science* 20:387–430.

Ceruzzi, Paul. 2005. Moore's Law and Technological Determinism: Reflections on the History of Technology. *Technology and Culture* 46 (3):584–93.

Chandler, Alfred D., and James W. Cortada, eds. 2000. *A Nation Transformed by Information: How Information Has Shaped the United States from Colonial Times to the Present.* New York: Oxford University Press.

Chapman, Allen. 2005. *England's Leonardo: Robert Hooke and the Seventeenth-Century Scientific Revolution* Bristol: Institute of Physics.

Chase, George C. 1980. History of Mechanical Computing Technology. *Annals of the History of Computing* 2:198–226.

Chaulnes, Michel-Ferdinand d'Albert d'Ailly, duc de. 1768. *Nouvelle méthode pour diviser les instruments de mathématique et d'astronomie.* Paris.

Choi, Y., and E. E. Swartzlander Jr. 2005. Parallel Prefix Adder Design with Matrix Representation. Paper read at the 17th IEEE Symposium on Computer Arithmetic, June 27–29, 2005. Available from http://www.acsel-lab.com/arithmetic/arith17/papers/ARITH17_Choi.pdf.

Church, William Farr. 1972. *Richelieu and Reason of State.* Princeton: Princeton.

Clark, William. 2002. From Enlightenment to Romanticism: Lichtenberg and Göttingen Physics. In *Göttingen and the Development of the Natural Sciences,* edited by N. Rupke. Göttingen: Wallstein, 72–85.

Clement, Joseph. 1825. Stand for Drawing Boards of Large Area. *Transactions of the Society for the Encouragement of Arts, Manufactures, and Commerce* 43:138–42.

Cléro, Jean Pierre. 2001. *Les Pascal à Rouen 1640–1648.* Rouen: Université de Rouen.

Clifton, Gloria. 1996. *Directory of British Scientific Instrument Makers, 1550–1851.* London: Zwemmer in association with the National Maritime Museum.

Cohen, I. B. 1990. Notes on Babbage, Aiken, and Bowditch. *Annals of the History of Computing* 12 (1):70.

Cohen, Patricia Cline. 1982. *A Calculating People: The Spread of Numeracy in Early America*. Chicago: University of Chicago Press.

Colbert, Jean Baptiste, Pierre Clément, and Pierre de Brotonne. 1861. *Lettres, instructions et mémoires de Colbert, publiés d'après les ordres de l'empereur, sur la proposition de Son Excellence M. Magne, ministre secrétaire d'état des finances*. Paris: Imprimerie impériale.

Cole, Charles Woolsey. 1939. *Colbert and a Century of French Mercantilism*. 2 vols. New York: Columbia University Press.

Coleman, Gabriella. 2012. *Coding Freedom: The Ethics and Aesthetics of Hacking*. Princeton: Princeton University Press.

Collier, Bruce. 1990. *The Little Engines That Could've: The Calculating Machines of Charles Babbage*. New York: Garland. Original edition, 1970.

Collins, H. M. 1985. *Changing Order: Replication and Induction in Scientific Practice*. London: Sage.

———. 1990. *Artificial Experts: Social Knowledge and Intelligent Machines*. Cambridge, MA: MIT Press.

———. 2001. Tacit Knowledge, Trust and the Q of Sapphire. *Social Studies of Science* 31:71–85.

———. 2010. *Tacit and Explicit Knowledge*. Chicago: University of Chicago Press.

Collins, H. M., and Martin Kusch. 1998. *The Shape of Actions: What Humans and Machines Can Do*. Cambridge, MA: MIT Press.

Collins, James B. 1988. *Fiscal Limits of Absolutism: Direct Taxation in Early Seventeenth-Century France*. Berkeley: University of California Press.

Comrie, Leslie J. 1946. The Application of Commercial Calculating Machines to Scientific Computing. *Mathematical Tables and Other Aids to Computation* 2 (16):149–59.

Condorcet. 1847–49. *Oeuvres de Condorcet*. Edited by F. Arago and A. O'Connor. Paris.

Condren, Conal, and Ian Hunter. 2008. Introduction: The *Persona* of the Philosopher in the Eighteenth Century. *Intellectual History Review* 18 (3):315–317.

Constant, Edward, II. 1999. Reliable Knowledge and Unreliable Stuff: On the Practical Role of Rational Beliefs. *Technology and Culture* 40 (3):324–57.

Cook, Harold J. 2007. *Matters of Exchange: Commerce, Medicine, and Science in the Dutch Golden Age*. New Haven: Yale University Press.

Cook, Simon. 2005. Minds, Machines and Economic Agents: Cambridge Receptions of Boole and Babbage. *Studies in History and Philosophy of Science Part A* 36 (2):331–50.

Copeland, B. Jack. 2006. Colossus and the Rise of the Modern Computer. In *Colossus: The Secrets of Bletchley Park's Codebreaking Computers*, edited by B. J. Copeland. Oxford: Oxford University Press, 101–15.

Cortada, James W. 1996. Commercial Applications of the Digital Computer in American Corporations, 1945–1995. *Annals of the History of Computing* 18 (2):18–29.

———. 2000. *Before the Computer: IBM, NCR, Burroughs, and Remington Rand and the Industry They Created, 1865–1956*. Princeton: Princeton University Press.

Coward, William. 1702. *Second thoughts concerning human soul: Demonstrating the Notion of Human Soul, As believ'd to be a Spiritual and Immaterial Substance, united to Human Body, To be an invention of the Heathens, And not Consonant to the Principles of Philosophy, Reason, or Religion*. London: A. Basset.

Crinò, Anna Maria. 1955. Sir Samuel Morland nei suoi rapporti col granduca Cosimo III di Toscana. In *English Miscellany, a Symposium of History, Literature and the Arts*. Rome: Edizioni di storia e letteratura, 237–46.

———. 1957. I rapporti di Sir Samuel Morland con Cosimo III dei Medici. In *Fatti e figure del Seicento anglo-toscano: Documenti inediti sui rapporti letterari, diplomatici e culturali fra Toscana e Inghilterra*. Florence: L. S. Olschki, 213–61.

Crosnier, Jules. 1910. *La Société des Arts et ses collections*. Geneva: "Nos anciens et leurs oeuvres."

Daston, Lorraine. 1994. Enlightenment Calculations. *Critical Inquiry* 21 (1):182–202.

Daston, Lorraine, and H. Otto Sibum. 2003. Introduction: Scientific Personae and Their Histories. *Science in Context* 16 (1–2):1–8.

Daumas, Maurice. 1953. *Les instruments scientifiques aux XVIIe et XVIIIe siècles*. Paris: PUF.

Davillé, L. 1920–22. Le séjour de Leibniz à Paris. *Archiv für Geschichte der Philosophie* 32:142–9, 33:67–78, 34:14–40, 35:50–62.

Davis, Martin. 2001. *Engines of Logic: Mathematicians and the Origin of the Computer*. New York: W. W. Norton.

De Fouchy, J. P. G. 1763. Eloge de M. Le Marquis Poleni. *Histoire et Mémoires de l'Académie Royale des Sciences*. Histoire, 151–63.

Derham, W. 1700. *The Artificial Clock-Maker. a Treatise of Watch & Clock-Work, Wherein the Art of Calculating Numbers for Most Sorts of Movements Is Explained, to the Capacity of the Unlearned.* . . . 2nd enlarged ed. London: J. Knapton.

Des Chene, Dennis. 2006. "Animal" as Category: Bayle's "Rorarius." In *The Problem of Animal Generation in Early Modern Philosophy*, edited by J. Smith. Cambridge: Cambridge University Press, 215–32.

Descotes, Dominique. 1986. "Usage de la machine": Manuscrit acquis par le Centre international Blaise Pascal. *Courrier du Centre international Blaise Pascal*, no. 8:4–23.

———. 1989. Pascal et le marketing. In *Mélanges offerts au professeur Maurice Descotes*. Pau: Université de Pau, 141–60.

Dickinson, H. W. 1970. *Sir Samuel Morland, Diplomat and Inventor, 1625–1695*. Cambridge: Published for the Newcomen Society, by W. Heffer and Sons Limited.

Diderot, Denis. 1975–. *Oeuvres complètes. Édition critique et annotée*. Edited by H. Dieckmann, J. Fabre, J. Proust, and J. Varloot. Paris: Hermann.

Diderot, Denis, and Jean Le Rond d'Alembert. 1751–72. *Encyclopédie, ou, Dictionnaire raisonné des sciences, des arts et des métiers*. Geneva: Chez Briasson.

Diehl, Wilhelm. 1930. Lebensbeschreibung des Obristen und Oberbaudirektors, auch Direktors des Oberbaukollegs Johann Helfich von Müller. *Hessische Chronik* 17:1–21.

Domat, Jean. 1989. *Les loix civiles dans leur ordre naturel*. Edited by S. G. Fabre and J. Remy. Caen: Centre de Philosophie politique et juridique.

dom Calmet. 1751. *Bibliothèque lorraine, ou Histoire des hommes illustres qui ont fleuri en Lorraine, dans les Trois Évêchés, dans l'archevêché de Trèves, dans le duché de Luxembourg, etc.* Nancy: A. Leseure.

Duchesneau, François. 2003. Leibniz's Model for Analyzing Organic Phenomena. *Perspectives on Science* 11 (4):378–409.

Duff, William. 1767. *An Essay on Original Genius; and Its Various Modes of Exertion in Philosophy and the Fine Arts, Particularly in Poetry*. London.

Du Marsais, César Chesneau. 1743. Le Philosophe. In *Nouvelles libertés de penser*. Amsterdam.

Durand-Richard, Marie-José. 1996. L'École algébrique anglaise: Les conditions conceptuelles et institutionnelles d'un calcul symbolique comme fondement de la connaissance. In *L'Europe mathématique-Mathematical Europe*, edited by C. Goldstein, J. Gray, and J. Ritter. Paris: Editions de la Maison des sciences de l'homme, 445–77.

Dutton, H. I. 1984. *The Patent System and Inventive Activity during the Industrial Revolution, 1750–1852*. Manchester: Manchester University Press.

Eagleton, Catherine, and Boris Jardine. 2005. Collections and Projections: Henry Sutton's Paper Instruments. *Journal of the History of Collections* 17:1–13.

Easton, Patricia, ed. 1997. *Logic and the Workings of the Mind: The Logic of Ideas and Faculty Psychology in Early Modern Philosophy*. Vol 5. of *North American Kant Society Studies in Philosophy*. Atascadero, CA: Ridgeview.

Edgerton, David. 1999. From Innovation to Use: Ten Eclectic Theses on the Historiography of Technology. *History and Technology* 16 (2):111–36.

Edmondson, Joseph. 1885. Summary of Lecture on Calculating Machines, Delivered before the Physical Society of London, March 28, 1885. *London, Edinburgh, and Dublin Philosophical Magazine*, 5th ser., vol. 20 (July).

Edwards, M. R. 2002. *Pushing Gravity: New Perspectives on Le Sage's Theory of Gravitation*. Vol. Montreal: C. Roy Keys Inc.

Elster, Jon. 1975. *Leibniz et la formation de l'esprit capitaliste*. Paris: Aubier Montaigne.

Emigh, Rebecca J. 2002. Numeracy or Enumeration? The Uses of Numbers by States and Societies. *Social Science History* 26 (4):653–98.

Engel, Claire-Elaine. 1946. Genève et l'Angleterre: Les De Luc, 1727–1817. *Zeitschrift für Schweizerische Geschichte* 26:479–504.

Ensmenger, Nathan. 2010. *The Computer Boys Take Over: Computers, Programmers, and the Politics of Technical Expertise*. Cambridge, MA: MIT Press.

Erickson, P., J. L. Klein, L. Daston, R. Lemov, T. Sturm, and M. D. Gordin. 2013. *How Reason Almost Lost Its Mind: The Strange Career of Cold War Rationality*. Chicago: University of Chicago Press.

Favaro, A. 1896. Intorno alla vita ed ai lavori di Tito Livio Burattini. *Memorie del Reale Instituto Veneto di Scienze, Lettere ed Arti* 25 (8):1–140.

Felt, D. E. 1916. *Mechanical Arithmetic, or, the History of the Counting Machine*. Chicago: Washington Institute.

Ferguson, Eugene S. 1971. Leupold's "Theatrum Machinarum": A Need and an Opportunity. *Technology and Culture* 12 (1):64–68.

———. 1977. The Mind's Eye: Nonverbal Thought in Technology. *Science* (26 August):827–36.

Foisil, Madeleine. 1970. *La révolte des nu-pieds et les révoltes normandes de 1639*. Paris: Presses Universitaires de France.

Forbes, Eric Gray. 1970. Tobias Mayer (1723–62): A Case of Forgotten Genius. *British Journal for the History of Science* 5 (1):1–20.

Fors, Hjalmar. 2014. *The Limits of Matter: Chemistry, Mining, and Enlightenment*. Chicago: University of Chicago Press.

Fox, Celina. 2009. *The Arts of Industry in the Age of Enlightenment*. New Haven: Paul Mellon Centre for Studies in British Art and Yale University Press.

Francoeur. 1822. Rapport fait par M. Francoeur sur la machine à calculer de M. le Chevalier Thomas, de Colmar. . . . *Bulletin de la Société d'Encouragement pour l'Industrie Nationale*. 21(February):33–36.

Fréchet, Michel. 2004. Machines d'Arithmétique. In *Instruments scientifiques à travers l'histoire*, edited by E. Hébert. Paris: Ellipses, 337–63.

Fries, Jakob Friedrich. 1803. *Reinhold, Fichte, Schelling*. Leipzig: A. L. Reinicke.

Frumkin, Maximilian. 1947–48. The Early History of Patents for Invention. *Transactions of the Chartered Institute for Patent Agents* 66:20–69.

Fuegi, J., and J. Francis. 2003. Lovelace & Babbage and the Creation of the 1843 "Notes." *Annals of the History of Computing* 25 (4):16–26.

Gallon, Jean-Gaffin. 1735–77. *Machines et inventions approuvées par l'Académie royale des sciences depuis son établissement . . . (jusqu'en 1754) avec leur description*. 7 vols. Paris: G. Martin: J. B. Coignard fils: H. L. Guérin.

Gammon, Martin. 1997. "Exemplary Originality": Kant on Genius and Imitation. *Journal of the History of Philosophy* 35 (4):563–92.

Gardner, Martin. 1982. *Logic Machines and Diagrams*. 2nd ed. Chicago: University of Chicago Press.

Garrett, Aaron. 2013. Mind and Matter. In *The Oxford Handbook of British Philosophy in the Eighteenth Century*, edited by J. A. Harris. Oxford: Oxford University Press, 171–96.

Gaukroger, Stephen. 1997. Descartes' Early Doctrine of Clear and Distinct Ideas. In *The Genealogy of Knowledge: Analytical Essays in the History of Philosophy and Science*. Aldershot: Ashgate, 131–52.

Gauvin, Jean-François. 2006a. Artisans, Machines, and Descartes's *Organon*. *History of Science* 44:187–216.

———. 2006b. Le cabinet de physique du château de Cirey et la philosophie naturelle du Mme Du Châtelet et de Voltaire. In *Emilie Du Châtelet: Rewriting Enlightenment Philosophy and Science*, edited by J. P. Zinsser and J. C. Hayes. Oxford: Voltaire Foundation, 165–202.

———. 2008. Habits of Knowledge: Artisans, Theory and Mechanical Devices in Seventeenth-Century France. PhD diss., History of Science, Harvard University, Cambridge, MA.

Gérard, Alexander. 1774. *An Essay on Genius*. London: Printed for W. Strahan; T. Cadell.

Gersten, Christian-Ludovicus. 1735. The Description and Use of an Arithmetical Machine Invented by Christian-Ludovicus Gersten, F. R. S. Professor of Mathematicks in the University of Giessen. Inscribed to Sir Hans Sloane, Bart. President of the Royal Society. *Philosophical Transactions of the Royal Society of London* 39:79–97.

Gill, Thomas. 1822. On Various Cements. In *The Technical Repository*, edited by T. Gill. London: T. Cadell, 412–19.

Gingerich, Owen. 1986. Instruments in the Beijing National Palace. *Bulletin of the Scientific Instrument Society*, no. 10:8.

Ginn, William Thomas. 1991. Philosophers and Artisans: The Relationship between Men of Science and Instrument Makers in London, 1820–1860. PhD diss., Unit for the History of Science, University of Kent at Canterbury.

Glassman, Robert B., and Hugh W. Buckingham. 2007. David Hartley's Neural Vibrations and Psychological Associations. In *Brain, Mind and Medicine: Essays in Eighteenth-Century Neuroscience*, edited by H. Whitaker, C. U. M. Smith, and S. Finger. New York: Springer, 177–90.

Godin, Benoît. 2008. Innovation: The History of a Category. *Project on the Intellectual History of Innovation*, Working Paper No. 1. Available from http://www.csiic.ca/PDF/Intellectual No1.pdf.

Goldstine, Herman H. 1972. *The Computer from Pascal to von Neumann*. Princeton: Princeton University Press.

————. 1977. A Brief History of the Computer. *Proceedings of the American Philosophical Society* 121:339–45.

Goldstine, Herman H., and Adele Goldstine. 1946. The Electronic Numerical Integrator and Computer (ENIAC). *Mathematical Tables and Other Aids to Computation* 2 (15):97–110.

Gomme, Arthur Allan. 1946. *Patents of Invention: Origin and Growth of the Patent System in Britain, Science in Britain.* London: Published for the British Council by Longmans, Green, and Co.

Gordon, Robert B. 1988. Who Turned the Mechanical Ideal into Mechanical Reality? *Technology and Culture* 29 (4):744–78.

Gouhier, Henri. 1978. *Cartésianisme et augustinisme au XVIIe siècle.* Paris: J. Vrin.

Grabiner, Judith V. 1998. "Some Disputes of Consequence": Maclaurin among the Molasses Barrels. *Social Studies of Science* 28:139–68.

Graf, Klaus-Dieter. 1994. Calculating Machines in China and Europe in the 17th Century: The Western View. In *Proceedings of the Fifth Five Nations Conference on Mathematics Education,* edited by K. Yokochi and H. Okamori. Osaka. Available from http://www.inf.fu-berlin.de/users/graf/cmframe.html.

Grattan-Guinness, I. 1990. Work for the Hairdressers: The Production of de Prony's Logarithmic and Trigonometric Tables. *Annals of the History of Computing* 12 (3):177–85.

Gravatt, William F. 1857. *Specimens of Tables, Calculated, Stereomoulded and Printed by Machinery.* London: Longman, Brown, Green, Longmans, and Roberts.

Gray, Peter. 1873. On the Arithmometer of M. Thomas (de Colmar), and Its Application to the Construction of Life Contingency Tables. *Journal of the Institute of Actuaries and Assurance Magazine* 17 (4):249–66.

Green, Christopher D. 2005. Was Babbage's Analytical Engine Intended to Be a Mechanical Model of the Mind? *History of Psychology* 8 (1):35–45.

Gregory, Frederick. 1983. Neo-Kantian Foundations of Geometry in the German Romantic Period. *Historia Mathematica* 10 (2):184–201.

Grier, David Alan. 2005. *When Computers Were Human.* Princeton: Princeton University Press.

Grillet, René. 1673. *Curiositez mathématiques de l'invention du Sr Grillet, horlogeur à Paris.* Paris: Jean Baptiste Coignard.

Grua, Gaston. 1956. *La justice humaine selon Leibniz.* Paris: Presses Universitaires de France.

Guiffrey, Jules. 1881. *Comptes des Bâtiments du Roi sous le règne de Louis XIV: Tome Premier, Colbert: 1664–1680.* Paris: Imprimerie nationale.

Guyer, Paul. 1996. *Kant and the Experience of Freedom: Essays on Aesthetics and Morality.* Cambridge: Cambridge University Press.

H., J. 1780. Letter of 6.2.1780. *Monthly Review* 62 (February):176.

Hahn, Philipp Matthäus. 1779. Beschreibung einer Rechnungs-Maschine, wodurch man ohne Mühe, durch bloße Herumführung eines Triebels, die vier gewöhnlichen Rechnungs-Arten verrichten kann. *Der Teutsche Merkur* 2.Viertelj.:137–54.

————. 1785. Vertheidigung der Hahnschen Rechnungs-Maschine gegen einige Misverständnisse. *Der Teutsche Merkur* 2.Viertelj.:86–95.

————. 1979. *Die Kornwestheimer Tagebücher, 1772–1777.* Edited by M. Brecht and R. F. Paulus. Berlin: De Gruyter.

————. 1983. *Die Echterdinger Tagebücher, 1780–1790.* Edited by M. Brecht and R. F. Paulus. Berlin: De Gruyter.

————. 1987. *Werkstattbuch.* 4 vols. Stuttgart: Württembergisches Landesmuseum Stuttgart.

Hamann, J. G. 1894. *Briefwechsel zwischen Hamann und Lavater*. Edited by Heinrich Funck. Königsberg: R. Leupold.

Hamscher, Albert N. 1987. The Conseil Privé and the Parlements in the Age of Louis XIV: A Study in French Absolutism. *Transactions of the American Philosophical Society* 77 (2): i-162.

Hanley, Sarah. 1997. Social Sites of Political Practice in France: Lawsuits, Civil Rights, and the Separation of Powers in Domestic and State Government, 1500–1800. *The American Historical Review* 102 (1):27–52.

Hannyngton, J. 1865. On the Adaptation of Assurance Formulae to the Arithmometer of M. Thomas. *The Assurance Magazine, and Journal of the Institute of Actuaries* 12 (3):184.

———. 1871. On the Use of M. Thomas de Colmar's Arithmometer in Actuarial and Other Computations. *Journal of the Institute of Actuaries and Assurance Magazine* 16 (4):244–53.

Harley, Robert. 1879. The Stanhope Demonstrator. *Mind* 4 (14):192–210.

———. 1890. On the Stanhope Logical and Arithmetical Machines. *Journal and Proceedings of the Royal Society of New South Wales* 24:121–24.

Harris, John R. 1998. *Industrial Espionage and Technology Transfer: Britain and France in the Eighteenth Century*. Aldershot: Ashgate.

Hartley, David. 1791. *Observations on Man, His Frame, His Duty, and His Expectations*. London: J. Johnson.

Hayden-Roy, Priscilla A. 1994. *"A Foretaste of Heaven": Friedrich Hölderlin in the Context of Württenberg Pietism*. Amsterdam: Rodopi.

Hegel, G. W. F. 1977. *The Difference between Fichte's and Schelling's System of Philosophy*. Edited by H. S. Harris and W. Cerf. Albany: State University of New York Press.

———. 1986. *La Différence entre les systèmes philosophiques de Fichte et de Schelling*. Edited by B. Gilson. Paris: Librairie Philosophique J. Vrin.

Heide, Lars. 2009. *Punched-Card Systems and the Early Information Explosion, 1880–1945*. Baltimore: Johns Hopkins University Press.

Heilbron, J. L. 1993. *Weighing Imponderables and Other Quantitative Science around 1800*. Berkeley: University of California Press.

———. 2011. Natural Philosophy. In *Wrestling with Nature: From Omens to Science*, edited by P. Harrison, R. L. Numbers, and M. H. Shank. Chicago: University of Chicago Press, 173–99.

Heinich, Nathalie. 1993. *Du peintre à l'artiste: Artisans et académiciens à l'âge classique*. Paris: Minuit.

Herder, Johann Gottfried. 1881–1913. *Herders Sämmtliche Werke*. Edited by B. Suphan. Berlin: Weidmannsche Buchhandlung.

———. 1940. *God: Some Conversations*. Translated by F. H. Burkhardt. New York: Veritas Press.

Hilaire-Pérez, Liliane. 1991. Invention and the State in 18th-Century France. *Technology and Culture* 32 (4):911–31.

———. 2006. *L'invention au siècle des lumières*. Paris: Albin Michel.

———. 2007. Technology as a Public Culture in the Eighteenth Century: The Artisans' Legacy. *History of Science* 45:135–53.

———. 2013. *La pièce et le geste: Artisans, marchands et savoir technique à Londres au XVIIIe siècle*. Paris: A. Michel.

Hilaire-Pérez, Liliane, and Catherine Verna. 2006. Dissemination of Technical Knowledge in the Middle Ages and the Early Modern Era. *Technology and Culture* 47:536–65.

Hoart, M. 1870. *Advertisement for Arithmomètre of Thomas de Colmar*. Paris.

Hobbes, Thomas. 1981. *Computatio sive logica*. Translated by A. Martinich. Edited by I. C. Hungerland and G. R. Vick. New York: Abaris Books.

———. 1999. *De corpore: Elementorum philosophiae sectio prima*. Edited by K. Schuhman. Paris: Vrin.

Hont, Istvan. 2005. *Jealousy of Trade: International Competition and the Nation-State in Historical Perspective*. Cambridge, MA: Belknap Press of Harvard University Press.

Hooke, Robert. 1935. *The Diary of Robert Hooke, 1672–1680, Transcribed from the Original in the Possession of the Corporation of the City of London (Guildhall Library)*. Edited by H. W. Robinson and W. Adams. London: Taylor and Francis.

Hooke, Robert, and Richard Waller. 1705. *The Posthumous Works of Robert Hooke Containing His Cutlerian Lectures, and Other Discourses, Read at the Meetings of the Illustrious Royal Society*. London: Printed by S. Smith and B. Walford.

Howard, Nicole. 2008. Marketing Longitude: Clocks, Kings, Courtiers, and Christiaan Huygens. *Book History* 11:59–88.

Hoyau. 1822. Description d'une machine à calculer nommée Arithmomètre, de l'invention de M. le Chevalier Thomas, de Colmar. *Bulletin de la Société d'Encouragement pour l'Industrie Nationale* 21(November):356–68.

Hughes, T. P. 1971. *Elmer Sperry: Inventor and Engineer*. Baltimore: Johns Hopkins University Press.

Hulme, E. Wyndham. 1917. Privy Council Law and Practice of Letters Patent for Invention from the Restoration to 1794. *Law Quarterly Review* 33:63–75, 180–95.

Hume, David. 1987. *Essays: Moral, Political, Literary*. Edited by E. F. Miller. Indianapolis: Liberty Fund.

Hunter, Michael, and Edward B. Davis, eds. 1999–2000. *The Works of Robert Boyle*. 14 vols. London: Pickering & Chatto.

Huskey, Velma R., and Harry D. Huskey. 1980. Lady Lovelace and Charles Babbage. *Annals of the History of Computing* 2 (4):299–329.

Huygens, Christiaan. 1888. *Oeuvres complètes*. 22 vols. The Hague: M. Nijhoff.

Hyman, Anthony. 1982. *Charles Babbage, Pioneer of the Computer*. Princeton: Princeton University Press.

Iliffe, Rob. 1992. "In the Warehouse": Privacy, Property and Priority in the Early Royal Society. *History of Science* 30:29–68.

———. 1995. Material Doubts: Hooke, Artisan Culture and the Exchange of Information in 1670s London. *British Journal for the History of Science* 28:285–318.

Ingold, Tim. 2007. Materials against Materiality. *Archaeological Dialogues* 14 (1):1–16.

———. 2010. The Textility of Making. *Cambridge Journal of Economics* 34 (1):91–102.

Inkster, Ian. 1990. Mental Capital: Transfers of Knowledge and Technique in Eighteenth Century Europe. *Journal of European Economic History* 19:403–41.

Isoré, Jacques. 1937. De l'existence des brevets de l'invention en droit français avant 1791. *Revue historique de droit français et etranger*, 4th ser., vol. 16:94–130.

Ivernois, François d'. 1782. *Tableau historique et politique des revolutions de Genève*. Geneva.

Jackson, Myles. 2000. *Spectrum of Belief: Jospeh von Fraunhofer and the Craft of Precision Optics*. Cambridge, MA: MIT Press.

Jacob, Margaret C. 2007. Mechanical Science on the Factory Floor: The Early Industrial Revolution in Leeds. *History of Science* 45 (2):197–221.

Jacob, Margaret C., and Larry Stewart. 2004. *Practical Matter: Newton's Science in the Service of Industry and Empire, 1687–1851*. Cambridge, MA: Harvard University Press.

Jaffee, David. 2010. *A New Nation of Goods: The Material Culture of Early America, Early American Studies*. Philadelphia: University of Pennsylvania Press.

Janetschek, Hellmut. 1988. Die Mechaniker- und Uhrmacherfamilie Braun und die braunschen Rechenmaschinen: Ergänzungen und Richtigstellungen zur Bisherigen Forschung. *Blätter für Technikgeschichte* 50:165–79.

Jansen, P. 1951. Une tractation commerciale au XVIIe siècle. *Revue d'histoire des sciences et de leurs applications* 4 (2):173–76.

Jennings, Louis J. 1885. *The Croker Papers: The Correspondence and Diaries of the Late Right Honourable John Wilson Croker, Secretary to the Admiralty from 1809 to 1830*. London: John Murray.

Johns, Adrian. 2009. *Piracy: The Intellectual Property Wars from Gutenberg to Gates*. Chicago: University of Chicago Press.

Johnson, Ann. 2009. *Hitting the Brakes: Engineering Design and the Production of Knowledge*. Durham: Duke University Press.

Johnston, Stephen. 1997. Making the Arithmometer Count. *Bulletin of the Scientific Instrument Society* 52:12–21.

Jones, Matthew L. 2006. *The Good Life in the Scientific Revolution: Descartes, Pascal, Leibniz and the Cultivation of Virtue*. Chicago: University of Chicago Press.

———. 2014. Space, Evidence, and the Authority of Mathematics in the Eighteenth Century. In *The Routledge Companion to Eighteenth-Century Philosophy*, edited by A. Garrett. London: Routledge, 203–31.

———. 2016. Calculating Devices and Computers. In *A Companion to the History of Science*, edited by B. Lightman. Malden, MA: Wiley-Blackwell, 472–87.

Jones, Peter. 2008. *Industrial Enlightenment: Science, Technology and Culture in Birmingham and the West Midlands, 1760–1820*. Manchester: Manchester University Press.

Kant, Immanuel. 1986. *Philosophical Correspondence, 1759–1799*. Translated by A. Zweig. Chicago: University of Chicago Press.

———. 1992a. *Lectures on Logic*. Edited by J. M. Young. Cambridge: Cambridge University Press.

———. 1992b. *Theoretical Philosophy, 1755–1770*. Edited by D. Walford and R. Meerbote. Cambridge: Cambridge University Press.

———. 1996. *Critique of Pure Reason*. Translated by W. S. Pluhar. Indianapolis: Hackett.

———. 2000. *Critique of the Power of Judgment*. Translated by P. Guyer. New York: Cambridge University Press.

———. 2005. *Notes and Fragments*. Translated by C. Bowman, P. Guyer and F. Rauscher. Edited by P. Guyer. Cambridge: Cambridge University Press.

Karpinski, Louis Charles. 1933. *The Story of Figures*. Detroit: Burroughs Adding Maching Co.

Kästner, Abraham G. 1751. Réflexions sur l'origine du plaisir, où l'on tâche de prouver l'idée de Des-Cartes: Qu'il naît toujours du sentiment de la perfection de nous-mêmes. *Histoire de l'Académie Royale des sciences et belle lettres (Berlin)*. 1749:478–88.

———. 1756. *De eo quod studium matheseos facit ad virtutem, oratio inaug*. Göttingen.

———. 1783. *Vermischte Schriften*. 3rd ed. Vol. 2. Altenburg: Richter.

———. 1784. [Report on Müller's Presentation of Calculating Machine]. *Göttingische Anzeigen von gelehrten Sachen* 2. Heft:1201–6.

Keller, Vera. 2008. Cornelis Drebbel (1572–1633): Fame and the Making of Modernity. PhD diss., History, Princeton University, Princeton.

———. 2015. *Knowledge and the Public Interest, 1575–1725*. Cambridge: Cambridge University Press.

Kelty, Christopher M. 2008. *Two Bits: The Cultural Significance of Free Software*. Durham: Duke University Press.

Kemp, Martin. 1977. From "Mimesis" to "Fantasia": The Quattrocento Vocabulary of Creation, Inspiration and Genius in the Visual Arts. *Viator* 8:347–98.

———. 2004. *Leonardo*. Oxford: Oxford University Press.

Kerviler, René Pocard du Cosquer de. 1874. *Le chancelier Pierre Séguier. Etudes sur sa vie privée, politique et litteraire et sur le groupe academique de ses familiers et commensaux*.

Keyssler, J. G. 1751. *Johann Georg Keysslers . . . Neueste Reisen durch Deutschland, Boehmen, Ungarn, die Schweiz, Italien und Lothringen*. Hanover: N. Foerster.

Kidwell, Peggy A. 1990. American Scientists and Calculating Machines: From Novelty to Commonplace. *Annals of the History of Computing* 12 (1):31–40.

King, Henry C., and John R. Millburn. 1978. *Geared to the Stars: The Evolution of Planetariums, Orreries, and Astronomical Clocks*. Toronto: University of Toronto Press.

Kistermann, Friedrich W. 1995. Die Rechentechnik um 1600 und Wilhelm Schickards Rechenmaschine. In *Zum 400. Geburtstag von Wilhelm Schickard*, edited by F. Seck. Sigmaringen: Jan Thorbecke Verlag, 241–72.

———. 1998. Blaise Pascal's Adding Machine: New Findings and Conclusions. *Annals of the History of Computing* 20:69–76.

———. 1999. When Could Anyone Have Seen Leibniz's Stepped Wheel? *Annals of the History of Computing* 21:68–72.

———. 2001. How to Use the Schickard Calculator. *Annals of the History of Computing* 23:80–85.

Kollman, A. F. C. 1828. Letter on Müller's Calculating Machine, Oct. 27. *Gentleman's Magazine* 98 (July–December):412.

Kratzenstein, Christian. 1779. Auszüge aus Briefen des Hrn. Prof. Kratzenstein an Hrn. Bernoulli. In *Astronomisches Jahrbuch oder Ephemeriden für das Jahr 1782, nebst einer Sammlung der neuesten in die astronomischen Wissenschaften einschlagenden Beobachtunger, Nachrichten, Bemerken und Abhandlungen*, edited by Akademie der Wissenschaften zu Berlin. Berlin, 135–46.

Landes, David S. 1979. Watchmaking: A Case Study in Enterprise and Change. *Business History Review* 53 (1):1–39.

Lange, Werner. 1981. Kurze Erläuterung zur Darstellung des Funktionsablaufes der Rechenmaschine des Joh.-Helfreich[sic]-Müller. Typescript. Available from http://www.rechenmaschinen -illustrated.com/pdf/MuellerRM_Lange.pdf.

Lavoisy, Olivier. 2004. Illustration and Technical Know-How in Eighteenth-Century France. *Journal of Design History* 17 (2):141–62.

Le Cerf, Mr., and Lord Viscount Mahon. 1778. Account of the Advantages of a Newly-Invented Machine Much Varied in Its Effects, and Very Useful for Determining the Perfect Proportion between Different Moveables Acting by Levers and Wheel and Pinion. By Mr. Le Cerf, Watch-Maker at Geneva; Communicated by Lord Viscount Mahon, F. R. S. *Philosophical Transactions of the Royal Society of London* 68:950–98.

Lécuyer, Christophe. 2007. *Making Silicon Valley: Innovation and the Growth of High Tech, 1930–1970*. Cambridge, MA: MIT Press.

Lécuyer, Christophe, and David C. Brock. 2006. The Materiality of Microelectronics. *History and Technology* 22 (3):301–25.

Lefèvre, Wolfgang, ed. 2004. *Picturing Machines 1400–1700*. Cambridge, MA: MIT Press.

Lehmann, N. Joachim. 1987. Leibniz' Ideenskizze zum "Sprossenrad." *NTM: Zeitschrift für Geschichte der Naturwissenschaft, Technik und Medizin* 24 (1):83–89.

———. 1993. Neue Erfahrungen zur Funktionsfähigkeit von Leibniz' Rechenmaschine. *Studia Leibnitiana* 25:174–88.

Leibniz, Gottfried Wilhelm. 1845. *Epistolae XLVI ad Teuberum concionatorem aulae Cizensis*. Edited by C. F. Nobbe. Leipzig: Staritz.

———. 1875. *Die philosophischen Schriften*. Edited by K. Gerhardt. Berlin: Weidmann.

———. 1906. *Leibnizens nachgelassene Schriften physikalischen, mechanischen und technischen Inhalts*. Edited by E. Gerland. Leipzig: B. G. Teubner.

———. 1966. *Geschichtliche Aufsätze und Gedichte*. Edited by G. H. Pertz. Hildesheim: G. Olms. Original edition, 1847.

———. 1969. *Gottfried Wilhelm Leibniz. Philosophical Papers and Letters*. Translated by L. E. Loemker. Dordrecht: Kluwer Academic Publishers.

Leopold, J. H. 1980. Christiaan Huygens and His Instrument Makers. In *Studies on Christiaan Huygens*, edited by H. J. M. Bos et al. Lisse: Swets, 221–33.

Lepage, Henri. 1854. Notes sur le Mécanicien Philippe Vayringe. *Journal de la Société d'archéologie et du comité du Musée lorraine et du Musée historique lorraine* 3:79–82.

Lesaulnier, Jean. 1992. *Port-Royal insolite: Edition critique du Receuil de choses diverses*. Paris: Klincksieck.

Leupold, Jacob. 1727. *Theatrum arithmetico-geometricum: Das ist: Schau-Platz der Rechen -und Mess-Kunst, darinnen enthalten dieser beyden Wissenschaften nöthige Grund-Regeln und Handgriffe so wohl, als auch die unterschiedene Instrumente und Machinen, welche theils in der Ausübung auf den Papier theils auch im Felde besonderen Vortheil geben können, in sonderheit wird hierinnen erklärt*. Leipzig: C. Zunkel.

Lichtenberg, Georg Christoph. 1966. *Sudelbücher*. Munich: Carl Hanser Verlag.

———. 1983–2004. *Briefwechsel*. Edited by U. Joost, A. Schöne, and J. Hoffmann. 5 v. in 6 vols. Munich: Beck.

———. 2003. *"Ist es ein Traum, so is es der größte und erhabenste der je ist geträumt worden . . . ": Aufzeichnungen über die Theorie der Schwere von George-Louis Le Sage*. Edited by H. Zehe and W. Hinrichs, *Nachrichten der Akademie der Wissenschaften in Göttingen. II. Mathematische-Physicalische Klasse*. No. 1. Göttingen: Vandenhoek & Ruprecht.

Li Di, Bai Shangshu, and Michael R. Williams. 1992. Chinese Calculators Made during the Kangxi Reign in the Qing Dynasty. *Annals of the History of Computing* 14 (4):63–67.

Lindgren, Michael. 1990. *Glory and Failure: The Difference Engines of Johann Müller, Charles Babbage and Georg and Edvard Scheutz*. Translated by C. G. McKay. Cambridge, MA: MIT Press.

Linebaugh, Peter. 2003. *The London Hanged: Crime and Civil Society in the Eighteenth Century*. 2nd ed. London: Verso.

Liu, Lu (刘潞), ed. 1998. *Qinggong Xiyang yiqi* 请宫西洋仪器 *(Western Armillary Spheres in the Qing Court)*. Hong Kong: Commercial Press.

Long, Pamela O. 1991. Invention, Authorship, "Intellectual Property," and the Origin of Patents: Notes toward a Conceptual History. *Technology and Culture* 32:846–84.

———. 2001. *Openness, Secrecy, Authorship: Technical Arts and the Culture of Knowledge from Antiquity to the Renaissance.* Baltimore: Johns Hopkins University Press.

Lubar, Steven. 1995. Representation and Power. *Technology and Culture* 36:S54–S82.

MacLeod, Christine. 1988. *Inventing the Industrial Revolution: The English Patent System, 1660–1800.* Cambridge: Cambridge University Press.

———. 2004. The European Origins of British Technological Predominance. In *Exceptionalism and Industrialization: Britain and Its European Rivals, 1688–1815,* edited by L. Prados de la Escosura. Cambridge: Cambridge University Press, 111–26.

———. 2007. *Heroes of Invention: Technology, Liberalism and British Identity, 1750–1914.* Cambridge: Cambridge University Press.

MacLeod, Roy M. 1971. Of Medals and Men: A Reward System in Victorian Science, 1826–1914. *Notes and Records of the Royal Society of London* 26 (1):81–105.

Macsorley, O. L. 1961. High-Speed Arithmetic in Binary Computers. *Proceedings of the IRE* 49 (1):67–91.

Magalotti, Lorenzo. 1980. *Lorenzo Magalotti at the Court of Charles II: His "Relazione d'Inghilterra" of 1668.* Translated by W. E. K. Middleton. Waterloo, Ont.: Wilfrid Laurier University Press.

Magnin, Charles, and Marco Marcacci. 2001. Le projet de réforme du Collège (1774): Entre instruction publique, politique et économie. In *H.B. De Saussure (1740–1799): Un régard sur la terre,* edited by J.-D. Candaux and R. Sigrist. Geneva: Georg Editeur, 409–29.

Mahon, Charles Lord Viscount. 1778. Description of a Most Effectual Method of Securing Buildings against Fire, Invented by Charles Lord Viscount Mahon, F. R. S. *Philosophical Transactions of the Royal Society of London* 68:884–94.

Mahoney, Michael S. 1988. The History of Computing in the History of Technology. *Annals of the History of Computing* 10:113–25.

M'Leod, H. 1885. International Inventions Exhibition. Group XXVIII. Philosophical Instruments. *Journal of the Society of Arts* 33 (1707):942–46.

Manning, A. 1860. *Valentine Duval: An Autobiography of the Last Century.* London: R. Bentley.

Margócsy, Dániel. 2014. *Commercial Visions: Science, Trade, and Visual Culture in the Dutch Golden Age.* Chicago: University of Chicago Press.

Marguin, Jean. 1993. Le reporteur et la naissance du calcul mécanique. *La Revue du Musée des arts et métiers* (2):26–32.

———. 1994. *Histoire des instruments et machines à calculer: Trois siècles de mécanique pensante, 1642–1942.* Paris: Hermann.

Martin, Ernst. 1992. *The Calculating Machines.* Translated by P. A. Kidwell and M. R. Williams. Cambridge, MA: MIT Press. Original edition, 1925.

Mathias, Peter. 1969. Who Unbound Prometheus? Science and Technical Change 1600–1800. *Bulletin of Economic Research* 21 (1):3–16.

Matschoss, C. 1939. *Great Engineers.* Freeport, NY: Books for Libraries Press.

Matthews, Bruce. 2011. *Schelling's Organic Form of Philosophy: Life as the Schema of Freedom.* Albany: State University of New York Press.

McConnell, Anita. 2007. *Jesse Ramsden (1735–1800): London's Leading Scientific Instrument Maker.* Aldershot: Ashgate.

McGee, David. 1995. Making Up Mind: The Early Sociology of Invention. *Technology and Culture* 36 (4):773–801.

———. 1999. From Craftsmanship to Draftsmanship: Naval Architecture and the Three Traditions of Early Modern Design. *Technology and Culture* 40:209–36.

————. 2004. The Origins of Early Modern Machine Design. In *Picturing Machines 1400–1700*, edited by W. Lefèvre. Cambridge, MA: MIT Press, 53–84.

McKenna, Antony. 1990. L'Histoire du brochet et de la grenouille: Pascal et Izaac Walton. *Courrier du Centre International Blaise Pascal*, no. 12:18–19.

Meisenzahl, Ralf, and Joel Mokyr. 2011. The Rate and Direction of Invention in the British Industrial Revolution: Incentives and Institutions. *National Bureau of Economic Research Working Paper Series* No. 16993.

Mellot, Jean-Dominique. 1984. Le régime des privilèges et permissions d'imprimer à Rouen au XVIIe siècle. *Bibliothèque de l'école des chartes* 142 (1):137–52.

Ménard, Jean-Louis. 2005. *La révolte des nu-pieds en Normandie au XVIIème siècle*. Paris: Dittmar.

Merck, Johann Heinrich. 1784. Nachricht von einer neuen Rechen-Maschine, welche Herr Ingenieur-Hauptmann Müller zu Darmstadt in abgewichnem Jahre erfunden hat. *Der Teutsche Merkur* 1.Viertelj.:269–75.

————. 2007. *Briefwechsel*. Edited by U. Leuschner. Göttingen: Wallstein.

Merzbach, Uta C. 1977. *Georg Scheutz and the First Printing Calculator*. Washington, DC: Smithsonian Institution Press.

Mesnard, Jean. 1965. *Pascal et les Roannez*. Paris: Desclée De Brouwer.

————. 1985. La mécénat scientifique avant l'Académie des sciences. In *L'âge d'or du mécénat*, edited by R. Mousnier. Paris: Editions de CNRS, 107–17.

Meurillon, Christian. 1982. La Machine arithmétique à la genèse des ordres pascaliens. *Revue des Sciences Humaines* (186–87):147–58.

————. 2001. Le Chancelier, les Nu-pieds, et la machine: Pascal père et fils à Rouen. In *Les Pascal à Rouen 1640–1648*, edited by J. P. Cléro. Rouen: Université de Rouen, 89–105.

Michaux, Bernard. 2001. Pascal, Descartes et les Artisans. In *Les Pascal à Rouen 1640–1648*, edited by J. P. Cléro. Rouen: Université de Rouen, 197–215.

Mill, John Stuart. 1981. *The Collected Works of John Stuart Mill, Volume I—Autobiography and Literary Essays*. Edited by J. M. Robson and J. Stillinger. Toronto: University of Toronto Press.

Miller, David Philip. 1999. The Usefulness of Natural Philosophy: The Royal Society and the Culture of Practical Utility in the Later Eighteenth Century. *British Journal for the History of Science* 32 (2):185–201.

Mindell, David A. 2002. *Between Human and Machine: Feedback, Control, and Computing before Cybernetics*. Baltimore: Johns Hopkins University Press.

Miniati, Mara. 1993. Le savoir scientifique et ses instruments en Italie au XVIIe siècle. In *Diffusion du savoir et affrontement des idées, 1600–1770*. Montbrison: Association du Centre Culturel de la Ville de Montbrison, 147–60.

Mody, Cyrus C. M. 2011. *Instrumental Community: Probe Microscopy and the Path to Nanotechnology*. Cambridge, MA: MIT Press.

Mokyr, Joel. 2002. *The Gifts of Athena: Historical Origins of the Knowledge Economy*. Princeton: Princeton University Press.

————. 2005a. The Intellectual Origins of Modern Economic Growth. *Journal of Economic History* 65 (2):285–351.

————. 2005b. Long-Term Economic Growth and the History of Technology. In *Handbook of Economic Growth*, edited by A. Philippe and N. D. Steven. Amsterdam: Elsevier, 1113–80.

————. 2007. Knowledge, Enlightenment, and the Industrial Revolution: Reflections on *The Gifts of Athena*. *History of Science* 45:185–96.

———. 2009. Intellectual Property Rights, the Industrial Revolution, and the Beginnings of Modern Economic Growth. *American Economic Review* 99 (2):349–55.

Moore, Jonas. 1650. *Moores Arithmetick Discovering the Secrets of That Art in Numbers and Species: In Two Bookes:*. . . . London: Printed by Thomas Harper for Nathani[el] Brookes.

———. 1681. *A Mathematical Compendium;*. . . . 2nd ed. London: Printed for Robert Harford.

Morar, Florin-Stefan. 2009. Leibniz's Calculating Machine and the Social Meaning of Invention in the Context of Early Modern Science. Master's thesis, Fakultät für Soziologie, Universität Bielefeld, Bielefeld.

———. 2014. Reinventing Machines: The Transmission History of the Leibniz Calculator. *British Journal for the History of Science* 48 (1): 123–46.

Morland, Samuel. 1673. *The Description and Use of Two Arithmetick Instruments, Together with a Short Treatise.* . . . London: Moses Pitt.

———. 1695. *The Urim of Conscience* . . . London: Printed by J. M. for A. Roper.

Mourlevat, Guy. 1988. *Les machines arithmétiques de Blaise Pascal.* Vol. 51, *Mémoires de l'Académie des sciences, belles-lettres et arts de Clermont-Ferrand.* Clermont-Ferrand: La Francaise d'Edition et d'Imprimerie.

Mourlevat, Guy, Alain Jean-Baptiste, and Yvon Formento. 1981. *La machine arithmétique ou Pascaline.* Vol. 6, *Annales du Centre Regional de documentation pédagogique de Clermont-Ferrand.* Clermont-Ferrand: Centre Regional de documentation pédagogique de Clermont-Ferrand.

Mukerji, Chandra. 2002. Cartography, Entrepreneurialism, and Power in the Reign of Louis XIV. In *Merchants and Marvels: Commerce, Science and Art in Early Modern Europe,* edited by P. H. Smith and P. Findlen. New York: Routledge, 248–76.

———. 2006. Tacit Knowledge and Classical Technique in Seventeenth-Century France: Hydraulic Cement as a Living Practice among Masons and Military Engineers. *Technology and Culture* 47:713–33.

———. 2009. *Impossible Engineering: Technology and Territoriality on the Canal du Midi.* Princeton: Princeton University Press.

Müller, Conrad H. 1904. *Studien zur Geschichte der Mathematik: Inbesondere des mathematischen Unterrichts an der Universität Göttingen im 18. Jh.* Leipzig: Teubner.

Müller, Johann Helfrich. 1783. Schreiben an Hrn. Prof. Lichtenberg, eine neue von ihm erfundene Rechenmaschine betreffend. *Göttingisches Magazin der Wissenschaften und Litteratur* 3 (5):774–83.

Müller, Johann Helfrich, and P. E. Klipstein. 1786. *Beschreibung seiner neu erfundenen Rechenmaschine: Nach ihrer Gestalt, ihrem Gebrauch & Nutzen.* Frankfurt: Varrentrapp & Wenner.

Murphy, John Joseph. 1966. Entrepreneurship in the Establishment of the American Clock Industry. *Journal of Economic History* 26 (2):169–86.

Musson, Albert E. 1975. Joseph Whitworth and the Growth of Mass Production Engineering. *Business History* 17:109–49.

Nagase, Haruo. 1998. Rhétorique de la machine arithmétique: Signification de son invention dans la pensée de Pascal. *Etudes de Langue et Littérature Françaises,* no. 72:17–30.

Nagel, Josef. 1960. Beschreibung der Rechenmaschine des Antonius Braun. *Blätter für Technikgeschichte* 22:81–87.

Neher-Bernheim, R. 1983. Un savant juif engagé: Jacob Rodrigue Péreire (1715–1780). *Revue des Etudes Juives* 142 (3–4):373–51.

Netz, Raviel. 2002. Counter Culture: Towards a History of Greek Numeracy. *History of Science* 40:321–52.

Newman, Aubrey. 1969. *The Stanhopes of Chevening: A Family Biography*. London: Macmillan.

Nielsen, Michael. 2012. *Reinventing Discovery: The New Era of Networked Science*. Princeton: Princeton University Press.

Nummedal, Tara E. 2007. *Alchemy and Authority in the Holy Roman Empire*. Chicago: University of Chicago Press.

Nuvolari, Alessandro. 2004. Collective Invention during the British Industrial Revolution: The Case of the Cornish Pumping Engine. *Cambridge Journal of Economics* 28 (3):347–63.

———. 2005. Open Source Software Development: Some Historical Perspectives. *First Monday* 10 (10). Available from http://dx.doi.org/10.5210/fm.v10i10.1284.

Ocagne, Maurice d'. 1986. *Le Calcul simplifé: Graphical and Mechanical Methods for Simplifying Calculation*. Translated by J. Howlett and M. R. Williams. Cambridge, MA: MIT Press.

Oklobdzija, Vojin G., Bart R. Zeydel, Hoang Dao, Sanu Mathew, and Ram Krishnamurthy. 2003. Energy-Delay Estimation Technique for High-Performance Microprocessor VLSI adders. *Proceedings of the 16th IEEE Symposium on Computer Arithmetic*:272–79.

Oldenburg, Henry. 1965–86. *The Correspondence of Henry Oldenburg*. Edited by A. R. Hall and M. B. Hall. 13 vols. Madison: University of Wisconsin Press.

Orcibal, J. 1950. Descartes et sa philosophie jugés à l'hôtel Liancourt (1669–1674). In *Descartes et le cartésianisme hollandais. Etudes et documents*. Paris: PUF, 87–107.

Oudshoorn, Nelly, and Trevor Pinch, eds. 2003. *How Users Matter: The Co-construction of Users and Technology*. Cambridge, MA: MIT Press.

Owen, D. 2002. *Hume's Reason*. Oxford: Oxford University Press.

Owens, Larry. 1986. Vannevar Bush and the Differential Analyzer: The Text and Context of an Early Computer. *Technology and Culture* 27 (1):63–95.

———. 1996. Where Are We Going, Phil Morse: Changing Agendas and the Rhetoric of Obviousness in the Transformation of Computing at MIT, 1939–1957. *Annals of the History of Computing* 18 (4):34–41.

Pagani, Catherine. 2001. *"Eastern Magnificence & European Ingenuity": Clocks of Late Imperial China*. Ann Arbor: University of Michigan Press.

Pancaldi, Giuliano. 2003. *Volta: Science and Culture in the Age of Enlightenment*. Princeton: Princeton University Press.

Pannabecker, John R. 1998. Representing Mechanical Arts in Diderot's *Encyclopédie*. *Technology and Culture* 39 (1):33–73.

Pantalony, David. 2004. Seeing a Voice: Rudolph Koenig's Instruments for Studying Vowel Sounds. *American Journal of Psychology* 117:425–42.

Parker, Charles Stuart. 1891. *Sir Robert Peel . . . from his Private Correspondence*. London: J. Murray.

Parker, David S. 1989. Sovereignty, Absolutism and the Function of the Law in Seventheenth-Century France. *Past & Present* (122):36–74.

Pascal, Blaise. 2000. *Pensées*. Edited by G. Ferreyrolles and P. Sellier. Paris: Livre de Poche.

Pastorino, Cesare. 2009. The Mine and the Furnace: Francis Bacon, Thomas Russell, and Early Stuart Mining Culture. *Early Science and Medicine* 14 (6):630–60.

Paulus, Rudolf F. 1975. Die Briefen von Philipp Matthäus Hahn an Johann Caspar Lavater. *Blätter für württembergische Kirchengeschichte* 75:61–84.

Pelaez, Eloina. 1999. The Stored-Program Computer: Two Conceptions. *Social Studies of Science* 29 (3):359–89.

Pérez-Ramos, Antonio. 1988. *Francis Bacon's Idea of Science and the Maker's Knowledge Tradition.* Oxford: Clarendon Press.

Perkins, Franklin. 2004. *Leibniz and China: A Commerce of Light.* Cambridge: Cambridge University Press.

Phillips, C. J. 2014. *The New Math: A Political History.* Chicago: University of Chicago Press.

Pictet, Marc-Auguste. 1802. *Voyage de trois mois, en Angleterre, en Ecosse, et en Irlande pendant l'été de l'an IX. (1801 v. st.).* Geneva: Impr. de la Bibliothèque Britannique.

Pigman, G. W. 1980. Versions of Imitation in the Renaissance. *Renaissance Quarterly* 33 (1):1–32.

Pinch, Trevor J., and Wiebe E. Bijker. 1987. The Social Construction of Facts and Artifacts: Or How the Sociology of Science and the Sociology of Technology Might Benefit Each Other. In *The Social Construction of Technological Systems,* edited by W. E. Bijker, T. P. Hughes, and T. J. Pinch. Cambridge, MA: MIT Press, 17–51.

Playfair, John, and Nevil Maskelyne. 1778. On the Arithmetic of Impossible Quantities. *Philosophical Transactions of the Royal Society of London* 68:318–43.

Polanyi, Michael. 1962. *Personal Knowledge: Toward a Post-Critical Philosophy.* Corrected ed. Chicago: University of Chicago Press.

Poleni, Giovanni. 1709. *Joannis Poleni Miscellanea: Hoc est I. Dissertatio de barometris, & thermometris, II. Machinae aritmeticae ejusque usus descriptio, III. De sectionibus conicis parallelorum in horologiis solaribus tractatus.* Venice: Apud Aloysium Pavinum.

Poni, C. 1993. The Craftsman and the Good Engineer: Technical Practice and Theoretical Mechanics in JT Desaguliers. *History and Technology* 10 (4):215–32.

Popplow, Marcus. 1998. Protection and Promotion: Privileges for Inventions and Books of Machines in the Early Modern Period. *History of Technology* 20:103–24.

———. 2002. *Models of Machines: A "Missing Link" between Early Modern Engineering and Mechanics.* Max Planck Institut for the History of Science Preprint 225.

———. 2004. Why Draw Pictures of Machines? The Social Context of Early Modern Machine Drawings. In *Picturing Machines 1400–1700,* edited by W. Lefèvre. Cambridge, MA: MIT Press, 17–48.

Pottage, Alain, and Brad Sherman. 2010. *Figures of Invention: A History of Modern Patent Law.* Oxford: Oxford University Press.

Pouzet, Régine. 2001. *Chronique des Pascal: "Les affaires du monde" d'Etienne Pascal à Marguerite Périer (1588–1733).* Paris: Honoré Champion.

Prager, Frank D. 1964. Examination of Inventions from the Middle Ages to 1836. *Journal of the Patent and Trademark Office Society* 46:268–91.

Preud'homme, Louis-Baptiste. 1778. Considerations pratiques sur les engrenages des roues & pignons en Horlogerie. *Mémoires de la Société établie à Genève pour l'encouragement des Arts & de l'Agriculture* 1, part 2:81–132.

Prévost, P. 1805. *Notice de la Vie et des Ecrits de George Louis Le Sage.* Geneva: J. J. Paschoud.

Prévost, Pierre, and George-Louis Le Sage. 1804. *Essais de philosophie, ou étude de l'esprit humain . . . Suivis de quelques opuscules de feu G. L. Le Sage, Tome Second.* Geneva: J. J. Paschoud.

Priestley, Joseph. 1775. *The History and Present State of Electricity, with Original Experiments.* 3rd ed. 2 vols. London: Bathurst &c.

Priestley, Mark. 2011. *A Science of Operations.* London: Springer.

Pullan, J. M. 1969. *The History of the Abacus*. New York: Praeger.

Pütter, Johann Stephan. 1765. *Versuch einer Academischen Gelehrten-Geschichte von der Georg-Augustus-Universität zu Göttingen*. Göttingen: Wittwe Vandenhoek.

Pye, David. 1968. *The Nature and Art of Workmanship*. Cambridge: Cambridge University Press.

Raffaelli, Tiziano. 1994. The Early Philosophical Writings of Alfred Marshall. *Research in the History of Economic Thought and Methodology*. Archival Supplement 4:53–156.

Raggio, Olga. 1958. The Myth of Prometheus: Its Survival and Metamorphoses up to the Eighteenth Century. *Journal of the Warburg and Courtauld Institutes* 21 (1/2):44–62.

Ratcliff, Jessica R. 2007. Samuel Morland and His Calculating Machines c. 1666: The Early Career of a Courtier-Inventor in Restoration London. *British Journal for the History of Science* 40:159–79.

Reinert, Sophus A. 2011. *Translating Empire: Emulation and the Origins of Political Economy*. Cambridge, MA: Harvard University Press.

Rescher, Nicholas. 1992. Leibniz Finds a Niche: Settling in at the Court of Hanover 1676–1677. *Studia Leibnitiana* 24:25–48.

———. 2014. Leibniz's Machina Deciphratoria: A Seventeenth-Century Proto-Enigma. *Cryptologia* 38 (2):103–15.

Reynolds, Joshua 1997. *Discourses on Art*. Edited by R. R. Wark. New Haven: Paul Mellon Centre for Studies in British Art and Yale University Press.

Richards, Joan L. 2006. Historical Mathematics in the French Eighteenth Century. *Isis* 97 (4): 700–713.

Roberts, C. J. D. 1987. Babbage Difference Engine No-1 and the Production of Sine Tables. *Annals of the History of Computing* 9 (2):210–12.

———. 1990a. *Babbage's First Difference Engine: How It Was Intended to Work*. Available from https://web.archive.org/web/20150907215540/https://sites.google.com/site/babbagedifference engine/howitwasintendedtowork.

———. 1990b. *History of Babbage's Difference Engine No. 1*. Available from https://web.archive .org/web/20061128143214/http://babbage.bravehost.com/histde1.htm.

———. 1990c. *Analysis of the Costs of and Government Expenditure on Babbage's 1st Difference Engine*. Available from https://web.archive.org/web/20071010133237/http://babbage.brave host.com/Expenditure.htm.

Roberts, Lissa, and Ian Inkster. 2009. Special Issue: Mindful Hand. *History of Technology* 29:103–211.

Roberts, Lissa, Simon Schaffer, and Peter Dear, eds. 2007. *The Mindful Hand: Inquiry and Invention from the Late Renaissance to Early Industrialisation*. Amsterdam: Koninklijke Nederlandse Akademie van Wetenschappen.

Robinet, André. 1994. *G. W. Leibniz: Le meilleur des mondes par la balance de l'Europe*. Paris: Presses Universitaires de France.

Rojas, Raúl, and Ulf Hashagen, eds. 2000. *The First Computers: History and Architectures*. Cambridge, MA: MIT Press.

Ronfort, Jean-Nérée. 1989. Science and Luxury: Two Acquisitions by the J. Paul Getty Museum. *The J. Paul Getty Museum Journal* 17:47–82.

Rosenband, Leonard N. 2000. The Competitive Cosmopolitanism of an Old Regime Craft. *French Historical Studies* 23:455–76.

———. 2007. Becoming Competitive: England's Papermaking Apprenticeship, 1700–1800. In *The Mindful Hand: Inquiry and Invention from the late Renaissance to Early Industrialisation*, edited by L. Roberts, S. Schaffer, and P. Dear. Amsterdam: Koninklijke Nederlandse Akademie van Wetenschappen, 379–401.

Rossi, Paolo. 1970. *Philosophy, Technology, and the Arts in the Early Modern Era*. New York: Harper.

Rule, John. 1987. The Property of Skill in the Period of Manufacture. In *The Historical Meanings of Work*, edited by P. Joyce. Cambridge: Cambridge University Press, 99–118.

S. P. 1804. To the Editor. *Christian Observer, Conducted by Members of the Established Church* 3:149–51.

Salomon-Bayet, Claire. 1978. Les Académies scientifiques: Leibniz et l'Académie Royale des Sciences. In *Leibniz à Paris (1672–1676)*. Wiesbaden: Franz Steiner Verlag, 155–70.

Sambrook, James. 2005. The Psychology of Literary Creation and Literary Response. In *The Cambridge History of Literary Criticism*. Vol 4, *The Eighteenth Century*, edited by B. H. Nisbet and C. Rawson. Cambridge: Cambridge University Press.

Sander, H. 1784. *Beschreibung seiner Reisen durch Frankreich, die Niederlande, Holland, Deutschland und Italien: In Beziehung auf Menschenkenntnis, Industrie, Litteratur und Naturkunde insonderheit. Zweiter Teil*. Leipzig: Friedrich Gotthold Jacobäer und Sohn.

Saussure, H.-B. 1774. *Eclaircissements sur le projet de réforme pour le Collège de Genève*. Geneva.

Schaffer, Simon. 1990. Genius in Romantic Natural Philosophy. In *Romanticism and the Sciences*, edited by A. Cunningham and N. Jardine. Cambridge: Cambridge University Press, 82–98.

———. 1994. Babbage's Intelligence: Calculating Engines and the Factory System. *Critical Inquiry* 21:203–27.

———. 1999. Enlightened Automata. In *The Sciences in Enlightened Europe*, edited by W. Clark, J. Golinski, and S. Schaffer. Chicago: University of Chicago Press, 126–65.

———. 2002. Golden Means: Assay Instruments and the Geography of Precision in the Guinea Trade. In *Instruments, Travel and Science*, edited by M.-N. Bourguet, C. Licoppe, and H. O. Sibum. London: Routledge, 20–50.

Scheel, Günter. 2001. Helmstedt als Werkstatt für die Vervollkommmnung der von Leibniz erfundenen und konstruierten Rechenmaschine. *Braunschweigisches Jahrbüch für Landesgeschichte* 82:105–18.

Schickard, Wilhelm. 2002. *Briefwechsel*. Edited by Friedrich Seck. Stuttgart-Bad Cannstatt: Frommann-Holzboog.

Schmiedecke, Adolf. 1969. Leibniz' Beziehungen zu Zeitz. *Studia Leibnitiana* 1 (2):137–44.

Schubring, Gert. 2005. *Conflicts between Generalization, Rigor, and Intuition: Number Concepts Underlying the Development of Analysis in 17–19th Century France and Germany*. New York: Springer Science.

Schuster, John. 1984. Methodologies as Mythic Structures: A Preface to the Future Historiography of Method. *Metascience* 1/2:15–36.

Scranton, Philip. 2000. Missing the Target? A Comment on Edward Constant's "Reliable Knowledge and Unreliable Stuff." *Technology and Culture* 41 (4):752–64.

———. 2006. Technology-led Innovation: The Non-Linearity of US Jet Propulsion Development. *History and Technology* 22:337–67.

Sennett, Richard. 2008. *The Craftsman*. New Haven: Yale University Press.

Séris, Jean-Pierre. 1987. *Machine et communication: Du théâtre des machines à la mécanique industrielle*. Paris: Vrin.

Sewell, William H. 2010. The Empire of Fashion and the Rise of Capitalism in Eighteenth-Century France. *Past & Present* 206 (1):81–120.

———. 2012. Reply. *Past & Present* 216 (1):259–67.

Shaftesbury, Anthony Ashley Cooper, Earl of. 2001. *Characteristicks of Men, Manners, Opinions, Times*. Edited by Douglas den Uyl. 3 vols. Indianapolis: Liberty Fund.

Shapin, Steven. 1994. *A Social History of Truth: Civility and Science in Seventeenth-Century England*. Chicago: University of Chicago Press.

Sherry, David. 1991. The Logic of Impossible Quantities. *Studies in History and Philosophy of Science Part A* 22 (1):37–62.

Shimpi, Anand. 2012. *Intel's Haswell Architecture Analyzed: Building a New PC and a New Intel*, October 5. Available from http://www.anandtech.com/show/6355/intels-haswell-architecture/.

Shovlin, John. 2003. Emulation in Eighteenth-Century French Economic Thought. *Eighteenth-Century Studies* 36 (2):224–30.

Sibum, H. Otto. 1994. Reworking the Mechanical Value of Heat: Instruments of Precision and Gestures of Accuracy in Early Victorian England. *Studies in History and Philosophy of Science* 26:73–106.

———. 2003. Experimentalists in the Republic of Letters. *Science in Context* 16 (1–2):89–120.

Silberstein, Marcel. 1961. *Erfindungsschutz und merkantilistische Gewerbeprivilegien*. Zürich: Polygraphischer Verlag.

Simpson, A. D. C. 1989. Robert Hooke and Practical Optics: Technical Support at a Scientific Frontier. In *Robert Hooke: New Studies*, edited by M. Hunter and S. Schaffer. Woodbridge: Boydell Press, 33–62.

Smiles, Samuel. 1864. *Industrial Biography: Iron-Workers and Tool-Makers*. Boston: Ticknor and Fields.

Smith, C. U. M. 1987. David Hartley's Newtonian Neuropsychology. *Journal of the History of the Behavioral Sciences* 23 (2):123–36.

Smith, David Kammerling. 1995. "*Au bien du commerce*": Economic Discourse and Visions of Society in France. PhD diss., History, University of Pennsylvania.

Smith, Justin. 2011. *Divine Machines: Leibniz and the Sciences of Life*. Princeton: Princeton University Press.

Smith, Pamela H. 1994. *The Business of Alchemy: Science and Culture in the Holy Roman Empire*. Princeton: Princeton University Press.

———. 2004a. *The Body of the Artisan: Art and Experience in the Scientific Revolution*. Chicago: University of Chicago Press.

Smith, Pamela H., and Paula Findlen, eds. 2002. *Merchants & Marvels: Commerce, Science, and Art in Early Modern Europe*. New York: Routledge.

Smith, Roger. 2004b. The Swiss Connection: International Networks in some Eighteenth-Century Luxury Trades. *Journal of Design History* 17 (2):123–39.

Soll, Jacob. 2009. *The Information Master: Jean-Baptiste Colbert's Secret State Intelligence System*. Ann Arbor: University of Michigan Press.

Solon, M. L. 1905. The Rouen Porcelain. *The Burlington Magazine for Connoisseurs* 7 (26):116–24.

Sonenscher, Michael. 1989. *Work and Wages: Natural Law, Politics and the Eighteenth-Century French Trades*. Cambridge: Cambridge University Press.

———. 2012. The Empire of Fashion and the Rise of Capitalism in Eighteenth-Century France. *Past & Present* 216 (1):247–58.

Splinter, Susan. 2005. Eine unbekannte Rechenmaschine von Leibniz? In *Physica et historia: Festschrift für Andreas Kleinert zum 65. Geburtstag*, edited by S. Splinter. Halle/Saale: Deutsche Akademie der Naturforscher Leopoldina, 147–54.

Stäbler, Walter. 1992. *Pietistische Theologie im Verhör: Das System Philipp Matthäus Hahns und seine Beanstandung durch das würtembergische Konsistorium.* Stuttgart: Calwer.

Stalnaker, Joanna. 2010. *The Unfinished Enlightenment: Description in the Age of the Encyclopedia.* Ithaca: Cornell University Press.

Stanhope, Ghita, and G. P. Gooch. 1914. *The Life of Charles Third Earl Stanhope.* London: Longmans, Green, and Co.

Starobinski, Jean. 1989. Dieu observable: Les commencements de la science genevoise. In *Table d'orientation: L'auteur et son autorité.* Lausanne: Editions l'âge d'homme, 13–34.

Stein, Erwin, Franz Otto Kopp, Karin Wiechmann, and Gerhard Weber. 2006. Neue Forschungsergebnisse und Nachbauter zur Vier Spezies-Rechenmaschine und zur dyadischen Rechenmaschine nach Leibniz. In *VIII. Internationaler Leibniz-Kongress: Einheit in der Vielheit,* edited by H. Berger. Hannover: Gottfried-Wilhelm-Leibniz-Gesellschaft, 1018–25.

Stein, Erwin, and Franz Otto Kopp. 2010. Konstruktion und Theorie der Leibnizschen Rechenmaschinen im Kontext der Vorläufer, Weiterentwicklungen und Nachbauten: Mit einem Uberblick zur Geschichte der Zahlensysteme und Rechenhilfsmittel. *Studia Leibnitiana* 42(1), 1–128.

———. 2014. Konstruktiv-mathematische Erforschung der Dezimalmaschine und funktionsoptimierter Hannoverscher Nachbau für vollständige Zehnerüberträge. In *Das letzte Original. Die Leibniz-Rechenmaschine der Gottfried Wilhelm Leibniz Bibliothek,* edited by A. Walsdorf, K. Badur, E. Stein, and F. O. Kopp. Hannover: Gottfried Wilhelm Leibniz Bibliothek-Niedersächsische Landesbibliothek, 173–253.

Stein, Ludwig. 1888. Die in Halle aufgefundenen Leibnitz-Briefe im Auszug mitgetheilt. *Archiv für die Geschichte der Philosophie* 1:78–91.

Stern, J. P. 1959. *Lichtenberg: A Doctrine of Scattered Occasions.* Bloomington: Indiana University Press.

Stewart, Larry. 1999. Other Centres of Calculation, or, Where the Royal Society Didn't Count: Commerce, Coffee-Houses and Natural Philosophy in Early Modern London. *British Journal for the History of Science* 32 (2):133–53.

———. 2007. Experimental Spaces and the Knowledge Economy. *History of Science* 45 (2):155–77.

Sturdy, D. J. 1995. *Science and Social Status: The Members of the Académie des Sciences 1666–1750.* Woodbridge: Boydell Press.

Sulzer, Johann Georg. 1773–75. *Allgemeine Theorie der schönen Künste.* Leipzig: Weidmann und Reich.

Summers, David. 1987. *The Judgment of Sense: Renaissance Naturalism and the Rise of Aesthetics.* Cambridge: Cambridge University Press.

Sunder, Madhavi. 2007. The Invention of Traditional Knowledge. *Law and Contemporary Problems* 70:99–124.

Swade, Doron. 1995. *Charles Babbage's Difference Engine No. 2 Technical Description, Science Museum Papers in the History of Technology,* no. 4. London: Science Museum.

———. 2001. *The Difference Engine: Charles Babbage and the Quest to Build the First Computer.* 1st American ed. New York: Viking.

———. 2003. Calculation and Tabulation in the Nineteenth Century: Airy versus Babbage. PhD diss., University College London.

———. 2005. The Construction of Charles Babbage's Difference Engine No. 2. *Annals of the History of Computing* 27 (3):70–88.

————. 2010. Automatic Computation: Charles Babbage and Computational Method. *The Rutherford Journal*. Available from http://www.rutherfordjournal.org/article030106.html.

————. 2011. Pre-electronic Computing. In *Festschrift Randell*, edited by C. B. Jones and J. L. Lloyd. Berlin: Springer, 53–83.

Swartzlander, Earl. 1990. *Computer Arithmetic*. 2 vols. Los Alamitos, CA: IEEE Computer Society Press.

————. 1995. Generations of Calculators. *Annals of the History of Computing* 17 (3):75–77.

Targosz, Karolina. 1977. "Le dragon volant" de Tito Livio Burattini. *Annali dell'Istituto e Museo di Storia della Scienza di Firenze* II, fasc 2.:67–85.

————. 1982. *La cour savante de Louise-Marie de Gonzague et ses liens scientifiques avec la France*. Translated by V. Dimov. Warsaw: Wydawnictwo Polskiej Akademii Nauk.

————. 1995. La cour royale de Pologne au XVII siècle: Centre préacadémique. In *Lieux du pouvoir au Moyen Age et à l'époque moderne*, edited by M. Tymowski. Warsaw: Wydawnictwa Uniwersytetu Warszawskiego, 215–37.

Taton, René. 1982. Noveaux Documents sur le "Dragon Volant" de Burattini. *Annali dell'Istituto e Museo di Storia della Scienza di Firenze* 7 (2):161–68.

Taylor, E. G. R. 1954. *The Mathematical Practitioners of Tudor and Stuart England*. Cambridge: Cambridge University Press.

Taylor, N. K. 1992. Charles Babbage's Mini-Computer: Difference Engine No. 0. *Bulletin of the Institute of Mathematics and Its Applications* 28 (6–8):112–14.

Tee, G. J. 1994. More about Charles Babbage's Difference Engine No. 0. *Bulletin of the Institute of Mathematics and Its Applications* 30 (9–10):134–37.

Thirsk, Joan. 1978. *Economic Policy and Projects: The Development of a Consumer Society in Early Modern England*. Oxford: Clarendon Press.

Thomas, Keith. 1987. Numeracy in Early Modern England. *Transactions of the Royal Historical Society*, 5th ser., vol. 37:103–32.

Thomson, Ann. 2008. *Bodies of Thought: Science, Religion, and the Soul in the Early Enlightenment*. Oxford: Oxford University Press.

————. 2010. Animals, Humans, Machines and Thinking Matter, 1690–1707. *Early Science and Medicine* 15 (1–2):3–37.

Tonelli, Giorgio. 1959. Der Streit über die Mathematische Methode in der Philosophie in der ersten Hälfte des 18. Jahrhunderts und die Entstehung von Kants Schrift über die "Deutlichkeit." *Archiv für Philosophie* 9:37–66.

Tresch, John. 2012. *The Romantic Machine: Utopian Science and Technology after Napoleon*. Chicago: University Of Chicago Press.

Turck, J. A. V. 1921. *Origin of Modern Calculating Machines: A Chronicle of the Evolution of the Principles That Form the Generic Make up of the Modern Calculating Machine*. Chicago: Western Society of Engineers.

Turing, Alan M. 1950. Computing Machinery and Intelligence. *Mind* 59 (236):433–60.

————. 2004a. Lecture to L. M. S. Feb. 20, 1947. In *The Essential Turing: The Ideas That Gave Birth to the Computer Age*, edited by B. J. Copeland. Oxford Clarendon Press, 378–94.

————. 2004b. Systems of Logic Based on Ordinals (1939). In *The Essential Turing: The Ideas That Gave Birth to the Computer Age*, edited by B. J. Copeland. Oxford: Clarendon Press, 146–208.

Turner, Anthony. 1998. Mathematical Instrument-Making in Early Modern Paris. In *Luxury*

Trades and Consumerism in Ancien Régime Paris: Studies in the History of the Skilled Workforce, edited by R. Fox and A. Turner. Aldershot: Ashgate, 63–96.

———. 2008. "Not to Hurt of Trade": Guilds and Innovation in Horology and Precision Instrument Making. In *Guilds, Innovation, and the European Economy, 1400–1800*, edited by S. R. Epstein and M. R. Prak. Cambridge: Cambridge University Press, 264–87.

Turvey, Peter J. 1991. Sir John Herschel and the Abandonment of Charles Babbage's Difference Engine No. 1. *Notes and Records of the Royal Society of London* 45 (2):165–76.

Umbach, Maiken. 2000. *Federalism and Enlightenment in Germany, 1640–1806*. London: Hambledon Press.

Varley, Cornelius. 1825. Copying Screws by the Lathe. *Transactions of the Society for the Encouragement of Arts, Manufactures, and Commerce* 43:90–93.

———. 1832. Planing Machine. *Transactions of the Society for the Encouragement of Arts, Manufactures, and Commerce* 49:157–85.

Varley, Samuel. 1820. An Account of a Telescope of a New and Singular Construction, Invented by the Right Hon. the Late Earl Stanhope. *London Journal of Arts and Sciences* 1:31–51, 109–29.

Vayringe, Philippe. 1732. *Cours de philosophie mecanique et experimentale . . .* Lunéville: Chez Nicolas Galland.

Vérin, Hélène. 1993. *La gloire des ingénieurs: L'intelligence technique du XVIe au XVIIIe siècle*. Paris: Albin Michel.

Verthamont, François de, and Amable Floquet. 1842. *Diaire ou journal du voyage du chancelier Séguier en Normandie après la sedition des Nu-pieds (1639–1640) et documents relatifs à ce voyage et à la sédition*. Rouen: E. Frère.

von Freytag Löringhoff, Bruno Baron. 1978. Die Rechenmaschine. In *Wilhelm Schickard, 1592–1635: Astronom, Geograph, Orientalist, Erfinder d. Rechenmaschine*, edited by F. Seck. Tübingen: Mohr, 288–307.

von Hippel, Eric. 2005. *Democratizing Innovation*. Cambridge, MA: MIT Press.

von Mackensen, Ludolf. 1968a. Die Vorgeschichte und Entstehung der ersten digitalen 4-Spezies-Rechenmaschine von Gottfried Wilhelm Leibniz nach bisher unerschlossenen Manuskripten und Zeichnungen mit einem Quellenanhang der Hauptdokumente. Dr. rer. nat., Fakultät für allgemeine Wissenschaften, Technischen Hochschule München, Munich.

———. 1968b. Les origines de la machine arithmétique de Leibniz. In *XIIe Congrès International d'Histoire des Sciences. Paris 1968. Tome XA Histoire des Instruments Scientifiques*. Paris: Albert Blanchard, 75–78.

———. 1969. Zur Vorgeschichte und Entstehung der ersten digitalen 4-Spezies-Rechenmaschine von Gottfried Wilhelm Leibniz. In *Akten des (ersten) Internationalen Leibnizkongresses, Hannover 1966*. Wiesbaden: F. Steiner, 34–68.

Voskuhl, Adelheid. 2013. *Androids in the Enlightenment: Mechanics, Artisans, and Cultures of the Self*. Chicago: Chicago University Press.

Wakefield, André. 2010. Leibniz in the Mines. *Osiris* 25:171–88.

Walker, R. B. 1973. Advertising in London Newspapers, 1650–1750. *Business History* 15:112–30.

Walsdorf, Ariane. 2014. Biographie einer Rechenmaschine: Die Entwicklungs- und Herstellungsgeschichte der Leibniz-Rechenmaschine. In *Das letzte Original. Die Leibniz-Rechenmaschine der Gottfried Wilhelm Leibniz Bibliothek*, edited by Klaus Badur, Ariane Walsdorf, Erwin Stein, and Franz Otto Kopp. Hannover: Gottfried Wilhelm Leibniz Bibliothek-Niedersächsische Landesbibliothek, 15–118.

Walton, Izaak. 1655. *The Compleat Angler, or, The Contemplative Man's Recreation*. 2nd ed., much enlarged. London: Printed by T. M. for Rich. Marriot.

Warwick, Andrew. 1995. The Laboratory of Theory, or, What's Exact about the Exact Sciences. In *Values of Precision*, edited by M. N. Wise. Princeton: Princeton University Press, 135–72.

Weber, Otto. 1980. Ein "Computer" des 18. Jahrhunderts: Die Rechenmaschine des Landbaumeisters Müller. *Photorin: Mitteilungen der Lichtenberg-Gesellschaft*. 3:13–23.

Webster, Charles. 1975. *The Great Instauration: Science, Medicine and Reform, 1626–1660*. London: Duckworth.

Weinbrot, Howard D. 1985. "An Ambition to Excell": The Aesthetics of Emulation in the Seventeenth and Eighteenth Centuries. *Huntington Library Quarterly* 48 (2):121–39.

Weiss, Stephan. 2001–. Beiträge zur Geschichte des mechanischen Rechnens. Available from http://www.mechrech.de.

Wess, Jane. 1997. The Logic Demonstrators of the 3rd Earl Stanhope (1753–1816). *Annals of Science* 54 (4):375–95.

West, Frank. 1889. Arithmometer Calculating Machine. *English Mechanic and World of Science* 49 (1256):159.

Wilberg, Ernst-Eberhard. 1977. *Die Leibniz'sche Rechenmaschine und die Julius-Universität in Helmstedt*. Edited by A. Kuhlenkamp. *Beiträge zur Geschichte der Carolo-Wilhelmina*, Band 5. Braunschweig: Braunschweigischer Hochschulbund.

Williams, Michael R. 1981. The Scientific Library of Charles Babbage. *Annals of the History of Computing* 3:235–40.

———. 1983. From Napier to Lucas: The Use of Napier's Bones in Calculating Machinery, 1617–1900. *Annals of the History of Computing* 5 (3):279–96.

———. 1985. *A History of Computing Technology*. Englewood Cliffs, NJ: Prentice Hall.

———. 1992. Joseph Clement: The First Computer Engineer. *Annals of the History of Computing* 14 (3):69–76.

Williams, Rosalind. 2000. "All That Is Solid Melts into Air": Historians of Science in the Information Revolution. *Technology and Culture* 41:641–68.

Willmoth, Frances. 1993. *Sir Jonas Moore: Practical Mathematics and Restoration Science*. Woodbridge: Boydell Press.

Wolfe, Charles T. 2010. Why Was There No Controversy over Life in the Scientific Revolution? In *Controversies in the Scientific Revolution*, edited by V. Boantza and M. Dascal. Amsterdam: John Benjamins, 187–219.

———. 2014. Materialism. In *The Routledge Companion to Eighteenth-Century Philosophy*, edited by A. Garrett. London: Routledge, 91–118.

Wolfe, Charles T., and Motoichi Terada. 2008. The Animal Economy as Object and Program in Montpellier Vitalism. *Science in Context* 21 (4):537–79.

Woodmansee, Martha. 1994. *The Author, Art, and the Market: Rereading the History of Aesthetics*. New York: Columbia University Press.

Wright, M. T. 1994. Stanhope Calculating Machine, Techfile 1905-148. Science Museum, London.

Yates, JoAnne. 2000. Business Use of Information and Technology during the Industrial Age. In *Nation Transformed by Information: How Information Has Shaped the United States from Colonial Times to the Present*, edited by A. D. Chandler and J. W. Cortada. New York: Oxford University Press, 107–36.

———. 2005. *Structuring the Information Age: Life Insurance and Technology in the Twentieth Century*. Baltimore: Johns Hopkins University Press.

Yolton, John W. 1991. *Locke and French Materialism*. Oxford: Clarendon Press.

Young, Edward. 1759. *Conjectures on Original Composition*. 2nd ed. London: A. Millar and R. and J. Dodsley.

Zahn, Manfred. 1965a. Fichtes, Schellings und Hegels Auseinandersetzung mit dem Logischen Realismus Christoph Gottfried Bardilis. *Zeitschrift für philosophische Forschung* 19 (2): 201–23.

———. 1965b. Fichtes, Schellings und Hegels Auseinandersetzungen mit dem Logischen Realismus Chr. G. Bardilis (Fortsetzung und Schluß). *Zeitschrift für philosophische Forschung* 19 (3): 453–79.

Zeydel, B. R., D. Baran, and V. G. Oklobdzija. 2010. Energy-Efficient Design Methodologies: High-Performance VLSI Adders. *IEEE Journal of Solid-State Circuits* 45 (6):1220–33.

Zöller, Günter. 2000. The Unpopularity of Transcendental Philosophy: Fichte's Controversy with Reinhold (1799–1801). *Pli* 10:50–76.

Index

Page numbers in italics refer to figures.

abacus, 3, 5
Abeille, Louis-Paul, 232–33
absolutism: contestation over, 107–10, 120–21;
    ideology of, 108–9, 119–22; Stuart kings and,
    108, 112–14
Académie des sciences (France): Leibniz and,
    78, 97–98, 100, 114–20; *Mémoires* of, 37;
    Stanhope and, 191
Académie of Lunéville, Vayringe and, 135–38
Academy of Sciences (Copenhagen), Stanhope
    and, 191
Academy of Sciences (Göttingen), Müller at, 151, 158
actuarial work, calculating machines and, 2, 42,
    242, 244–45
Adamson, Humphrey, Morland machine and,
    15, 21
adequacy, patents and, 93, 20–23
Airy, George Biddell, on Difference Engines, 95
Alder, Ken, 184, 263n70
American Arithmometer Corporation, 245. *See
    also* Burroughs adding machine
analog: anachronism of term, 262n41; computers,
    246–47
Analytical Engine: anticipating carry and, 44,
    53–54; drawings and, 50; philosophical
    implications of, 212–13, 234–35
analytical machine: Leibniz's, 51; Stanhope's, 197–98
animals: Leibniz on, 81, 83; Pascal on, 42–43, 221
anthropomorphism: Babbage on, 54; Pascal on,
    42–43, 221
Archimedes, as inventor, 85–86
arithmetic: computer, 240–41, 248; history of
    computing and, 247; Hobbes on, 4–5; low
    status of, 4–5; Petit on, 13; primers, 3–4, 21,
64, 111; reason as more than, 216–21; tacit
    knowledge in, 29, 262n47, 272n66. *See also*
    carry; multiplication
Arithmometer, 242, 243
Arnauld, Antoine: Leibniz and, 57; Pascal and,
    35–36
artisans
—coordination of: in Beijing, 56–57; challenge
    of, 5; Hahn and, 147; Hooke and, 64–65;
    Jarvis on, 207; Leibniz and, 58; Morland
    and, 20–21; Müller and, 147–49; necessity
    for, 56–57; Pascal and, 33–34; Stanhope and,
    162–71, 191–96. *See also* Clement; Hansen;
    Herschel; Ollivier
—cosmopolitanism of, 135
—creativity of: Clement and, 50–53, 125; Mor-
    land and, 19–21; Ollivier and, 74–75, 78;
    Vayringe and, 135–39
—dynamism of, 9, 164
—education of, 135–38, 205–6
—freedom of, 86, 180–81, 210–11
—German artisans, 78–79
—hierarchy among, 20, 34, 83–84, 139, 181,
    189–90, 231
—historiography on, 8–9
—limits of, 31–35, 207
—ordinary artisans: Hahn and, 147; Müller on,
    152; patents and, 109; privileges and, 102;
    Stanhope and, 194. *See also* craftsmanship
    of certainty; manufacture; reduction to
    practice
—vanity of, 32–33
—*See also* Clement; emulation; Jarvis; knowl-
    edge: artisanal; Ollivier; Vayringe

associational psychology. *See* psychology, materialist
authorship: Babbage and, 160–61; collective, 7; as deceptive category, 7–8, 99, 160–61; need to hide, 121; romantic view of, 99, 124, 153. *See also* emulation; imitation; vegetative model
Automatic Computing Engine (Turing), 237, 239

Babbage, Charles: carry and, 44–45, 47–48, 50–54; circular gears and, 94–95; contracts and, 206–67; crediting Clement, 52; criticism of, 91–92, 95–96; design problems, 208; on drawing, 49–50, 53, 124–25; as elite artisan, 122–23; envy of, 96; on God, 234–35; on human nature, 234; manufacture and, 54, 89, 122–23, 125; on miracles, 234–35; novelty of Difference Engine and, 157–61; property in Difference Engine and, 50, 94, 123–25; reduction to practice and, 94, 125; rewards of invention, 96; rights to invention and, 122; on risks of invention, 89, 96, 124; Royal Society of London and, 89–96, 157–59, 202–6; Scheutz engines and, 209–10; state support for, 88–96; on toolmaking, 205; vindication of, 208–9
Babbage, Herschel, donation of calculating machines, 160
Bacon, Francis, 1, 10, 82
Baily, Francis, criticism of Babbage, 158
Bardili, Christoph Gottfried, on thinking and calculating, 218
Bayle, Pierre, on organization of complex beings, 222–23
Belfanti, Carlo Marco, on privileges, 104
Bentley, Richard, on thinking matter, 225
Berg, Maxine, 133, 180
Berkeley, George, on evidence of mathematics, 4, 219–20
Berström, Johan Wilhelm, difference engines and, 209
Berthoud, Ferdinand, Stanhope visits, 191
Biagioli, Mario, on intellectual property, 268n44, 268n55, 274n15, 282n114
Bibliothéque Publique du Roi, Stanhope visits, 191
bill, Morland's parliamentary, 112–13
Bischoff, Johann Paul, visit to Hahn, 152–53
Blondeau, Pierre, Morland machine and, 18–20
Boistissandeau, Hillerin de, calculating machine of, 37
Bonnet, Charles, 217
Boyle, Robert, 10, 41, 223
Braun, Anton, the elder, 130, 132, 135–36
Braun, Anton, the younger, 138–39, 142
Brewster, David: on Babbage's originality, 161; on patent system, 122
Brock, David: on materiality of microelectronics, 241; on oversocialized history, 260n18, 266n13

Brunel, Isambard Kingdom, on patents, 233
Buchta, Christoph Enoch, 80
Buffon, Georges-Louis Leclerc, Comte de, on mathematics, 219
Bullock, James, Stanhope machines and, 165, 192
Burattini, Tito Livio, as emulator of inventions, 37
Burroughs adding machine, 242–43
Bush, Vannevar, 246

calculating aids (other than machines), superiority of, 2
calculating machines. *See names of inventors and makers*
calculating machines, philosophical implications of: Babbage on, 234–36; *Christian Observer* on, 199; Hahn on, 153–56; Leibniz on, 222–24; Lichtenberg on, 226; Marshall on, 236–37; Morland on, 40–41; Pascal on, 41–43, 221; reason and, 4–5, 215–17, 221; Stanhope on, 197–99, 215–16, 221, 233–34; Turing, 212–13, 237–38
calculation, not mathematics, 218–21
Carcavy, Pierre de, invigilator of inventors and philosophers, 114–17
carry
—approaches to: anticipating, 53–54, 240–41; Babbage's Difference Engine and, 44–45, 47–48; contemporary, 239–41; Hahn's, 146; by hand, 3; Leibniz's, 62–63, 68–73, 75, 80, 145; look-ahead, 240–41; Moore's, 21; Morland's, 23–24, 117; Müller's, 149; Pascal's, 24, 27–31, 100; Roberval on, 27–28; Schickard's, 26–27; Stanhope's, 48, 171, 175, 178, 182–88. *See also* keeping-it-digital problem; spiral wheel; sufficient-force problem
—problems of: propagation of, 25–26, 183; simultaneous addition, 70, 183. *See also* keeping-it-digital problem; sufficient-force problem
Cartesian, 43, 217
chancellor of the exchequer, 88, 91–92
Charles II, as patron of Morland, 110–14
Charles Stuart, Duke of Lennox, Morland and, 16
Chevening (estate of Stanhope), 163, 166, 172, 174–78, 184–88, 190, 194–95
Christina (queen of Sweden), 107
Church, Dr., novelty of Difference Engine and, 157
circular calculating machine: Hahn, 150–51; Leupold, 143–44; Müller, 147–48; Stanhope, 175
Clapham Sect, 199
Clarke, Samuel, 234
Clement, Joseph: Babbage's views on, 50–52, 123; carry and, 50–53; creativity of, 50–53, 125; danger of losing, 52–53; drawing and, 53, 201–5; "drawing board of large area," *204*, 205; disclosure and, 200–201; employees of,

205–6; exclusivity and, 124; Herschel and, 50–53, 94, 201–5; Jarvis on, 10; property in Difference Engine project, 50, 123–25; reduction to practice and, 94; Royal Society report and, 94; security mechanisms and, 51–52, 94; tools and, 50, 200–201, 205–6; workshop of, 50, 95, 201–6

clockmakers: counterfeiters and, 32, 191; interchangeable parts and, 196; ordinary, 149, 152; resistance to privileges, 108; Stanhope and, 162, 167, 171, 178, 191–92. *See also* Hahn; Ollivier; Ramsden; Vayringe; Vuillamy

clockmaking: in China, 56; in Geneva, 133–34, 167–71, 191; Hahn and, 147; historiography of, 196; Leibniz on, 74; in London, 192–93; in Paris, 66, 108, 191; Stanhope's, 171, 191; theory and practice in, 169

Colbert, Jean-Baptiste: industrial policy, 66–67, 114–15; Leibniz and, 74, 78, 100, 114–21. *See also* Carcavy

Collier, Bruce, on Difference Engines, 206, 265n16, 265n38, 271n1, 272n16, 286n24

Colmar, Thomas de, calculating machine of, 1, 242–45

Colwall, Daniel, 64

communication, technical: Abeille on, 232–33; among artisans, 65, 134, 137; barriers to, 40, 69; between Leibniz and Ollivier, 69–70, 75; need for, 170; through ostension, 38–41; Stanhope and, 65–66, 162. *See also* disclosure; drawing; open technique

competition: in early twentieth century, 245; intellectual property and, 6; virtuousness of, 127–28, 135, 139–40. *See also* emulation

Compleat Angler, Pascal and, 42

Comptometer, 245

computation: Hobbes on, 4–5; low status of, 4–5; reasoning and, 4–5, 216–21; Stanhope on, 197–99, 215–16, 221. *See also* arithmetic

computer arithmetic, field of, 240–41, 248

Computer History Museum, Mountain View, Calif., Difference Engine at, 48

computers, general purpose, calculation and, 246–47

computing, history of: computational, 11, 247; idealist, 11, 238, 247; Mahoney on, 11–12

Comrie, Leslie, scientific computation and, 246–47

Condillac, Étienne Bonnot de, on mathematics, 216

Condorcet, Marie Jean Antoine Nicolas de Caritat, marquis de, 218

contracts: Babbage and, 94–95, 206–7; Leibniz and, 59, 67, 76–80, 97, 117–18

coordinating. *See under* artisans

Copernican planisphere, 137, 138

counterfeiting: calculating machines, 32, 171; as industrial development, 133–34; reduction to practice and, 103

Coward, William, 224–25

Cox, Christopher, 64

craftsmanship of certainty, 14, 75, 152, 244

craftsmanship of risk, 51, 74, 152, 244

craftspeople. *See* artisans

creativity. *See* artisans: creativity of; freedom; originality

credit: Babbage on, 88–89, 92, 95–96; for disclosure, 120–21, 151–52; division of, 160–61; financial, 88–89, 97–98, 115, 120, 136–38; for improvement of invention, 142; multiple invention and, 126, 128, 161; originality and, 214; philosophical, 96, 157–58; reasonable advantage, 97, 115, 120; Sonenscher on, 58. *See also* economy of glory; novelty; priority

Cromwell, Oliver, 19

cylindrical calculating machine. *See* circular calculating machine

Daer, Basil, Lord, 158

d'Albert d'Ailly, Michel Ferdinand, duc de Chaulnes, knowledge dissemination and, 38–40

Dalencé, Jérôme, privilege broker, 117–19, 277n97

d'Alibray, Charles Vion, poem on Pascal's machine, 41

Darmstadt, 37, 126–27, 147

Davy, Sir Humphry, 89–90

Desaguliers, J. T., Vayringe and, 137–38

Descartes, René, 4, 10, 86, 153, 216, 220

Deutsches Museum, 135

diagrams: Diderot's, on Pascal, 27, 39–40; Leibniz's, 61–62, 71–74; Leupold's disclosure through, *143*, *144*; Stanhope's, 166, 193–94. *See also* drawing

Diderot, Denis: artisanal knowledge and, 39–40; on mathematics, 219; on Pascal's machine, 27, 39–40

Difference Engine: carry in, 44–45, 47–48, 50–53; definition of, 45–46; "Difference Engine 0," 47, 89; "Difference Engine 1," 47, 95; "Difference Engine 2," 47, 52, 160, 208–9; novelty of, 157–61; property in Difference Engine, 50, 94, 123–25; security of, 48, 94; tool making and, 206. *See also* Babbage; Clement; Herschel; Jarvis

difference engines, Scheutz, 209–10

digital: anachronism of term, 262n41; electronic computers, 245, 247. *See also* keeping-it-digital problem

discernment, mechanical: Hahn on, 149–50; Hooke and, 64–65, 68, 149; Leibniz on, 84–86; Ramsden, 194; as tacit knowledge, 15–16 disclosure: Carcavy on, 116–17; Clement and, 200–201; Diderot and, 39–40; *Encyclopédie* and, 39–40; Hahn and, 144–45, 149; hylomorphic, 40; Leibniz and, 120–21, 129, 144; Leupold and, *143*, 144; mental concept of invention and, 116–17; Morland and, 112–14; Müller and, 151–52; through ostension, 38–41; specification and, 109–10, 113–14. See also specification

disinterestedness, invention and, 89–90. See also freedom

Doctorow, Cory, 7

Domat, Jean, on hierarchy among artisans, 34

draftsman, 39, 47, 51, 160, 200, 201, 206

drawing: adequacy and, 93; Alder on, 189; anticipating carry and, 54; artisan hierarchy and, 189; Babbage on, 49–50, 53, 124–25; Clement and, 50–53, 93, 123–25, 201–5; in *Encyclopédie*, 39–40; formal, 10, 39–40; Jarvis and, 47, 207; Leibniz and, 61–62, 71–74; McGee on, 189; Ollivier, use of, 68, 73–74; property in, 123–25; Stanhope and, 162–66, 185–86, 188–90, 193–94; tools for, 10, 205

duc de Chevreuse, Charles-Honoré d'Albert de Luynes, Leibniz machine and, 117, 119

duc de Roannez, Artus Gouffier, Paris bus system and, 106

duc of Liancourt, Pascal on, 42–43

Duke of Lennox, 16, 18, 20

Duke of Lorraine. *See* Francis Stephan

Duke of Wellington, Babbage and, 93, 122

Earl of Shaftesbury, 3rd (Anthony Ashley Cooper), on artistic unity, 227–28

economy of glory: absolutism and, 109; Babbage and, 96, 161; Carcavy on, 115–16; Leibniz and, 98, 115–17; Leibniz's failure and, 130, 133; Merck on, 126; Pascal and, 103, 106–7; Poleni within, 130, 132; specification and, 109–10. *See also* invention: multiple; novelty; priority

Edison, Thomas, 6, 180

elasticity, inadequate knowledge of, 31, 84

Electronic Numerical Integrator and Computer (ENIAC), 11, 246

emulation: affective power of, 125, 144; Aristotle on, 125; Burattini exemplifies, 37; Clement and, 200–201, 205–6, 210; Colmar and, 242; Difference Engine 2 and, 208–9; in *Encyclopédie*, 127, 139; freedom within, 210; Geneva and, 169; industrial espionage and, 140; Leupold and, 144; manufacture and, 128, 169; natural philosophy and, 170–71; political economy and, 128, 137, 141, 168; programming and, 248; proof of competence and,

37, 129, 134–35; Sweden and, 209–10; tension with genius and, 141–42; Vayringe exemplifies, 135–39; virtuousness of competition in, 127–28, 135, 139–40; Young on, 135, 140

Encyclopédie (of Diderot and d'Alembert): "emulation," 127, 139; Le Sage in, 167; Pascal's machine, *28*, 39–40; technical description in, 39–40; "tooth" (mechanical), 168

engineers: French royal, 32; Morland's theology and, 40–41

enlightenment: high versus industrial, 40, 170–71, 180–81, 283n30; information in, 133

Euler, Leonhard: as mathematician, not physicist, 218; Stanhope and, 171

evidence, mathematical: Berkeley on, 219–20; Kant on, 220–21; Kästner on, 230

exclusivity: Clement and, 124; locality of privileges and, 104; Ollivier and, 79

Exposition of 1851, calculating machines at, 205

failure: as inspiration, 35–38, 129–30, 133; Leibniz's, 70, 80–81, 119, 129–30, 133, 145; Leupold on, 142–44; methodological virtues of, 5–6; Pascal's, 13, 35–37, 102–4

Fairchild, B. H., "The Art of the Lathe," 200

Favre, Antoine, pinned barrel and, 191

Felt, D. E., on the invention of calculating machines, 245

Felt and Tarrant Corporation, 245

Ferguson, Eugene, 188, 284n56

Ferrier, Jean, Descartes and, 261n11

Fichte, Johann Gottlieb, 218

Florence, 16, 19, 37, 136

Forbidden Palace, calculating machines produced in, 56

foresight, mechanical, Babbage on, 53–54

Francis Stephan, Duke of Lorraine, later Grand Duke of Tuscany and Holy Roman emperor, 135–37

freedom: of artisans, 86, 180–81, 210–11; to create, 227–28, 231–32; despite financial interest, 210–11; from financial interest, 210–21, 232; Kant and, 210–11, 230–32; materialist psychology and, 227–29, 236; in privileges, 102; in programming, 247–48

Friedrich, Johann, Duke, patron of Leibniz, 87, 98, 120

Fries, Jacob Friedrich, 217–18

frog, Pascal on, 42

Fromantle, John, Morland machine and, 18, 20

Galileo, Galilei, 16, 19, 85–86

Gallois, Jean, 117

Gauvin, Jean-François, on Pascal's machine, 32

gears, optimal shape of. *See* teeth

Geneva: clockmaking, 133–34, 167–71; politics of, 167; Stanhope and, 166–71, 191–92
genius: artistic unity and, 229, 231–32; financial interest and, 211, 232; history of technology and, 7; Kant on, 231–32; mathematics as exercises for, 230; mathematics excluded from, 232; tension with emulation, 141–42; tension with imitation, 128
Gérard, Alexander, on unified forms, 229
Gerbier, Sir Balthasar, on Pascal's machine, 13
Gersten, Christian Ludwig: calculating machine of, 132–34, 144; Leibniz's failure as inspiration, 132
gifts: Hooke and, 65; machines as, 37, 56, 96, 107, 137; Pascal and, 103–5, 107; privileges as, 101, 103–5, 116, 121–22; sovereigns and, 108
Gilbert, Davies, 90–91
glory. *See* economy of glory
God: as artisan, 222–24, 231–32, 234; Babbage on, 234–35; Bayle on, 223; Boyle on, 223; Coward on, 225; Hahn on, 153–56; Kant on, 227–28, 231–32; Leibniz on, 81–83, 121, 222–24
Gödel, Kurt, Turing on, 237–38
goldsmiths, as counterfeiters, 32, 106
Gonzague, Louise-Marie de, calculating machines and, 37
Göttingen, Leibniz machine in, 174, 225, 278nn16–17, 279n19
Göttingen Academy, Müller and, 151–52, 158
Gould, Benjamin, 209
Grillet, René, machine of, 35–37, 60
Grimaldi, Claudio, SJ, attempted purchase of Leibniz machine, 56

Haas from Augsburg, 80
Hahn, Philipp Matthäus: account of invention, 144–47; Bischoff and, 152–54; calculating machine of, 144–47; coordination of artisans, 147; disclosure and, 144–45, 149; on God as creator, 154–55; on Herder and Spinoza, 153–55; on iterative implementation, 150, 155; on Leibniz's failure, 145; manufacture and, 147; on materialism, 154–55; against Müller, 149–51; nineteenth-century view of, 242–43; on ordinary artisans, 147; second invention and, 126–28, 142; springs and, 148–49; stepped drum and, 145–46; workshop of, 147
Hamann, Johann Georg, letter to Kant, 141
Hanover, 58, 61–62, 67–68, 71–72, 76, 79–80, 87, 98, 120
Hansen, Friedrich A.: Ollivier and, 76–80; poor understanding of machine, 76
Harris, John, 180
Harrison, John, 191

Hartley, David, associationist psychology of, 217, 229
Harz mountains, Leibniz mining effort in, 120
Haux, Mr., 64
Hegel, Georg Wilhelm Friedrich, 153, 218
Hell, Father, on Braun's machine, 138
Herder, Johann: Hahn reads, 153–55; on Spinoza, 153–55
Herschel, J. W. F.: on Clement, 50–53, 94, 201–5; defense of Babbage and, 92; on novelty of Difference Engine, 158–59; Royal Society report and, 93–94; on security of Difference Engine, 51–52; on tool making, 205–6; translation of Müller, 159
Hevelius, Johannes, 65
Hilaire-Pérez, Liliane, 137, 180
Hobbes, Thomas, on reasoning, 4–5, 217, 219
Hollerith punch card machines, 246
Holy Roman emperor, 114, 130, 137, 146
Hont, Istvan, on emulation, 140
Hooke, Robert: calculating machine of, 63–66; impropriety of, 64, 68; knowledge of London, 64–66; Leibniz's machine and, 63–68; mechanical discernment, 64–65, 68, 149; model of machine and, 64–65, 267n28; on Morland, 65–66; Royal Society and, 63–66
Horace (Quintus Horatius Flaccus), 228, 230
Horatian conception of unity, 33
horny piece, 162, 165–66, 178, 183–84, 187–88, 194. *See also* spiral wheel
hubris, philosophical, 1, 5, 9, 213
humility, Stanhope on, 199, 233, 235, 238
Huygens, Christiaan, 20, 108, 115–16, 171
hylomorphism: defined, 9; drawing and, 40, 49–50, 193; Hahn's theology and, 152–56; intellectual property and, 99, 101; rules and, 231; techniques permitting, 10, 40, 193

Iliffe, Rob, on Hooke, 64–65
imitation: Asian goods and, 133–34; creativity and, 228–32; Diderot and, 39–40; Hahn and, 150–51; innovation and, 7–8, 128–29; Kant on, 139, 141–42, 231–32; Leibniz on, 85, 222, 224; manufacture and, 201; Marshall on, 236–37; nineteenth-century critique, 3, 214, 236; of Pascal machines, 35–39; not plagiary, 138; Reynolds on, 139; Sulzer and, 139, 181; Turing and, 237–38; Vayringe and, 136–37; Young on, 135, 140, 227. *See also* emulation
import-substitution, 133–34
industrial espionage, 12, 37, 98, 128, 133–37, 140, 153, 169
Industrial Revolution, 180
Ingold, Tim, 9, 283n5
innovation. *See* artisans: creativity of; emulation; imitation; novelty; originality

Institution of Civil Engineers, 201
intellectual property: Abeille rejects, 232–33;
     emergence of, 99, 107–10, 114, 121; litigation
     and, 145; Pascal and, 101; significance of,
     6. *See also* authorship; disclosure; patents;
     privileges; property; specification
invention
—accounts of: Hahn, 144–47; Leupold, 144;
     Müller, 147–48; Pascal, 31–35
—collective, 7, 128, 232–33. *See also* emulation;
     imitation
—multiple: Babbage and precursors, 160–61;
     celebration of, 152; credit for, 126, 128, 161;
     Leibniz on Hooke's, 68. *See also* emulation;
     Hahn; Müller
—risks of: Babbage on, 89, 96, 124; Leibniz
     on, 74–75, 78–79, 97–98, 118; Ollivier on, 74;
     Stanhope on, 192. *See also* craftsmanship
     of risk
—state support for: Babbage, 88–96; Leibniz,
     97–98, 114–20; Pascal, 31–35, 100–107;
     Vayringe, 138
—theory of: Babbage's, 1, 248; Bacon's, 1; clas-
     sical rhetorical theory and, 227; Hilaire-
     Pérez's, 180; Leibniz's, 83–86; Le Sage's, 167;
     Mokyr's, 181; Stanhope's, 172, 178–82, 197–98

James I, 112
Jarvis, Charles, 47, 160, 206–8, 210
Johnston, Stephan, 244
Journal des Savants, 37, 38, 117
journeymen, 81, 130, 147–49, 152, 167, 192

Kangxi emperor, 56
Kant, Immanuel: on freedom, 210–11, 230–33;
     on genius, 231–32; on God, 227–28, 231–32;
     Hamann and, 141; on imitation, 139, 141–42,
     231–32; on mathematics, 220–21, 232; on
     Newton, 141, 232; on physico-theology, 231;
     profit not philosophical, 210–11
Karl Eugen, Duke of Württemberg, 144
Kästner, Abraham: on Leibniz's machine, 279n19;
     on mathematical unity, 230
Kearns, Michael, 212
keeping-it-digital problem, 29–30, 49, 51, 63, 70,
     94, 100, 239
Kepler, Johannes, Schickard and, 26
Keyssler, Johann Georg, 137
Knibb, Samuel, maker of Morland machine, 261n28
knowledge
—artisanal knowledge: Diderot and, 39–40;
     exemplary breakdown of, 18; formalization
     of, 39–40, 84–85; hierarchy within, 20, 181;
     Hooke and, 63–66; Leibniz's account of, 82–
     86; Pascal's account of, 31–35, 78, 82; political
     economy and, 81; propositional knowledge

within, 9, 15, 39–40; of social world of skill,
     15, 20–21, 58–59, 63–66; unity impossible
     with, 33–34; worth stealing, 67. *See also*
     craftsmanship of certainty; craftsmanship
     of risk; tacit knowledge
—propositional knowledge: convergence of
     philosophy and artisanal practice and, 169–
     71, 178; conversion of artisanal to, 39–40,
     84–86; inadequacy of, 31–32, 83, 171; making
     practical, 169; Mokyr on, 181; social relations
     and, 9, 15; tacit knowledge and, 15
Kölbing, Georg Heinrich (clockmaker), 79
Kollman, A. F. C., on Müller machine, 158
Kratzenstein, Christian, calculating machine of,
     133–34

Lardner, Dionysius: on Babbage's machines, 48,
     160, 209; on spiral wheel, 160
lathes: Clement's work on, 50, 124, 193, 200–201;
     Fairchild on, 200; Pascal on, 33
Lavater, Johann Kaspar, on Hahn's machine, 145
Le Cerf, Mr., 168
Lécuyer, Christophe: on materiality of micro-
     electronics, 241; on oversocialized history,
     260n18, 266n13
Leibniz, Gottfried Wilhelm: Académie des sci-
     ences and, 78, 97–98, 100, 114–20; addition,
     simultaneous, 70, 172; analytic machine, 57;
     calculus of, 57, 87; carry mechanism and,
     62–63, 68–73, 75, 80, 145; Colbert and, 74,
     78, 100, 114–21; on confused and distinct
     knowledge, 85; contracts and, 59, 67, 76–80,
     97, 117–18; credit for machine and, 79; criti-
     cism of, 115–16; drawings and, 61–62, 71–74;
     emblem of machine and, 55; failure of, 70,
     80–81, 119, 129–30, 133, 145; on God, 81–83,
     121, 222–24; on hierarchy of artisans, 83–84;
     on Hooke, 64, 68; on knowledge of artisans,
     82–86; on limits to materialism, 222–24; on
     machinelike action, 85–86; manufacture
     and, 66, 87; *Memoire* for Ollivier, 68–78;
     mining and, 120–21; monads of, 154; on
     Morland, 21, 117; move to Hanover, 76, 120–
     21; nondisclosure and, 120–21, 129, 144; ob-
     taining secrets and, 67, 115, 268n50; Ollivier
     and, 68–80; on organic machines, 82; Paris
     and, 66–67, 82–88, 114–20; on Pascal, 21, 117;
     patronage and, 87, 97–98, 114–20; privileges
     and, 97–98, 114–20; as projector, 87, 116–18;
     risk of invention and, 74–75, 78–79, 97–98,
     118; Royal Society and, 59–66, 68, 117; state
     support, 97–98, 114–20; stepped drum and,
     60–62, 75, 144, 266–67n19, 278n16; technical
     competence and, 86–87; unconscious influ-
     ence and, 229; variable cogwheel of, 60–61.
     See also Ollivier

Leibnizian-Wolffians, 216
Le Philosophe, 216
Lepine (Epine), calculating machine of, 37
Le Roy, Pierre, Stanhope visit to, 191
Le Sage, Georges Louis, 166–67, 171, 181, 191–92, 226
Lessig, Laurence, 7
Lessing, Gotthold Ephraim, 153
Leupold, Jacob: cylindrical calculating machine of, 143–44; disclosure and, 144; Hahn and, 150–51; history of calculating machines and, 142–44
Levi (calculating machine maker), 38
liberal arts, 227, 231, 232, 247
Lichtenberg, Georg Christoph: on Euler, 218; on mathematics, 218–19; on mechanism, 225–27; on Newton, 218; on Stanhope machines, 152
Locke, John: associative psychology and, 229; on logic, 216, 220; on thinking matter, 224
logic, formal: carry and, 240–41; history of computing and, 11, 247, 257; Le Sage on, 167; Stanhope on, 173, 178–80, 197–99; Turing on, 237–38. *See also* computation; invention: theory of; reasoning, as computation; symbolic reasoning
logical demonstrator, Stanhope's, 197–98
London: Hooke and, 64–66; Leibniz and, 59, 63, 98, 117; Müller and, 152; Stanhope and, 192–93
London Chemical and Philosophical Society, 193
London *Gazette*, Morland and, 110–11
Long, Pamela, on artisanal property, 105
Lord Derby, 206
Lord Kelvin, analog computer of, 246
Lorraine, Duchy of, 136–38
Louis XIV, potential patron of Leibniz, 114–15
Lovelace, Ada, on machines and originality, 212–13, 235
Lunéville, 129, 135–37

Mahon, Charles. *See* Stanhope, Charles
Mahoney, Michael, on history of computing, 11
manufacture: Babbage and, 54, 89, 122–23, 125; clockmaking and, 133; creativity and, 227; elite artisans and, 125; emulation and, 128, 169; God and, 228; Hahn and, 147; Leibniz on, 66, 87; of microprocessors, 248; Morland and, 110; Müller on, 151–52, 158; nineteenth-century, of machines, 5, 242–44; Pascal and, 33, 104; privileges and, 99–103; Stanhope and, 196; Vayringe and, 136–37. *See also* artisans: ordinary; craftsmanship of certainty; industrial espionage; privileges; reduction to practice
Mariotte, Edme, Leibniz and, 77

market, development of, 242, 245–47
Marshall, Alfred, association machine of, 236–37
Masham, Damaris, Leibniz and, 223
materialism: Coward on, 224–25; eighteenth-century challenge to, 153–54, 199, 214–15, 217, 225–27; Hahn on, 154–55; Kant on, 231; Leibniz on limits of, 222–24; Lichtenberg on, 225–27; Morland on limits of, 40–41; originality and, 227–30; Pascal on, 41–43
materialization: Babbage on, 207; Clement's tools and, 201–2; funding for, 119; Hahn and, 145, 155; microprocessors and, 240–41, 248; social framework for, 20–21, 58–59. *See also* hylomorphism
mathematics, not calculation, 218–21. *See also* arithmetic; calculation
mathematics, philosophy of, eighteenth century, 218–21. *See also* Berkeley; evidence; Kant; Lichtenberg; logic
Mayer, Tobias, 225
McGee, David, 189, 260n25, 284n62
mechanical analogies, 170, 178–80
mechanical art, 34, 135, 142, 151, 193, 210, 227
mechanism (philosophy). *See* materialism
Mémoires de l'Académie des sciences, 37
Merck, Johann Heinrich: Hahn's suspicions of, 281n92; on improvement of existing machine, 142; misunderstandings of, 151; on Müller's machine, 126–28, 151
microprocessors, 240–41, 248
Mill, John Stuart, machinelike upbringing, 236
models (of machine): Babbage's, 89, 91; Gersten's, 132; Hooke on value of, 66; Hooke's, 64–65, 267n28; Leibniz's, 59; Pascal's, 100–101; Popplow on, 267n27; Stanhope's, 162, 171–78, 194
Mokyr, Joel: on competence, 194; on propositional knowledge and invention, 180–81
monopoly: absolutism and, 110–13, 116; Leibniz seeks, 97–98; Morland seeks, 112–23; Pascal seeks, 34, 100–101, 103–6. *See also* absolutism; patents; privileges
monsters, Pascal on mechanical, 33–34, 74, 104
Moore, Sir Jonas, arithmetical primers, 21, 64
Moritz-Wilhelm of Zeitz, 80
Morland, Sir Samuel: bookselling, 111; carry and, 23–24, 117; Charles II and, 110–14; coordinative skill and, 20–21; engineering work of, 111–13; Hooke on, 65; Leibniz on, 21, 110, 117; London *Gazette*, and, 110–11; manufacture and, 110; on materialism, 40–41; monopoly and, 112–13; multiplying machine of, 16–21; publications, 22–23; pumps, 111–13; specification, 112–14
mountebanks. *See* projectors
Müller, Captain Johannes: account of invention, 147–48; calculating machine, superiority of,

Müller, Captain Johannes (*cont.*)
148, 151; carry mechanism and, 149; coordination of artisans, 147–49; criticism of Hahn, 149–51; debts to Hahn, 150–51; Difference Engine and, 158; disclosure, 151–52; manufacture and, 151–52, 158; nineteenth-century view of, 242–43; on ordinary artisans, 152; second invention and, 126–28, 142; stepped drum, 148–49; table printing, 158–59; translated by Herschel, 159; use of springs, 148–49; workshop of, 148
multiplication: mechanizing, 59–62; Morland's machine for, 16–21; Napier's bones and, 16

Napier's bones, 16, 60, 64, 144
Nasmyth, James, 206
natural philosophers. *See* philosophers
nescience: as legitimizing imitation, 130; about Leibniz's machine, 129; about technical practitioners, 10, 88–89, 134–35
Newton, Isaac: Kant on, 141, 232; Lichtenberg on, 218; simultaneous discovery and, 152; Stanhope on, 197–98, 215
Newtonianism, and materialism, 225–26
Normandy, Pascal family and, 105–7
North, John, correspondence with Stanhope, 162, 196–97
novelty: absolutism and, 109; Colmar and, 242; Difference Engine and, 157–61; global vs. local, 116, 130, 132; Leibniz on, 98, 116–17; materialism and, 227–30; Poleni and, 130; in privileges, 102. *See also* emulation; invention: multiple
numeracy, in early modern Europe, 3–4, 21

odometers, 26, 111
Oldenburg, Henry, and Leibniz machine, 64, 68, 119
Ollivier (Parisian artisan): carry, difficulties with, 68, 75; communication with Leibniz, 68–69, 77; contracts with, 76–80; credit and, 79; drawings and, 73; Hansen and, 76–80; inventiveness of, 70, 74–75, 78; Leibniz's choice of, 67–68; *Memoire* for, 68–78; ownership in machine, 78–79; payment to, 75, 118; resistance to work-discipline, 76–77, 79; risk of invention and, 74–75, 118
On the Economy of Machinery and Manufactures (Babbage), 49
open technique: historiography of, 180; Vayringe and, 137
organic: accounts of making, 228; Leibniz's conception of, 82, 228
originality: credit and, 214; Lovelace on, 212–14, 235; materialist psychology and, 227–30,

236–37; Turing on, 212–13, 237–38; vegetative model, 229–30
Oughtred Society, 259n5

papermaking, 133–34
Paris: artisanal center, 66–67; emulation in, 38; Leibniz and, 66–67; Stanhope in, 191–92
Parlement of Paris, 103, 108
Parliament, 89, 90, 112, 113
Pascal, Blaise: on animals, 42–43; on artisanal knowledge, 31–35, 78, 106; carry, approach to, 24, 27–31, 100; failure of, 13, 35–37; hylomorphism and, 104; on inadequacy of philosophers, 31–32; intellectual property and, 100; Leibniz on, 21, 100, 117; manufacture and, 33, 104; on materialism, 41–43; models of machine and, 100; on monsters, 33–34, 74, 104; patronage and, 105–6; quest for privilege, 34–35, 100–107; Séguier and, 105–6; state support for, 31–35
Pascal, Etienne, as royal official, 107, 275n35
Patent Novel Machine, 230
patents: Abeille on, 232–33; Brewster on, 122; intellectualization of, 109–10; lawsuits over, 245; privileges versus, 101–2, 104–5. *See also* intellectual property; privileges; specification
patronage: challenges of, 112–13; Leibniz and, 87, 97–98, 114–20; Morland and, 112–23; Pascal and, 105–6; Vayringe and, 136–39
Peel, Sir Robert, Difference Engine and, 90, 95–96
Pell, John: Hooke and, 65, 267n37; Morland and, 20
Pensées (Pascal), on calculating machines, 42
Pereira, Jacob Rodrigues, calculating machine of, 37–38
Perrault, Claude, calculating machine of, 37
Petit, Pierre, on failures of Pascal's machine, 13
philosophers, natural: as engineers or inventors, 31–32, 83–84; inadequacy of, 5, 10, 31–32, 168–70; sovereigns and, 107; transformation of, 10, 84–86, 170–71, 178. *See also* artisans: coordination of; knowledge: artisanal
philosophy: applied focus, need for, 10, 168–71; profit antithetical to, 210–11; profit compatible with, 98
Pictet, Marc-August, on Stanhope, 197–98
pike, Pascal on, 42
pinned barrel, 174, 190
plagiarism, 132, 139, 149, 161. *See also* imitation; novelty
Plan 28 (Babbage Analytical Engine), 47
Poleni, Giovanni: calculating machine, 130–34, 138, 144, 158; nescience and, 130; variable cogwheel, 130

political economy: Colbert and, 66–67, 114–15; emulation and, 128, 137, 141, 168; Leibniz and, 81; natural philosophers and, 6, 91, 170–71; superior artisans and, 67
precision instruments. *See* tools
Prince of Mecklenburg, 152
priority: Difference Engine and, 157–61; Hahn vs. Müller, 126–29, 149–51; Leibniz on predecessors, 21, 10, 117. *See also* novelty
privileges: Abeille on, 232–33; absolutism and, 107–10; as gifts, 101, 103–5, 116, 121–22; judicial bodies and, 106, 108; Leibniz's quest for, 114–20; manufacture and, 99–103; not intellectual property, 101–4; novelty and, 102; Pascal's quest for, 34–35, 100–107; versus patents, 101–2, 104–5
programming, as craft, 7, 247–48
projectors: Babbage and, 48, 88, 90, 93, 203; Burattini as one, 37; Carcavy and, 115; Gerbier as one, 13; Leibniz and, 84, 87, 116–17
Prometheus: second inventors and, 126–27, 151; Shaftesbury on, 228
property: Babbage and, 50, 94, 123–25; Clement and, 50, 123–25; Ollivier on, 78; Pascal and, 100; rights to invention, 122. *See also* intellectual property; monopoly; patents; privileges
propositional knowledge. *See under* knowledge
psychology, materialist: challenge of originality, 227–30, 236–37; Marshall machine and, 236–37
pumps: Leibniz's, 120; Morland's, 111–13
Punch (magazine), 212–13
Pye, David, on craftsmanship of risk and certainty, 14, 74

Ramsden, Jesse, 192, 194, 200
reasonable advantage, 97, 115, 120. *See also* credit
reasoning, as computation: rejection of, 4–5, 216–21; Stanhope on, 197–99, 215–16, 221
rectangular design: Hahn considers, 144–46, 175; Leibniz's, 62–63; Stanhope switches to, 175–78
reduction to practice: Babbage on, 94, 125; Braun and, 130; Clement and, 94; Colmar and, 244; corporate histories and, 245; Hahn and, 147; Leibniz's failure and, 119–20; Müller and, 150, 158–59; necessity for privileges, 102, 274n19; Pascal's failure and, 102; Royal Society and, 94; Scheutz difference engines and, 209. *See also* artisans: ordinary; craftsmanship of certainty
reinvention, 35, 39, 130, 133–34. *See also* emulation; imitation; invention: multiple
republic of letters: artisanal alternative to, 152; Leibniz within, 130; Müller's disclosure

within, 151; Poleni working in; 132–33. *See also* economy of glory
reward. *See* credit
Reynolds, Joshua, on imitation and competition, 139
Richelieu, Cardinal, 19, 106–7
rights: against monarchs, 106–7, 113–14; specification and, 100. *See also* patents; privileges
Roberval, Gilles Personne de: own machine and, 35; Pascal machine and, 15, 27, 30, 262n46
Rouen, Pascals in, 13, 103
Rousseau, Jean-Jacques: Bischoff and, 191; Stanhope and, 153
Rowley, John, 137
Roy, Pierre Le, 191
Royal Astronomical Society, 91
Royal Society Club, 192
Royal Society of London: adequacy and, 94; Babbage and, 89–96, 157–59, 202–6; Gertsen and, 132; Herschel and, 94; Hooke and, 63–66; Leibniz and, 59–66, 68, 117; Stanhope and, 171
Royal Technical Institute (Sweden), 209

Sander, Heinrich, on Hahn's machine, 146
Saussure, H. B., 168
sautoir (Pascal), 27–29, 262n44
Schaffer, Simon, 207, 261n10, 271n140, 278n12
Schelling, Friedrich Wilhelm Joseph, 218
Scheutz, Edvard and George, difference engines of, 209–10
Schickard, Wilhelm, calculating machines of, 21, 26–27, 30
Schlegel, Karl Wilhelm Friedrich, 153
Science Museum, London: Babbage's collection at, 158, 160; Difference Engine at, 47–49, 160, 201; Difference Engine 2 at, 208–9; Stanhope machines at, 162, 188, 190
scientific computation, 245–47
Second World War, history of computing and, 239, 246–47
secrets: getting from artisans, 67; Roberval's machine and, 35; state policies on, 115, 268n50; Vayringe and, 138. *See also* industrial espionage
security mechanisms (Difference Engine), 48, 51–53, 94, 158–59, 241
Séguier, Chancellor: Huygens and, 108; Pascal and, 31, 34, 100, 105, 275n36; tax rebellion and, 107
self-interest: Babbage and, 90; Carcavy on, 115; choice of artisan and, 67; Jarvis on, 210; Leibniz and, 67, 97–98, 120–21. *See also* disinterestedness; freedom
Sherp, Adam, 79

Shimpi, Anand, on microprocessor design, 248
Shortgrave, Richard, 65
skill. *See* knowledge: artisanal; tacit knowledge
Smithsonian Institution, Scheutz difference engine and, 209
Société d'Encouragment pour l'Industrie Nationale, Colmar and, 242
Society of Arts (Geneva): convergence of philosophy and artisanal practice in, 169–71, 178; emulation and, 140–41
Society of Arts (London): Clement and, 200–201; Colmar's machine and, 244–45; as international model, 90; nineteenth-century machines and, 244–45; publications, 193, 201, *204*; Royal society versus, 90–91
Sonenscher, Michael, 58, 280n58
South Sea Bubble, as cautionary tale, 88
sovereigns: gifts of, 121; limits of, 107–12, 121; Pascal on, 107, 109. *See also* absolutism; technical competence
specification: Morland and, 112–14; in patents, 101, 109–10, 114, 116, 122; Stanhope and, 194
Sperry, Elmer, and historiography of invention, 6
spiked wheel. *See* horny piece; spiral wheel
Spinoza, Baruch, Herder and Hahn on, 153–54
spiral wheel: Babbage and, 48, 160, 202, 284n52; Stanhope and, 48, 160, 166, 189–90, 202. *See also* horny piece
springs: artisans and knowledge of, 15, 31, 262n53; Braun's use of, 139; Clement's use of, 51; Colmar's use of, 244; Hahn's use of, 146–49; Leibniz's use of, 63, 75; Müller's use of, 149, 151; Pascal's use of, 29–30
Stanhope, Charles Mahon, 3rd Earl Stanhope: Académie des sciences (France), 191; Academy of Sciences (Copenhagen), 191; analytical machine, 197–98; artisans, coordination of, 162–71, 191–96; carry and, 48, 171, 175, 178, 182–88; convergence of philosophy and artisanal practice and, 169–71, 178; criticism of, 196–97; "Cylinder A," 171–78, 182–83; Difference Engines and, 159–60; drawing, invention through, 162–66, 185–86, 188–90, 193–94; education of, 166–71; freedom of artisans and, 180–81; Geneva and, 166–71, 191–92; humility and, 199, 233, 235, 238; invention, theory of, 172, 178–82, 197–98; logical demonstrator of, 197–98; logic of, 173, 178–80, 197–99; London and, 192–93; manufacture and, 196; manuscripts of, 164–65; mechanical reasoning and, 197–99, 215–16, 221; nineteenth-century view of, 242–43; rectangular machine, switch to, 175–78; Royal Society and, 171; stepped drums and, 173–78, 190; technical vocabulary of,

190–91; teeth (gears) and, 162, 168–71, 188; tools and, 196–97; trial device of, 183–84. *See also* horny piece; spiral wheel
Stanhope, Philip Henry, 4th Earl Stanhope: on Babbage's novelty, 159; on third Earl Stanhope's machines, 159
Stanton, Mr., 65
stepped drum (stepped reckoner, Staffelwalze): Hahn's, 145–46; Leibniz's, 60–62, 75, 144, 266–67n19, 278n16; Leupold on, 144; Müller's, 148–49; Stanhope's, 173–78, 190
St. Petersburg, 133, 152
Stuttgart, 129, 147
sufficient-force problem, 26–29, 63, 70, 100, 172
Sulzer, Johann Georg, on hierarchy of artisans, 139, 181
Sutton, Henry, Morland machine and, 18, 20
Swade, Doron, 264n2, 264n4, 264n6, 285n52, 287nn32–34, 287n36, 291n32
Sweden, 107, 209
Swift, Jonathan, on projectors, 90
symbolic reasoning, dangers of, 4–5, 218–21. *See also* logic; mathematics, philosophy of; reasoning, as computation

tacit knowledge, 15–16. *See also* knowledge: artisanal
Tate, Samuel, Colmar machines and, 244–45
technical competence: Carcavy and, 116; machines as proof of, 10–11, 37, 86–87, 88–89, 129, 134–35
technical illustration. *See* disclosure; drawing; specification
technology, history of: critique of originality, 7–8, 214; cumulative change in, 7–8, 178, 180–81, 211, 214; heroic narratives, 7, 211, 214; invention within, 6; linear model and, 9, 180, 189; Mahoney on, 11; overly socialized, 260n18, 266n13. *See also* computing, history of
teeth: epicycloidal shape, 169; Stanhope on, 162, 168–71, 188. *See also* stepped drum; variable cogwheel
Teuber, Gottfried, Leibniz machine and, 80
Teutsche Merkur, 126, 138, 145, 150, 281n92
theory: convergence of philosophy and artisanal practice and, 10, 169–71, 178; inadequacy of, 10, 31–32, 83. *See also* philosophers
thinking matter debate, 216, 224
Thompson, Isaac (engineer), 20
tools: Babbage on, 205; Clement and, 50, 124, 200–201, 205–6; duc du Chaulnes and, 38; materialization and, 8–9, 196–97, 201–2; in nineteenth century, 9, 244–45; Stanhope, 196–97; S. Varley and, 193; Vayringe and, 137–38

trade-offs: in Colmar machines, 244; in contemporary computer arithmetic units, 239, 244–45, 248
Treasury (UK), Babbage and, 49, 89, 90, 91, 93, 95, 202
Turing, Alan: on Automatic Computing Engine, 237, 239; on carry, 240; general purpose computers and, 212–13, 247; in history of computing, 11–12, 247; on Lovelace, 212–13; on originality, 212–13, 237–38
Turner, Anthony, on Pascal's machine, 30

unity, artistic: impossibility for artisans, 33–34, 104; genius and, 229, 231–32; Kant on, 231–32; Kästner on, 230; organic machines and, 82; Pascal on, 33–34; sovereign maker and, 227–28
University of Helmstedt, 80
University of Pennsylvania, 246
Unwin, Cathorn, 244
Urlin, Simon, 108
users' history, 244–46

variable cogwheel: Braun-Vayringe, 138–39; Leibniz, 60–61; Poleni, 130
Varley, Cornelius, Clement and, 193, 200–201
Varley, Samuel, Stanhope and, 193–94
Vauxhall, Morland displays at, 111
Vayringe, Philippe: Académie of Lunéville and, 135–38; calculating machine of, 136–73, 144, 146, 193, 209, 244; exemplar of emulation, 135–39; manufacture and, 136–37
vegetative model: divine creation and, 155, 229; originality and, 155, 229–30
Venice: Poleni and, 129, 131; privileges and, 101
Very Large Scale Integration (VLSI) processor, nonhylomorphic design and, 248
Vienna, 136–39
Voltaire, 135, 138, 153
Von Neumann, John, 247
Vulliamy, Benjamin, Stanhope hires, 192

wages: Difference Engine project and, 205; exemplary breakdown of, 18; Ollivier and tension over, 77–80, 118; risk in inventing and, 74, 78, 118
Wagner, Rudolf Christian, Leibniz machine and, 80
Wallastson, Dr., on early Difference Engine, 47
Whitworth, Joseph, importance of Babbage's project for, 200, 206
work-discipline, Ollivier's resistance to, 76–77, 79
workmen. See artisans
workshop: Bernström's, 209; Chinese, 56; Clement's, 50, 95, 201–8; Hahn's, 147; Jarvis and, 206–7; Ollivier's, 77; threat of moving, 115, 122. See also Chevening
Wright, Michael, on Stanhope machines, 190

Young, Edward: Conjectures on Original Composition, 135, 140, 155, 227; Night-thoughts, 225, 227